This card just may

Change YOUR Life

The Princeton Review is giving away a

FREE
SAT, GMAT, GRE, LSAT, or MCAT course every other month.

We'll choose one lucky winner who will be entitled to a free course. Interested? Simply fill out the card on the right and mail it back to us.

For more information about **Princeton Review** courses and products, call **(800) REVIEW-6.**

Entries must be received by February 15, 1999. You don't need to buy this book to enter. See official rules on reverse side.

MW00736486

○ Also, please send me FREE information
about paying for school (including student loan application/information.)

Name: _____

Address: _____

_____

City: _____ State: _____ Zip: _____

Is this address ○ School ○ Work ○ Home

Phone: _____

E-mail address: _____

School: _____ Graduation Year: _____

Cracking the AP Calculus AB   CACA9

↑ THE PRINCETON REVIEW
PO BOX 67
EAST MEADOW NY 11554-0067

OFFICIAL RULES:

We will conduct a random drawing on the fifteenth of every other month from all cards we've received between the last drawing and midnight of the day before. We'll hold these drawings through February 15, 1999. The winner of each drawing will receive (at no fee) an SAT, LSAT, GMAT, GRE or MCAT course at any Princeton Review location, each with an approximate value of $695. Your odds of winning depend upon the number of entries received. If you win, you must take your free course within six months of notification; the free course is not transferable except to immediate family members. This promotion is not open to employees of The Princeton Review or Random House and is, of course, void where prohibited by law. All taxes are the sole responsibility of the winners. No purchase is necessary: if the card that's supposed to be attached has already been ripped out, or if you're not buying this book (big mistake), you may enter by sending your own postcard with your name, address, and school to The Princeton Review, 2315 Broadway, New York, NY 10024-4332. You may also write us to get a list of prize winners. By the way, we're not responsible for lost, misdirected, illegible, or mutilated entries.

THE
PRINCETON
REVIEW

THE PRINCETON REVIEW

Cracking the AP
Calculus AB & BC

By David S. Kahn

THE PRINCETON REVIEW

Cracking the AP
Calculus AB & BC

By David S. Kahn

1998–99 EDITION
RANDOM HOUSE, INC.
NEW YORK 1997
WWW.RANDOMHOUSE.COM

Princeton Review Publishing, L.L.C.
2315 Broadway
New York, NY 10024
E-mail: info@review.com

ISSN: 1092-4086
ISBN: 0-375-75107-6

AP is a registered trademark of The Educational Testing Services.

Editor: Rachel Warren
Designer: Illeny Maaza
Production Editor: Amy Bryant

Manufactured in the United States of America on partially recycled paper.

9 8 7 6 5 4 3 2 1

ACKNOWLEDGMENTS

First of all, I would like to thank Arnold Feingold and Peter B. Kahn for doing every problem, reading every word, and otherwise lending their invaluable assistance. I also want to thank my editors, Doug French, Melanie Sponholz, and Rachel Warren; my producer Amy Bryant; and the rest of The Princeton Review for their assistance. Thanks also to Gary King for doing all of the test problems, and to Mike Fritz for reading the manuscript. Thanks to the production crew of Chris Thomas, Carmine Raspaolo, Adam Hurwitz, Patricia Acero, Mabel Villanueva, Robert "Precious" McCormack, John Bergdahl, Matt Reilly, John Pak, Effie Hadjiioannou, and Greta Englert. Thanks to Jeffrey and Miriam for moral support. Thanks Mom.

Finally, I would like to thank the people who really made all of this effort worthwhile—my students. Your support over the last two plus years truly helped make this book possible, so thank you:

Aaron, Abby, Aidan, Alex, Alex and Gabe, Alexes, Alexis, Ali, Alicia, Ally, Alyssa, Amanda B., Amanda C., Amanda M., Andrea T., Andrea V., Andrew B., Andrew and Allison, Andrew S., Andy and Allison, Anna L-W., Anna F., Anna M., April, Arya, Arthur, Asheley and Freddy, Ashley, Ashley and Lauren, Ben, Bethany and Lesley, Betsy and Jon, Blythe, Brett, Brooke, Butch, Caitlin, Caroline, Chad, Christine, Christine W., Claudia, Corinne, Courtney, Craig, Daniel, Danielle, Dara P., Dara M., Deborah, Devon, Dora, Eairinn, Elisa, Emily B., Emily C., Emily L., Emily S., Emily T. and Erica H., Eric, Erin, Frank, Gabby, Geoffrey, George, Gloria, Heather, Hilary, Holly K., Ingrid, Jacob, Jaclyn, Jan, Jason, Jay, Jenna, Jen, Jennifer B., Jennifer W., Jess, Jesse, John, John and Dan, Jon, Jordan, Josh, Josh M., Julie H., Julie and Dana, Julia, Kat, Kate D., Kate L., Kate S., Katie, Katrina, Kimberly, Kitty, Laura F., Laura G., Laura Z., Lauren R., Lauren T., Lauren W., Lee R., Lee S., Leigh, Liana, Lila, Lilaj, Lily, Lily H., Lindsay F., Lindsay R., Lindsay, Magnolia, Mara, Marietta, Marcia, Mariel, Marisa, Mary M., Matt, Matt B., Matt V., Matthew B., Matthew F., Maya and Rohit, Melissa, Meredith, Milton, Morgan, Nadia, Nathania, Nicole, Nikki, Nora, Oliver, Omar, Oren, Pam, Peter, Rachel A., Rachel B., Rachel F., Rachel M., Rachel R., Ramit, Rayna, Rebecca, Ricky, Ruthie, Sam C., Sam L., Sam W., Samuel C., Samar, Sara, Sarah L., Sarah M-D., Sarah S., Sascha, Sasha, Saya, Sonja and Talya, Sophia, Sophie, Stacey, Stacy, Stefanie, Stephanie, Tammy and Hayley, Taylor, Tenley, Terrence, Tripp, Victoria, Waleed, and Zach.

If I forgot anyone, I apologize. You will be in the next edition.

CONTENTS

INTRODUCTION

If you're reading this book, you're either about to take Advanced Placement (AP) calculus exam, or you're in the middle of it, getting ready for the exam. Either way, good luck! Calculus is a very difficult subject and one that a lot of students have trouble with. If it's any consolation, you should remember that, generally, only *strong* math students take AP calculus. If you weren't a math whiz, you wouldn't have made it this far! So be proud of yourself, and realize that you're not alone. Most people find calculus very hard.

One of the reasons why calculus is so difficult arises from a lack of understanding about the nature of the subject. You probably think that calculus is the end of a sequence of courses in mathematics that you arrive at after passing through algebra, geometry, trigonometry, etc. Wrong! It's the beginning of a whole new branch of mathematics. You must learn a whole new set of tools, and a whole new approach to problem solving. Calculus is going to teach you a new way of thinking and of looking at mathematics and at nature. Maybe it will even stick.

What do we mean by a whole new approach to problem solving? Well, before calculus, most of the problems that you had to do were one-step problems. You were asked a question, and you had one thing that you had to figure out. For example: *If you raise the price of an item that originally cost $50 to $80, by what percent did you raise the price*? This isn't terribly hard. You set up an equation, and you solve it. In fact, only in geometry do you have to do multi-stage thinking involving proofs, not problem solving. This means that up until now, your thinking has tended to be oriented towards the method that ought to be used to solve the problem, and how to execute the method.

In calculus, you are going to be solving multi-step problems. For example: *If two cars start at the same point and travel in directions 120 degrees apart, and if the first car is traveling at 60 miles per hour and the second car is traveling at 75 miles per hour, how fast is the distance between the two cars growing after three hours?* This problem requires you to figure out how far the cars have gone after three hours, set up an equation relating the speed of each car to the rate of change of the distance between them, and solve it. It's going to require some trigonometry, some algebra, some geometry, and some calculus. Much harder, isn't it?

The good news about AP calculus is that, because it is an introductory course, you won't delve too deeply into any of the topics (unless you have a sadistic teacher), but you will concentrate on learning a lot of techniques. There's a lot of memorization involved in calculus, and you'll be expected to look at a problem and immediately know which type of problem it is and which technique to use to solve it. This book will help you do that. It's broken down into many small units, each of which concentrates on a particular type of problem and the technique(s) you'll need to solve it.

The other good news is that this is not college calculus, which could be taught any number of different ways. This is AP calculus. This means that the curriculum and the test are standardized. You'll be expected to learn certain things, and you will *not* be expected to learn others. Furthermore, the exam tends to ask the same kinds of questions year after year. So once you learn how to do those types of problems, you'll be in good shape for the test.

The final bit of good news is that the curve on the test is *very* generous. If you can get 70 percent of the questions right, you'll get a very good score. You should understand, of course, that it's hard to get a lot of the questions right! If 70 percent is a very good score, then a lot of people will be getting under 50 percent right. But after you finish this book, you won't be one of them!

AP calculus is divided into two types: AB and BC. The former is supposed to be the equivalent of a semester of college calculus; the latter, a year. In truth, AB calculus covers closer to three quarters of a year of college calculus. In fact, the main difference between the two is that BC calculus tests some more theoretical aspects of calculus and it covers a few additional topics. In addition, BC calculus is harder than AB calculus. The AB exam usually tests straightforward problems in each topic. They're not too tricky, and they don't vary very much. The BC exam asks harder questions. But neither exam is tricky in the sense that the SAT is tricky. Nor do they test esoteric aspects of calculus. Rather, both tests tend to focus on testing whether you've learned the basics of differential and integral calculus. The tests are difficult because of the breadth of topics that they cover, not the depth.

Now, a word about the test itself. The AP exam comes in three parts. First, there is a section of multiple choice questions covering a variety of calculus topics. You're not permitted to use your calculator for this first section of the test. Second, there is a section of multiple choice questions similar to the first, except that you are permitted to use a calculator for these questions. These two sections contain approximately 40 questions.

Third, there's a section of six questions, each of which requires you to write out the solutions and the steps by which you solved it. You are permitted to use a calculator for these problems. Partial credit is given for various steps in the solution of each problem. You'll usually be required to sketch a graph in one of the questions. ETS does you a big favor here: you may use a graphing calculator. In fact, ETS recommends it! And they allow you to use programs as well. But here's the truth about calculus: most of the time, you don't need the calculator anyway. Remember: these are the people who bring you the SAT. Any gift from them should be regarded skeptically!

HOW TO USE THIS BOOK

This book consists of 20 chapters on calculus, each of which covers a topic or two on the AP exam. If you're just starting your course in calculus, use this to study each topic as your teacher presents it to you. We've tried to reduce each subject to its bare essentials, so you can teach yourself the subject. There's almost no theory in this book. Instead, we emphasize the mechanics of calculus. This is what we believe is most important at this stage, and, furthermore, this is all you'll need for the exam. There is only a tiny bit of theory on the AP exam. In fact, you could get every theoretical question wrong on the exam and still get a 5 on the test! Of course, we are covering the theory that you'll need to know, but we won't ask you to do proofs!

We wish we could show you a bunch of techniques like those for the SAT but, alas, plugging in and backsolving won't do you much good on the AP. After all, Joe Bloggs doesn't take AP calculus!! No, the sad truth is that you must actually learn the subject matter for this test. The good news is that you can learn this material just by going through each of the chapters in this book and by working with them assiduously.

The key to doing well on the exam is to memorize a variety of techniques for solving calculus problems, and to recognize when to use them. There's so much to learn in AP calculus that it's difficult to remember everything. Instead, you should be able to derive or figure out how to do certain things based on your mastery of a few essential techniques. In addition, you'll be expected to remember a lot of the math that you did before calculus—particularly trigonometry. If you're not strong in this area, take some time to review the appendix on prerequisite mathematics located at the back of this book.

Furthermore, if you can't derive certain formulae, you should memorize them! A lot of students don't bother to memorize the trigonometry special angles and formulae because they can do them on their calculators. This is a big mistake. You'll be expected to be very good with these in calculus, and if you can't recall them easily, you'll be slowed down and the problems will seem much harder. Make sure that you're also comfortable with analytic geometry. If you rely on your calculator to graph for you, you'll get a lot of questions wrong because you won't recognize the curves when you see them.

This advice is going to seem backward compared to what your teachers are telling you. In school you're often yelled at for memorizing things. Teachers tell you to *understand* the concepts, not just memorize the answers. Well, things are different here. The understanding will come later, after you're comfortable with the mechanics. In the meantime, you should learn techniques, practice them, and, through repetition, you will ingrain them in your memory.

Each chapter is divided into three types of problems: examples, solved problems, and practice problems. The first type are contained in the explanatory portion of the unit. They're designed to further your understanding of the subject and to show you how to get the problems right. Each step of the solution to the example is worked out, except for some simple algebraic and arithmetic steps that should come easily to you at this point.

The second type are solved problems. The solutions are worked out in approximately the same detail as the examples. Before you start work on each of these, cover the solution with an index card or something, then check the solution afterward. And you should read through the solution, not just assume that you knew what you were doing because your answer was correct.

The third type of problems are practice problems. Only the answers to them are given. We hope you'll find that each chapter offers you enough practice problems for you to be comfortable with the material. The topics that are emphasized on the exam have more problems, those that are de-emphasized have fewer. In other words, if a chapter has only a few practice problems, it's not an important topic on the AP and you shouldn't worry too much about it.

The recipe for success follows the following guidelines:

Do all of the problems! As with many other things in life, if you want to do well in calculus, you must practice! You should work carefully through the examples and solved problems, so that you really understand the method involved for getting the right answer. Often it won't be the calculus that you get wrong, it will be the algebra.

Know your algebra! One of the main reasons that students have trouble with calculus is that their algebra is weak. You must be good at factoring, reducing, simplifying, expanding, inverting, and so on. Otherwise, you won't be able to perform the proper steps that reduce the problem to following a simple technique. Many of the answer choices on the AP exam differ only by a constant or two, so you have to be strong at algebra to get those correct. Review the appendix on prerequisite mathematics if you're weak in this area.

Do the practice exams! After you've finished going through all of the chapters, there are practice exams for you to do. Each has step-by-step solutions to the problems. You should carefully go through the solutions to all of the problems you get wrong, and you might want to review the units on the ones that you got correct as well. (You might just have been lucky!) Each of the answers will tell you the chapter you should review if you got the question wrong.

Don't always trust your teachers or your textbooks! Some of the material that your teachers will show you is not on the AP. Furthermore, many classes use a college textbook for the AP calculus course. These tend to cover far more than you'll ever need for the exam. Although you need to know this extra material for class, don't worry about it for the exam. Just stick to what this book covers.

If the test is in three weeks, what do I do?! Some of you will be buying this book right before the AP exam. In that case, you should proceed directly to the practice exams. Once you know where your weaknesses lie, you can go back to the relevant chapters and improve your skills.

CHANGES TO THE AP EXAM

As you may have heard, the AP Exam was changed this year. The people in charge of the exam have said that it contains some of the biggest changes in a long time. This may be sending you into a panic, but don't worry, the new exam isn't that different from the old one. After all, <u>Calculus</u> hasn't changed, just the exam!

The changes are as follows:

◆ If you are in AB calculus, you no longer need to know Integration By Parts, Newton's Method, and Simpson's Rule.

◆ If you are in BC calculus, you no longer need to know Newton's Method, Simpson's Rule, and The Strict Definition of Limit.

◆ You now need to know Euler's Method and Slope Fields.

◆ For both exams, the format has changed.

◆ Part I used to contain 40 questions and now contains 45 questions.

◆ Part IA of the exam now contains 28 multiple choice questions where you ARE NOT allowed to use the calculator.

- Part IB contains 17 questions where you ARE allowed to use the calculator.

- Part I is now 15 minutes longer to accommodate the 5 extra questions.

- Part II is still 6 long format questions that will be more "calculator active" than before.

- BC students will now get an AB subscore. In other words, you will get two scores—one for the AB part of the test and one for the BC part of the test. So now, a student might get a score such as: 3 overall, 4 on the AB portion of the test.

As you can see, the changes are not really that great. The main difference is that the AP folks will be expecting you to use a calculator, so the answers to questions may be messier than they used to be. For example, on the old test, the answer to a question that asks you to find a local minimum might be nice, round numbers such as (3,10). On the new test, the answer to a question that asks you to find a local minimum might now be (2.817,9.634). This is because you will now be expected to use a calculator to find values. **All numbers should be rounded to 3 decimal places unless you are told otherwise.**

WHAT IS ON THE AP?

If you are taking AB Calculus

Limits

Continuity

Differential Calculus
    The Definition of the Derivative
    Differentiability
    Product Rule, Quotient Rule, Chain Rule
    Higher-Order Derivatives
    Implicit Differentiation
    Tangent and Normal Lines
    Mean Value Theorem for Derivatives and Rolle's Theorem
    Curve Sketching
    Maxima/Minima
    Related Rates
    Position, Velocity, and Acceleration
    Derivatives of Trig Functions
    Derivatives of Logarithmic and Exponential Functions
    Derivative of an Inverse

Integral Calculus
    Antiderivatives
    U-Substitution
    Area Under a Curve
    The Fundamental Theorems of Calculus
    The Trapezoid Rule
    The Mean Value Theorem for Integrals

Integrals of Logarithmic and Exponential Functions
The Area between Two Curves
Volume of a Solid of Revolution—Washers and Shells
Position, Velocity and Acceleration
Differential Equations
Separation of Variables

If you are taking the BC Exam

Topics in italics are not on the AB exam

Limits

Continuity

Differential Calculus
    The Definition of the Derivative
    Differentiability
    Product Rule, Quotient Rule, Chain Rule
    Higher-Order Derivatives
    Implicit Differentiation
    Tangent and Normal Lines
    Mean Value Theorem for Derivatives and Rolle's Theorem
    Curve Sketching
    Maxima/Minima
    Related Rates
    Position, Velocity, and Acceleration
    Derivatives of Trig Functions
    Derivatives of Logarithmic and Exponential Functions
    Derivative of an Inverse
    Logarithmic Differentiation
    L'Hopital's Rule
    Differentials
    Derivatives of Parametric Functions

Integral Calculus
    Antiderivatives
    U-Substitution
    Area Under a Curve
    The Fundamental Theorems of Calculus
    The Trapezoid Rule
    The Mean Value Theorem for Integrals
    Integrals of Logarithmic and Exponential Functions
    The Area between Two Curves
    Volume of a Solid of Revolution — Washers and Shells
    Position, Velocity and Acceleration
    Integration by Parts
    Integrals of Trig Functions and Inverse Trig Functions
    The Method of Partial Fractions
    Improper Integrals

Length of a Curve
Parametric Functions

Differential Equations
    Separation of Variables
    Euler's Method
    Slope Fields

Infinite Series
    Sequences
    Series
    Ratio and Integral Tests
    Power Series
    Taylor Series
    The Remainder of a Taylor Series

Some of the topics on the BC exam are only occasionally tested and thus we don't spend much time on them.

1

LIMITS

WHAT IS A LIMIT?

In order to understand calculus, you need to know what a "limit" is. A limit is the value a function (which usually is written "$f(x)$" on the AP exam) approaches as the variable within that function (usually "x") gets nearer and nearer to a particular value. In other words, when x is very close to a certain number, what is $f(x)$ very close to? As far as the AB test is concerned, that's all you have to know about evaluating limits. There's a more technical method that you BC students have to learn, but we won't discuss it until we have to—at the end of this chapter.

Let's look at an example of a limit: What is the limit of the function $f(x) = x^2$ as x approaches 2? In limit notation, the expression "the limit as x approaches 2" is written like this: $\lim_{x \to 2}$. In order to evaluate the limit, let's check out some values of $\lim_{x \to 2}$ as x increases and gets closer to 2 (without ever exactly getting there):

When $x = 1.9$, $f(x) = 3.61$.

When $x = 1.99$, $f(x) = 3.9601$.

When $x = 1.999$, $f(x) = 3.996001$.

When $x = 1.9999$, $f(x) = 3.99960001$.

As x increases and approaches 2, $f(x)$ gets closer and closer to 4. This is called the **left-hand limit** and is written $\lim\limits_{x \to 2^-}$. Notice the little minus sign!

What about when x is bigger than 2?

When $x = 2.1$, $f(x) = 4.41$.

When $x = 2.01$, $f(x) = 4.0401$.

When $x = 2.001$, $f(x) = 4.004001$.

When $x = 2.0001$, $f(x) = 4.00040001$.

As x decreases and approaches 2, $f(x)$ still approaches 4. This is called the **right-hand limit** and is written like this: $\lim\limits_{x \to 2^+}$. Notice the little plus sign!

We got the same answer when evaluating both the left and right-hand limits, which isn't exactly earth-shattering, because when x is 2, $f(x)$ **is** 4. You should always check both sides of the independent variable because, as you'll see shortly, sometimes you don't get the same answer. Therefore, we write that $\lim\limits_{x \to 2} x^2 = 4$.

We didn't really need to look at all of these decimal values to know what was going to happen when x got really close to 24. But it's important to go through the exercise because, typically, the answers get a lot more complicated. Let's do a few examples.

Example 1: Find $\lim\limits_{x \to 5} x^2$.

The approach is simple: plug in 5 for x, and you get 25.

Example 2: Find $\lim\limits_{x \to 3} x^3$.

Here the answer is 27.

There are some simple algebraic rules of limits that you should know. These are:

$$\lim_{x \to a} kf(x) = k \lim_{x \to a} f(x)$$

Example: $\lim\limits_{x \to 5} 3x^2 = 3 \lim\limits_{x \to 5} x^2 = 75$

$$\lim_{x \to a}[f(x) + g(x)] = \lim_{x \to a} f(x) + \lim_{x \to a} g(x)$$

Example: $\lim_{x \to 5}[x^2 + x^3] = \lim_{x \to 5} x^2 + \lim_{x \to 5} x^3 = 150$.

$$\lim_{x \to a}[f(x) \cdot g(x)] = \lim_{x \to a} f(x) \cdot \lim_{x \to a} g(x)$$

Example: $\lim_{x \to 5}\left[(x^2 + 1)\sqrt{x - 1}\right] = \lim_{x \to 5}(x^2 + 1)\lim_{x \to 5} \sqrt{x - 1} = 52$

Example 3: Find $\lim_{x \to 0}(x^2 + 5x)$.

Still nothing new. Plug in 0, and you get 0.

So far, so good. All you do to find the limit of a simple polynomial is plug in the number that the variable is approaching (the little number at the end of the arrow) and see what the answer is. Naturally, the process can get messier—especially if x approaches zero.

Example 4: Find $\lim_{x \to 0} \dfrac{1}{x^2}$.

If you plug in some very small values for x, you'll see that this function approaches ∞. And it doesn't matter whether x is positive or negative, you still get ∞. Look at the graph of $y = \dfrac{1}{x^2}$:

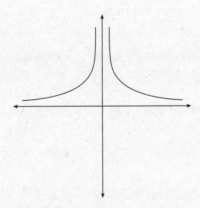

On either side of $x = 0$ (the y-axis), the curve approaches ∞.

Example 5: Find $\lim\limits_{x \to 0} \dfrac{1}{x}$.

Here you've got a problem. If you plug in some very small positive values for x (0.1, 0.01, 0.001, etc.), you approach ∞. In other words, $\lim\limits_{x \to 0^+} \dfrac{1}{x} = \infty$. But, if you plug in some very small negative values for x (–0.1, –0.01, –0.001, etc.) you approach $-\infty$. That is, $\lim\limits_{x \to 0^-} \dfrac{1}{x} = -\infty$. Because the right-hand limit is not equal to the left-hand limit, this limit does not exist.

Look at the graph of $\dfrac{1}{x}$:

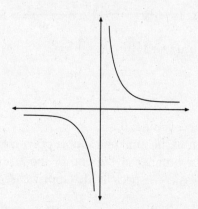

You can see that on the left side of $x = 0$, the curve approaches $-\infty$ (it heads downward forever), and on the right side of $x = 0$, the curve approaches ∞ (it heads upward forever). There are some very important points that we need to emphasize from the last two examples.

(1) If the left-hand limit of a function is not equal to the right-hand limit of the function, then the limit does not exist.

(2) A limit equal to infinity is not the same as a limit that does not exist.

(3) If k is a positive constant then $\lim\limits_{x \to 0^+} \dfrac{k}{x} = \infty$, $\lim\limits_{x \to 0^-} \dfrac{k}{x} = -\infty$, and $\lim\limits_{x \to 0} \dfrac{k}{x}$ does not exist.

(4) If k is a positive constant then $\lim\limits_{x \to 0^+} \dfrac{k}{x^2} = \infty$, $\lim\limits_{x \to 0^-} \dfrac{k}{x^2} = \infty$, and $\lim\limits_{x \to 0} \dfrac{k}{x^2} = \infty$.

Why does the limit exist in Example 4 but not for Example 5? Because when we have $\dfrac{k}{x^2}$, the function is always positive no matter what the sign of x is and thus the function has the same limit from the left and the right. But when we have $\dfrac{k}{x}$, the function's sign depends on the sign of x, and you get a different limit from each side.

Let's look at a few examples when the variable approaches infinity.

Example 6: Find $\displaystyle\lim_{x\to\infty}\frac{1}{x}$.

As x gets bigger and bigger, the value of the function gets smaller and smaller. Therefore, $\displaystyle\lim_{x\to\infty}\frac{1}{x}=0$.

Example 7: Find $\displaystyle\lim_{x\to-\infty}\frac{1}{x}$.

It's the same situation as the one in Example 6; as x decreases (it gets more negative), the value of the function also decreases. We write this:

$$\lim_{x\to-\infty}\frac{1}{x}=0$$

We don't have the same problem here that we did when x approached zero because "positive zero" is the same thing as "negative zero," whereas positive infinity is different from negative infinity.

Here's another rule:

If k and n are constants, $|x|>1$, and $n>0$ then $\displaystyle\lim_{x\to\infty}\frac{k}{x^n}=0$, and $\displaystyle\lim_{x\to-\infty}\frac{k}{x^n}=0$.

Example 8: Find $\displaystyle\lim_{x\to\infty}\frac{3x+5}{7x-2}$.

When you've got variables in both the top and the bottom, you can't just plug infinity into the expression. You'll get $\dfrac{\infty}{\infty}$. We solve this by using the following technique:

When an expression consists of a polynomial divided by another polynomial, divide each term of the numerator and the denominator by the highest power of x that appears in the expression.

The highest power of x in this case is x^1, so we divide every term in the expression (both top and bottom) by x, like so:

$$\lim_{x\to\infty}\frac{3x+5}{7x-2}=\lim_{x\to\infty}\frac{\dfrac{3x}{x}+\dfrac{5}{x}}{\dfrac{7x}{x}-\dfrac{2}{x}}=\lim_{x\to\infty}\frac{3+\dfrac{5}{x}}{7-\dfrac{2}{x}}.$$

Now when we take the limit, the two terms containing x approach zero. We're left with $\dfrac{3}{7}$.

Example 9: Find $\displaystyle\lim_{x\to\infty}\frac{8x^2-4x+1}{16x^2+7x-2}$.

Divide each term by (drum roll...) x^2. You get:

$$\lim_{x\to\infty}\frac{8-\dfrac{4}{x}+\dfrac{1}{x^2}}{16+\dfrac{7}{x}-\dfrac{2}{x^2}}=\frac{8}{16}=\frac{1}{2}.$$

Example 10: Find $\displaystyle\lim_{x\to\infty}\frac{-3x^{10}-70x^5+x^3}{33x^{10}+200x^8-1000x^4}$.

Here, divide each term by x^{10}:

$$\lim_{x\to\infty}\frac{-3x^{10}-70x^5+x^3}{33x^{10}+200x^8-1000x^4}=\lim_{x\to\infty}\frac{-3-\dfrac{70}{x^5}+\dfrac{1}{x^7}}{33+\dfrac{200}{x^2}-\dfrac{1000}{x^6}}=-\frac{3}{33}=-\frac{1}{11}$$

Focus your attention on the highest power of x. The other powers don't matter, because they're all going to disappear. Now we have three new rules for evaluating the limit of a rational expression as x approaches infinity:

> (1) If the highest power of x in a rational expression is in the numerator, then the limit as x approaches infinity is infinity.

Example: $\displaystyle\lim_{x\to\infty}\frac{5x^7-3x}{16x^6-3x^2}=\infty$.

> (2) If the highest power of x in a rational expression is in the denominator, then the limit as x approaches infinity is zero.

Example: $\displaystyle\lim_{x\to\infty}\frac{5x^6-3x}{16x^7-3x^2}=0$

> (3) If the highest power of x in a rational expression is the same in both the numerator and denominator, then the limit as x approaches infinity is the coefficient of the highest term in the numerator divided by the coefficient of the highest term in the denominator.

Example: $\displaystyle\lim_{x\to\infty}\frac{5x^7-3x}{16x^7-3x^2}=\frac{5}{16}$.

LIMITS OF TRIGONOMETRIC FUNCTIONS

At some point during the exam, you'll have to find the limit of certain trig expressions, usually as x approaches either zero or infinity. There are a four standard limits that you should memorize—with those, you can evaluate all of the trigonometric limits that appear on the test. As you'll see throughout this book, calculus requires that you remember all of your trig from previous years. If you're rusty, review the unit on Prerequisite Mathematics.

> Rule No 1: $\displaystyle\lim_{x\to 0}\frac{\sin x}{x}=1$.

This may seem strange, but if you look at the graphs of $f(x) = \sin x$ and $f(x) = x$, they have approximately the same slope near the origin (as x gets closer to zero). Since x and the sine of x are about the same as x approaches zero, their quotient will be very close to one. Furthermore, because $\lim\limits_{x \to 0} \cos x = 1$ (review cosine values if you don't get this!), we know that $\lim\limits_{x \to 0} \tan x = \lim\limits_{x \to 0} \dfrac{\sin x}{\cos x} = 0$.

Rule No. 2: $\lim\limits_{x \to 0} \dfrac{\cos x - 1}{x} = 0.$

Example 11: Find $\lim\limits_{x \to 0} \dfrac{\sin 3x}{x}$.

Use a simple trick: multiply the top and bottom of the fraction by 3. This gives us: $\lim\limits_{x \to 0} \dfrac{3 \sin 3x}{3x}$.

Next, substitute a letter for $3x$; say, a for example. Now, we get the following:

$$\lim\limits_{a \to 0} \frac{3 \sin a}{a} = 3 \lim\limits_{a \to 0} \frac{\sin a}{a} = 3(1) = 3$$

Example 12: Find $\lim\limits_{x \to 0} \dfrac{\sin 5x}{\sin 4x}$.

Now we get a bit more sophisticated. First, divide both the numerator and the denominator by x, like so:

$$\lim\limits_{x \to 0} \frac{\dfrac{\sin 5x}{x}}{\dfrac{\sin 4x}{x}}.$$

Next, multiply the top and bottom of the numerator by 5, and the top and bottom of the denominator by 4, which gives us:

$$\lim\limits_{x \to 0} \frac{\dfrac{5 \sin 5x}{5x}}{\dfrac{4 \sin 4x}{4x}}.$$

From the work we did in Example 11, we can see that this limit is $\dfrac{5}{4}$.

Guess what! You've got two more rules!

Rule No. 3: $\lim\limits_{x \to 0} \dfrac{\sin ax}{x} = a.$

Rule No. 4: $\lim\limits_{x \to 0} \dfrac{\sin ax}{\sin bx} = \dfrac{a}{b}.$

Example 13: Find $\lim\limits_{x \to 0} \dfrac{x^2}{1-\cos^2 x}$.

Using trigonometric identities (remember them?), you can replace $(1-\cos^2 x)$ with $\sin^2 x$:

$$\lim_{x \to 0} \frac{x^2}{1-\cos^2 x} = \lim_{x \to 0} \frac{x^2}{\sin^2 x} = 1$$

Here are other examples for you to try, with answers right beneath them. Give 'em a try, and check your work. (No peeking!)

PROBLEM 1. Find $\lim\limits_{x \to 3} \dfrac{x-3}{x+2}$.

Answer: If you plug in 3 for x, you get $\lim\limits_{x \to 3} \dfrac{x-3}{x+2} = \dfrac{0}{5} = 0$.

PROBLEM 2. Find $\lim\limits_{x \to 3} \dfrac{x+2}{x-3}$.

Answer: The left-hand limit is:

$$\lim_{x \to 3^-} \frac{x+2}{x-3} = -\infty$$

The right-hand limit is:

$$\lim_{x \to 3^+} \frac{x+2}{x-3} = \infty$$

These two limits are not the same. Therefore, the limit does not exist.

PROBLEM 3. Find $\lim\limits_{x \to 3} \dfrac{x+2}{(x-3)^2}$.

Answer: The left-hand limit is:

$$\lim_{x \to 3^-} \frac{x+2}{(x-3)^2} = \infty$$

The right-hand limit is:

$$\lim_{x \to 3^+} \frac{x+2}{(x-3)^2} = \infty$$

These two limits are the same, so the limit is ∞.

PROBLEM 4. Find $\lim\limits_{x \to -4} \dfrac{x^2 + 6x + 8}{x + 4}$.

Answer: If you plug –4 into the top and bottom, you get $\dfrac{0}{0}$. You have to factor the top into

$(x+2)(x+4)$ to get this:

$$\lim_{x \to -4} \frac{(x+2)(x+4)}{(x+4)}$$

Now, it's time to cancel like terms:

$$\lim_{x \to -4} \frac{(x+2)(x+4)}{(x+4)} = \lim_{x \to -4}(x+2) = -2$$

PROBLEM 5. Find $\lim\limits_{x \to \infty} \dfrac{15x^2 - 11x}{22x^2 + 4x}$.

Answer: Divide each term by x^2:

$$\lim_{x \to \infty} \frac{15x^2 - 11x}{22x^2 + 4x} = \lim_{x \to \infty} \frac{15 - \dfrac{11}{x}}{22 + \dfrac{4}{x}} = \frac{15}{22}$$

PROBLEM 6. Find $\lim\limits_{x \to 0} \dfrac{4x}{\tan x}$.

Answer: Replace $\tan x$ with $\dfrac{\sin x}{\cos x}$, which changes the expression into:

$$\lim_{x \to 0} \frac{4x}{\tan x} = \lim_{x \to 0} \frac{4x}{\dfrac{\sin x}{\cos x}} = \lim_{x \to 0} \frac{4x \cos x}{\sin x}$$

Since $\lim\limits_{x \to 0} \dfrac{x}{\sin x} = 1$ and $\lim\limits_{x \to 0} \cos x = 1$, the answer is 4.

Note: Pay careful attention to this next solved problem. It will be very important when you do work in chapter 4.

PROBLEM 7. Find $\lim\limits_{h \to 0} \dfrac{(5+h)^2 - 25}{h}$.

Answer: First, expand and simplify the numerator like this:

$$\lim_{h \to 0} \frac{(5+h)^2 - 25}{h} = \lim_{h \to 0} \frac{25 + 10h + h^2 - 25}{h} = \lim_{h \to 0} \frac{10h + h^2}{h}.$$

Next, factor h out of the numerator and the denominator like this:

$$\lim_{h \to 0} \frac{10h + h^2}{h} = \lim_{h \to 0} \frac{h(10 + h)}{h} = \lim_{h \to 0}(10 + h).$$

Taking the limit you get:

$$\lim_{h \to 0}(10 + h) = 10.$$

THE FORMAL DEFINITION OF A LIMIT

This section is for BC Calculus students only. (You AB students can talk amongst yourselves.) The official definition of a limit is very technical, mostly because it uses a bunch of Greek letters. It's a topic that's only covered in the BC curriculum, and it can be tough to grasp fully. Don't panic if this next concept eludes you, though; you can still get a 5 on the AP exam even if you get this particular question wrong!

Depending on the calculus book that you use, you may see the limit defined with slightly different terminology, but the essential idea is the same.

> **Definition:** The limit of $f(x)$ as x approaches x_0 is the number L if:
>
> Given any $\varepsilon > 0$, no matter how small, about L, there exists some $\delta > 0$ about x_0, such that for all x, $0 < |x - x_0| < \delta$ implies that $0 < |f(x) - L| < \varepsilon$.

We know what you're thinking: Could someone please translate that into English?

From a more simplistic perspective, the definition means this: L is the limit of a function at a value of x if, as x gets very close to x_0, $f(x)$ gets very close to L.

On the AP exam, sometimes all you'll need to know is this definition, so you should memorize the wording and understand the meaning. Sometimes, however, you'll have to work with the definition to prove a limit.

Example 14: Show that $\lim\limits_{x \to 3}(4x - 1) = 11$.

Using the definition, we let $x_0 = 3$, $L = 11$, and $f(x) = 4x - 1$. We need to show that for any $\varepsilon > 0$, there exists a number $\delta > 0$ such that $0 < |x - 3| < \delta$ implies that $0 < |(4x - 1) - 11| < \varepsilon$. If we solve this latter inequality we get:

$$|4x - 12| < \varepsilon$$

$$4|x - 3| < \varepsilon$$

$$|x - 3| < \frac{\varepsilon}{4}$$

This tells us that if $0 < \delta < \dfrac{\varepsilon}{4}$, the limit will hold. Because we can choose such a value of δ, the limit is 11.

Get it? What you're supposed to do with the definition of the limit is find a value of δ in terms of ε. Sometimes, you have to find a numerical value. Then all you do is replace the ε with a value, and solve for δ that way.

Example 15: Find δ such that the limit in the previous example is accurate to within 0.01.

This means that $\varepsilon = 0.01$. Then you have $0 < |(4x-1)-11| < 0.01$. Simplifying as you did above, you get:

$$|x-3| < \frac{0.01}{4} \text{ or } |x-3| < 0.0025.$$

Thus, any $0 < \delta < 0.0025$ gives us the requisite accuracy.

These are the only kinds of limit problems that you can expect from the BC exam, so we won't torture you any further.

PRACTICE PROBLEMS

Try these 30 problems to test your skill with limits. The answers are in chapter 21.

1. $\lim\limits_{x \to 8}\left(x^2 - 5x - 11\right) =$

2. $\lim\limits_{x \to 5}\left(\dfrac{x+3}{x^2 - 15}\right) =$

3. $\lim\limits_{x \to 0} \pi^2 =$

4. $\lim\limits_{x \to 3}\left(\dfrac{x^2 - 2x - 3}{x - 3}\right) =$

5. $\lim\limits_{x \to \infty}\left(\dfrac{10x^2 + 25x + 1}{x^4 - 8}\right) =$

6. $\lim\limits_{x \to \infty}\left(\dfrac{x^4 - 8}{10x^2 + 25x + 1}\right) =$

7. $\lim\limits_{x \to \infty}\left(\dfrac{x^4 - 8}{10x^4 + 25x + 1}\right) =$

8. $\lim\limits_{x \to \infty} \left(\dfrac{\sqrt{5x^4 + 2x}}{x^2} \right) =$

9. $\lim\limits_{x \to 6^+} \left(\dfrac{x + 2}{x^2 - 4x - 12} \right) =$

10. $\lim\limits_{x \to 6^-} \left(\dfrac{x + 2}{x^2 - 4x - 12} \right) =$

11. $\lim\limits_{x \to 6} \left(\dfrac{x + 2}{x^2 - 4x - 12} \right) =$

12. $\lim\limits_{x \to 0^+} \left(\dfrac{x}{|x|} \right) =$

13. $\lim\limits_{x \to 0^-} \left(\dfrac{x}{|x|} \right) =$

14. $\lim\limits_{x \to 7^+} \left(\dfrac{x}{x^2 - 49} \right) =$

15. $\lim\limits_{x \to 7} \left(\dfrac{x}{x^2 - 49} \right) =$

16. $\lim\limits_{x \to 7} \dfrac{x}{(x - 7)^2} =$

17. Let $f(x) = \begin{cases} x^2 - 5, x \le 3 \\ x + 2, x > 3 \end{cases}$.

    Find: (a) $\lim\limits_{x \to 3^-} f(x)$; (b) $\lim\limits_{x \to 3^+} f(x)$; and (c) $\lim\limits_{x \to 3} f(x)$

18. Let $f(x) = \begin{cases} x^2 - 5, x \le 3 \\ x + 1, x > 3 \end{cases}$.

    Find: (a) $\lim\limits_{x \to 3^-} f(x)$; (b) $\lim\limits_{x \to 3^+} f(x)$; and (c) $\lim\limits_{x \to 3} f(x)$

19. Find $\lim\limits_{x \to \frac{\pi}{4}} 3 \cos x$.

20. Find $\lim\limits_{x \to 0} 3\dfrac{x}{\cos x}$.

21. Find $\lim\limits_{x \to 0} 3\dfrac{x}{\sin x}$.

22. Find $\lim\limits_{x \to 0} \dfrac{\sin 3x}{\sin 8x}$.

23. Find $\lim\limits_{x \to 0} \dfrac{\tan 7x}{\sin 5x}$.

24. Find $\lim\limits_{x \to \infty} \sin x$.

25. Find $\lim\limits_{x \to \infty} \sin \dfrac{1}{x}$.

26. Find $\lim\limits_{x \to 0} \dfrac{x^2 \sin x}{1 - \cos^2 x}$.

27. Find $\lim\limits_{x \to 0} \dfrac{\sin^2 7x}{\sin^2 11x}$.

28. Find $\lim\limits_{h \to 0} \dfrac{(3+h)^2 - 9}{h}$.

29. Find $\lim\limits_{h \to 0} \dfrac{\sin(x+h) - \sin x}{h}$.

30. Find $\lim\limits_{h \to 0} \dfrac{\dfrac{1}{x+h} - \dfrac{1}{x}}{h}$.

2

CONTINUITY

Every AP exam has a few questions on continuity, so it's important to understand the basic idea of what it means for a function to be continuous. The concept is very simple: If the graph of the function doesn't have any breaks or holes in it within a certain interval, the function is continuous over that interval.

In general, simple polynomials are continuous everywhere; it's the wacky ones—trigonometric, exponential, rational, piecewise—that have continuity problems. Most of the test questions concern these last types of functions. In order to learn how to test whether a function is continuous, you'll need some more mathematical terminology.

THE DEFINITION OF CONTINUITY

In order for a function $f(x)$ to be continuous at a point $x = c$, it must fulfill *all three* of the following conditions:

Condition 1: $f(c)$ exists.

Condition 2: $\lim\limits_{x \to c} f(x)$ exists.

Condition 3: $\lim\limits_{x \to c} f(x) = f(c)$.

Let's look at a simple example of a continuous function. (Incidentally, you'll find that these functions are continuous almost everywhere, and the only possible difficulty will occur at a few specific values of x.)

Example 1: Is the function $f(x) = \begin{cases} x+1, & x < 2 \\ 2x-1, & x \geq 2 \end{cases}$ continuous at the point $x = 2$?

Condition 1: Does $f(2)$ exist?
Yup. It's equal to 2(2)-1=3.

Condition 2: Does $\lim\limits_{x \to 2} f(x)$ exist?

You need to look at the limit from both sides of 2. The left-hand limit is:

$$\lim_{x \to 2^-} f(x) = 2 + 1 = 3.$$

The right-hand limit is:

$$\lim_{x \to 2^+} f(x) = 2(2) - 1 = 3.$$

Because the two limits are the same, the limit exists.

Condition 3: Does $\lim\limits_{x \to 2} f(x) = f(2)$?
The two limits do equal each other, so yes; the function is continuous at $x = 2$.

A simple and important way to check whether a function is continuous is to sketch the function. If you can't sketch the function without lifting your pencil from the paper at some point, then the function is not continuous.

Now let's look at some examples of functions that are not continuous.

Example 2: Is the function $f(x) = \begin{cases} x+1, & x < 2 \\ 2x-1, & x > 2 \end{cases}$ continuous at $x = 2$?

Condition 1: Does $f(2)$ exist?

Nope. The function of x is defined if x is greater than or less than 2, but not if x is equal to 2. Therefore, the function is not continuous at $x = 2$. Notice that we don't have to bother with the other two conditions. Once you find a problem, the function is automatically not continuous, and you can stop.

Example 3: Is the function $f(x) = \begin{cases} x+1, & x < 2 \\ 2x+1, & x \geq 2 \end{cases}$ continuous at $x = 2$?

Condition 1: Does $f(x)$ exist?
Yes. It is equal to $2(2) + 1 = 5$.

Condition 2: Does $\lim\limits_{x \to 2} f(x)$ exist?
The left-hand limit is:

$$\lim_{x \to 2^-} f(x) = 2 + 1 = 3.$$

The right-hand limit is:

$$\lim_{x \to 2^+} f(x) = 2(2) + 1 = 5.$$

The two limits don't match, so the limit doesn't exist and the function is not continuous at $x = 2$.

Example 4: Is the function $f(x) = \begin{cases} x+1, & x < 2 \\ x^2, & x = 2 \\ 2x-1, & x > 2 \end{cases}$ continuous at $x = 2$?

Condition 1: Does $f(2)$ exist?
Yes. It's equal to $2^2 = 4$.

Condition 2: Does $\lim\limits_{x \to 2} f(x)$ exist?
The left-hand limit is:

$$\lim_{x \to 2^-} f(x) = 2 + 1 = 3.$$

The right-hand limit is:

$$\lim_{x \to 2^+} f(x) = 2(2) - 1 = 3.$$

Because the two limits are the same, the limit exists.
Condition 3: Does $\lim\limits_{x \to 2} f(x) = f(2)$?

Houston, we have a problem. The $\lim\limits_{x \to 2} f(x) = 3$, but $f(2) = 4$. Because these aren't equal, the answer is "no" and the function is not continuous at $x = 2$.

TYPES OF DISCONTINUITIES

There are four types of discontinuities you have to know: jump, point, essential, and removable.

A **jump** discontinuity occurs when the curve "breaks" at a particular place and starts somewhere else. In other words, $\lim\limits_{x \to a^-} f(x) \neq \lim\limits_{x \to a^+} f(x)$.

A sample looks like this:

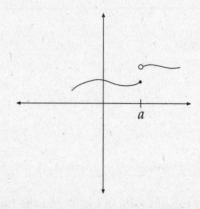

A **point** discontinuity occurs when the curve has a "hole" in it from a missing point because the function has a value at that point that's "off the curve". In other words, $\lim\limits_{x \to a} f(x) \neq f(a)$.

Here's what a point discontinuity looks like:

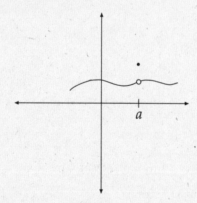

An **essential** discontinuity occurs when the curve has a vertical asymptote.

Like so:

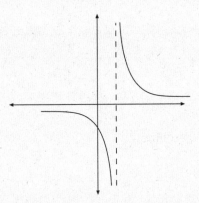

A **removable** discontinuity occurs when you have a rational expression with common factors in the numerator and denominator. Because these factors can be canceled, the discontinuity is "removable".

Here's an example:

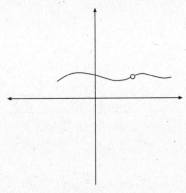

This curve looks very similar to a point discontinuity, but notice that with a removable discontinuity, $f(x)$ is not defined at the point. Whereas with a point discontinuity, $f(x)$ is defined there.

Now that you know what these four types of discontinuities look like, let's see what types of functions suffer from discontinuity.

Example 5: Consider the function:

$$f(x) = \begin{cases} x+3, & x \le 2 \\ x^2, & x > 2 \end{cases}$$

The left-hand limit is 5 as x approaches 2, and the right-hand limit is 4 as x approaches 2. Because the curve has different values on each side of 2, the curve is discontinuous at $x = 2$. We say that the curve "jumps" at $x = 2$ from the left-hand curve to the right-hand curve because the left and right-hand limits differ. It looks like this:

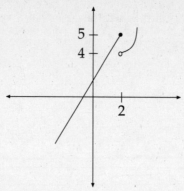

This is an example of a jump discontinuity.

Example 6: Consider the function:

$$f(x) = \begin{cases} x^2, & x \neq 2 \\ 5, & x = 2 \end{cases}.$$

This function violates the third condition, because $\lim\limits_{x \to 2} f(x) \neq f(2)$; the function is discontinuous at $x = 2$. The curve is continuous everywhere except at the point $x = 2$. It looks like this:

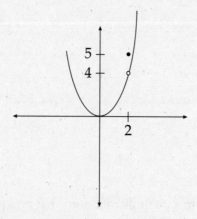

This is an example of a point discontinuity.

Example 7: Consider the function: $f(x) = \dfrac{5}{x-2}$.

The function is discontinuous because it's possible for the denominator to equal zero (at $x = 2$). This means that $f(2)$ doesn't exist, and the function has an asymptote at $x = 2$. In addition, $\lim\limits_{x \to 2^-} f(x) = -\infty$ and $\lim\limits_{x \to 2^+} f(x) = \infty$. The graph looks like this:

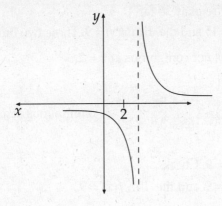

This is an example of an essential discontinuity.

Example 8: Consider the function:

$$f(x) = \frac{x^2 - 8x + 15}{x^2 - 6x + 5}.$$

If you factor the top and bottom, you can see where the discontinuities are:

$$f(x) = \frac{x^2 - 8x + 15}{x^2 - 6x + 5} = \frac{(x-3)(x-5)}{(x-1)(x-5)}.$$

The function has a zero in the denominator when $x = 1$ or $x = 5$, so the function is discontinuous at those two points. But, you can cancel the term $(x - 5)$ from both the numerator and the denominator, leaving you with:

$$f(x) = \frac{x-3}{x-1}.$$

Now, the reduced function *is* continuous at $x = 5$. Thus the original function has a removable discontinuity at $x = 5$. Furthermore, if you now plug $x = 5$ into the reduced function, you get:

$$f(5) = \frac{2}{4} = \frac{1}{2}.$$

The discontinuity is at $x = 5$, and there's a hole at $\left(5, \frac{1}{2}\right)$. In other words, if the original function were continuous at $x = 5$, it would have the value $\frac{1}{2}$. Notice that this is the same as: $\lim\limits_{x \to 5} f(x)$.

These are the types of discontinuities that you can expect to encounter on the AP examination. Here are some sample problems and their solutions. Cover the answers as you work, then check your results.

PROBLEM 1. Is the function $f(x) = \begin{cases} 2x^3 - 1, & x < 2 \\ 6x - 3, & x \geq 2 \end{cases}$ continuous at $x = 2$?

Answer: Test the conditions necessary for continuity.

Condition 1: $f(2) = 9$, so we're okay so far.

Condition 2: The $\lim_{x \to 2^-} f(x) = 15$ and the $\lim_{x \to 2^+} f(x) = 9$. These two limits don't agree, so the $\lim_{x \to 2} f(x)$ doesn't exist and the function is not continuous at $x = 2$.

PROBLEM 2. Is the function $f(x) = \begin{cases} x^2 + 3x + 5, & x < 1 \\ 6x + 3, & x \geq 1 \end{cases}$ continuous at $x = 1$?

Answer: Condition 1: $f(1) = 9$. Check.

Condition 2: The $\lim_{x \to 1^-} f(x) = 9$ and the $\lim_{x \to 1^+} f(x) = 9$.

Therefore, the $\lim_{x \to 1} f(x)$ exists and is equal to 9.

Condition 3: $\lim_{x \to 1} f(x) = f(1) = 9$.

The function satisfies all three conditions so it is continuous at $x = 1$.

PROBLEM 3. For what value of a is the function $f(x) = \begin{cases} ax + 5, & x < 4 \\ x^2 - x, & x \geq 4 \end{cases}$ continuous at $x = 4$?

Answer: Since $f(4) = 12$, the function passes the first condition.

For Condition 2 to be satisfied, the $\lim_{x \to 4^-} f(x) = 4a + 5$ must equal the $\lim_{x \to 4^+} f(x) = 12$. So, set $4a + 5 = 12$. If $a = \dfrac{7}{4}$, the limit will exist at $x = 4$ and the other two conditions will also be fulfilled.

Therefore, the value $a = \dfrac{7}{4}$ makes the function continuous at $x = 4$.

PROBLEM 4. Where does the function

$$f(x) = \frac{2x^2 - 7x - 15}{x^2 - x - 20}$$

have: (a) an essential discontinuity; and (b) a removable discontinuity?

Answer: If you factor the top and bottom of this fraction, you get:

$$f(x) = \frac{2x^2 - 7x - 15}{x^2 - x - 20} = \frac{(2x + 3)(x - 5)}{(x + 4)(x - 5)}$$

Thus, the function has an essential discontinuity at $x = -4$. If we then cancel the term $(x - 5)$, and substitute $x = 5$ into the reduced expression, we get $f(5) = \dfrac{13}{9}$. Therefore, the function has a removable discontinuity at $\left(5, \dfrac{13}{9}\right)$.

PRACTICE PROBLEMS

Now try these problems. The answers are in chapter 21.

1. Is the function $f(x) = \begin{cases} x+7, & x<2 \\ 9, & x=2 \\ 3x+3, & x>2 \end{cases}$ continuous at $x = 2$?

2. Is the function $f(x) = \begin{cases} 4x^2-2x, & x<3 \\ 10x-1, & x=3 \\ 30, & x>3 \end{cases}$ continuous at $x = 3$?

3. Is the function $f(x) = \begin{cases} 5x+7, & x<3 \\ 7x+1, & x>3 \end{cases}$ continuous at $x = 3$?

4. Is the function $f(x) = \sec x$ continuous everywhere?

5. Is the function $f(x) = \sec x$ continuous on the interval $\left[-\dfrac{\pi}{2}, \dfrac{\pi}{2}\right]$?

6. Is the function $f(x) = \sec x$ continuous on the interval $\left(-\dfrac{\pi}{2}, \dfrac{\pi}{2}\right)$?

7. For what value(s) of k is the function $f(x) = \begin{cases} 3x^2-11x-4, & x \le 4 \\ kx^2-2x-1, & x>4 \end{cases}$ continuous at $x = 4$?

8. For what value(s) of k is the function $f(x) = \begin{cases} -6x-12, & x<-3 \\ k^2-5k, & x=-3 \\ 6, & x>-3 \end{cases}$ continuous at $x = -3$?

9. At what point is the removable discontinuity for the function $f(x) = \dfrac{x^2+5x-24}{x^2-x-6}$?

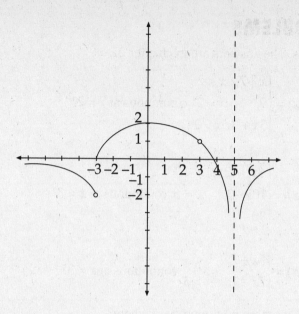

10. Given the graph of $f(x)$ above, find:

    (a) $\lim\limits_{x \to -\infty} f(x)$

    (b) $\lim\limits_{x \to \infty} f(x)$

    (c) $\lim\limits_{x \to 3^-} f(x)$

    (d) $\lim\limits_{x \to 3^+} f(x)$

    (e) $f(3)$

    (f)   Any other discontinuities.

3

THE DEFINITION OF THE DERIVATIVE

The main tool that you'll use in differential calculus is called **the derivative**. All of the problems that you'll encounter in differential calculus make use of the derivative, so your goal should be to become an expert at finding, or "taking", derivatives by the end of chapter 4. However, before you learn a simple way to take a derivative, your teacher will probably make you learn how derivatives are calculated by teaching you something called the "Definition of the Derivative."

DERIVING THE FORMULA

The best way to understand the definition of the derivative is to start by looking at the simplest continuous function: a line. As you should recall, you can determine the slope of a line by taking two points on that line and plugging them into the slope formula:

$$m = \frac{y_2 - y_1}{x_2 - x_1}$$ *m* stands for slope.

For example, suppose a line goes through the points (3,7) and (8,22). First, you subtract the *y*-coordinates (22 – 7) = 15. Next, subtract the corresponding *x*-coordinates (8 – 3) = 5. Finally, divide the first number by the second: $\frac{15}{5} = 3$. The result is the slope of the line: *m* = 3.

Notice that you can use the coordinates in reverse order and still get the same result. It doesn't matter in which order you do the subtraction as long as you're consistent (and you remember to put the *y*-coordinates in the numerator and the *x*-coordinates in the denominator).

Let's look at the graph of that line. The slope measures the steepness of the line, which looks like this:

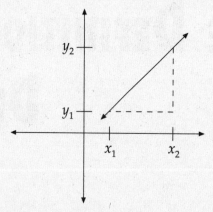

You probably remember your teachers referring to the slope as the "rise" over the "run." The rise is the difference between the *y*-coordinates (remember: "the rise is the *y*'s"), and the run is the difference between the *x*-coordinates. The slope is the ratio of the two.

Now for a few changes in notation. Instead of calling the *x*-coordinates x_1 and x_2, we're going to call them x_1 and $x_1 + h$, where *h* is the difference between the two *x*-coordinates. (Sometimes, instead of *h*, some books use Δx. Why? For the heck of it.) Second, instead of using y_1 and y_2, we use $f(x_1)$ and $f(x_1 + h)$. So now the graph looks like this:

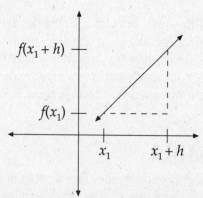

The picture is exactly the same—only the notation has changed.

THE SLOPE OF A CURVE

Suppose that instead of finding the slope of a line, we wanted to find the slope of a curve. Here, the slope formula no longer works because the distance from one point to the other is curved, not straight. But we could find the approximate slope if we took the slope of the line between the two points. This is called the secant line.

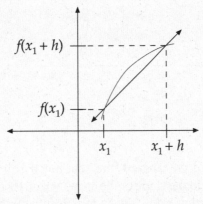

The equation for the slope of the secant line is:

$$\frac{f(x_1 + h) - f(x_1)}{h}.$$

Remember this formula!!

THE SECANT AND THE TANGENT

As you can see, the farther apart the two points are, the less the slope of the line corresponds to the slope of the curve.

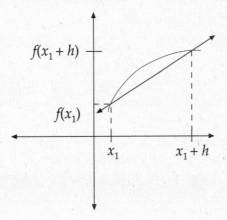

Conversely, the closer the two points are, the more accurate the approximation is.

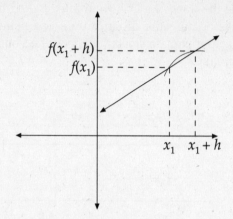

In fact, there is one line, called the **tangent line**, that touches the curve at exactly one point. The slope of the tangent line is equal to the slope of the curve at exactly this point. The object of using the above formula, therefore, is to shrink h down to an infinitesimally small amount. If we could do that, then the difference between $(x_1 + h)$ and x_1 would be a point. Graphically, it looks like this:

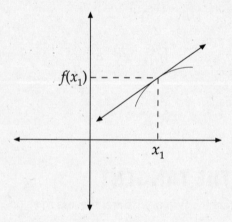

How do we perform this shrinking act? By using the limits we discussed in chapter 1. We set up a limit during which h approaches zero, like so:

$$\lim_{h \to 0} \frac{f(x_1 + h) - f(x_1)}{h}.$$

This is the definition of the derivative.

Notice that the equation is just a slightly modified version of the slope formula, with different notation. The only difference is that we're finding the slope between two points that are almost exactly next to each other.

Example 1: Find the slope of the curve $f(x) = y = x^2$ at the point (2,4).

This means that $x_1 = 2$ and $f(2) = 2^2 = 4$. If we can figure out $f(x_1 + h)$, then we can find the slope. Well, how did we find the value of $f(x)$? We plugged x_1 into the equation $f(x) = x^2$. To find $f(x_1 + h)$ we plug $x_1 + h$ into the equation, which now looks like this:

$$f(x_1 + h) = (2 + h)^2 = 4 + 4h + h^2.$$

Now plug this into the slope formula:

$$\lim_{h \to 0} \frac{f(x_1 + h) - f(x_1)}{h} = \lim_{h \to 0} \frac{4 + 4h + h^2 - 4}{h} = \lim_{h \to 0} \frac{4h + h^2}{h}.$$

Next, simplify by factoring h out of the top:

$$\lim_{h \to 0} \frac{4h + h^2}{h} = \lim_{h \to 0} \frac{h(4 + h)}{h} = \lim_{h \to 0} 4 + h.$$

Taking the limit as h approaches 0, we get 4. Therefore, the slope of the curve $y = x^2$ at the point (2,4) is 4. Now we've found the slope of a curve at a certain point, and the notation looks like this: $f'(2) = 4$. Remember this notation!

Example 2: Find the derivative of the equation in Example 1 at the point (5,25). This means that $x_1 = 5$ and $f(x) = 25$. This time,

$$(x_1 + h)^2 = (5 + h)^2 = 25 + 10h + h^2.$$

Now plug this into the formula for the derivative:

$$\lim_{h \to 0} \frac{f(x_1 + h) - f(x_1)}{h} = \lim_{h \to 0} \frac{25 + 10h + h^2 - 25}{h} = \lim_{h \to 0} \frac{10h + h^2}{h}$$

Once again, simplify by factoring h out of the top:

$$\lim_{h \to 0} \frac{10h + h^2}{h} = \lim_{h \to 0} \frac{h(10 + h)}{h} = \lim_{h \to 0} 10 + h.$$

Taking the limit as h goes to 0, you get 10. Therefore, the slope of the curve $y = x^2$ at the point (5,25) is 10, or: $f'(5) = 10$.

Using this pattern, let's forget about the arithmetic for a second and derive a formula.

Example 3: Find the slope of the equation $f(x) = x^2$ at the point $\left(x_1, x_1^2\right)$.

Follow the steps in the last two problems, but instead of using a number, use x_1. This means that $f(x_1) = x_1^2$ and $(x_1 + h)^2 = x_1^2 + 2x_1h + h^2$. Then the derivative is:

$$\lim_{h \to 0} \frac{x_1^2 + 2x_1h + h^2 - x_1^2}{h} = \lim_{h \to 0} \frac{2x_1h + h^2}{h}.$$

Factor h out of the top:

$$\lim_{h \to 0} \frac{h(2x_1 + h)}{h} = \lim_{h \to 0} 2x_1 + h.$$

Now take the limit as h goes to 0: you get $2x_1$. Therefore, $f'(x_1) = 2x_1$.

This example gives us a general formula for the derivative of this curve. Now we can pick any point, plug it into the formula, and determine the slope at that point. For example, the derivative at the point $x = 7$ is 14. At the point $x = \frac{7}{3}$, the derivative is $\frac{14}{3}$.

Easy, right?

DIFFERENTIABILITY

One of the important requirements for the differentiability of a function is that the function be continuous. Furthermore, even if a function is continuous at a point, the function is not necessarily differentiable there. Check out the graph below:

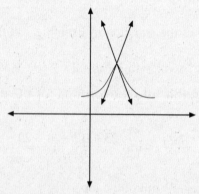

If a function has a cusp or a "sharp corner," you can draw more than one tangent line at that point. Therefore, the function is not differentiable there.

Another possible problem point occurs when the tangent line is vertical, because a vertical line has an infinite slope. For example, if the derivative of a function is $\frac{1}{x+1}$, it doesn't have a derivative at $x = -1$.

Try these problems on your own, then check your work against the right answers immediately beneath each problem.

PROBLEM 1. Find the derivative of $f(x) = 3x^2$ at (4,48).

Answer: $f(4 + h) = 3(4 + h)^2 = 48 + 24h + 3h^2$. Use the definition of the derivative

$$f'(4) = \lim_{h \to 0} \frac{48 + 24h + 3h^2 - 48}{h}.$$

Simplify:

$$\lim_{h \to 0} \frac{24h + 3h^2}{h} = \lim_{h \to 0} 24 + 3h = 24.$$

The slope of the curve at the point (4,48) is 24.

PROBLEM 2. Find the derivative of $f(x) = 3x^2$.

Answer: Since $f(x + h) = 3(x + h)^2 = 3x^2 + 6xh + 3h^2$, you can use the definition of the derivative:

$$f'(x) = \lim_{h \to 0} \frac{3x^2 + 6xh + 3h^2 - 3x^2}{h}.$$

Then simplify:

$$\lim_{h \to 0} \frac{6xh + 3h^2}{h} = \lim_{h \to 0} 6x + 3h = 6x. \text{ The derivative is } 6x.$$

PROBLEM 3. Find the derivative of $f(x) = x^3$.

Answer: $f(x + h) = (x + h)^3 = x^3 + 3x^2h + 3xh^2 + h^3$. By now, you should get the idea. First, use the definition of the derivative:

$$f'(x) = \lim_{h \to 0} \frac{x^3 + 3x^2h + 3xh^2 + h^3 - x^3}{h}.$$

And simplify:

$$\lim_{h \to 0} \frac{3x^2h + 3xh^2 + h^3}{h} = \lim_{h \to 0} 3x^2 + 3xh + h^2 = 3x^2.$$

The derivative is $3x^2$.
This next one will test your algebraic skills. Don't say we didn't warn you.

PROBLEM 4. Find the derivative of $f(x) = \sqrt{x}$.

Answer: $f(x + h) = \sqrt{x + h}$.
Step 1: Use the definition of the derivative:

$$f'(x) = \lim_{h \to 0} \frac{\sqrt{x + h} - \sqrt{x}}{h}.$$

Notice that this one doesn't cancel as conveniently as the other problems did. In order to simplify this expression, we have to multiply both the top and the bottom of the expression by $\sqrt{x+h}+\sqrt{x}$. (Your teacher might refer to this as the conjugate of the numerator.):

$$f'(x) = \lim_{h \to 0} \frac{\sqrt{x+h}-\sqrt{x}}{h}\left(\frac{\sqrt{x+h}+\sqrt{x}}{\sqrt{x+h}+\sqrt{x}}\right) = \lim_{h \to 0} \frac{x+h-x}{h\left(\sqrt{x+h}+\sqrt{x}\right)} = \lim_{h \to 0} \frac{h}{h\left(\sqrt{x+h}+\sqrt{x}\right)}$$

Simplify:

$$\lim_{h \to 0} \frac{1}{\left(\sqrt{x+h}+\sqrt{x}\right)} = \frac{1}{2\sqrt{x}}$$

The derivative is $\dfrac{1}{2\sqrt{x}}$.

Now, try a little trig.

PROBLEM 5. Find the derivative of $f(x) = \sin x$.

Answer: $f(x + h) = \sin(x + h) = \sin x \cos h + \cos x \sin h$.

Note: Don't confuse the sine of h with hyperbolic sine, which is abbreviated "sin$h$." The same is true for "cos$h$". The AP doesn't cover hyperbolic functions anyway, so you don't need to worry about them. Also, if you didn't remember the formula for the sine of the sum of two angles, review it!

Using the definition, you get this:

$$f'(x) = \lim_{h \to 0} \frac{\sin x \cos h + \cos x \sin h - \sin x}{h}.$$

Then the math gets icky. First, rearrange the numerator:

$$f'(x) = \lim_{h \to 0} \frac{\sin x(\cos h - 1) + \cos x \sin h}{h}.$$

Next, split the expression into two fractions:

$$\lim_{h \to 0} \frac{\sin x(\cos h - 1)}{h} + \lim_{h \to 0} \frac{\cos x \sin h}{h}.$$

As you learned in chapter 2, $\lim\limits_{h \to 0} \dfrac{\cos h - 1}{h} = 0$ and $\lim\limits_{x \to o} \dfrac{\sin h}{h} = 1$. Therefore, we now have:

$$\lim_{h \to 0} \sin x(0) + \lim_{h \to 0} \cos x(1) = \cos x.$$

The derivative of $\sin x$ is $\cos x$.

PRACTICE PROBLEMS

Now find the derivative of the following expressions. The answers are in chapter 21.

1. $f(x) = 5x$ at $x = 3$

2. $f(x) = 4x$ at $x = -8$

3. $f(x) = 2x^2$ at $x = 5$

4. $f(x) = 5x^2$ at $x = -1$

5. $f(x) = 8x^2$

6. $f(x) = -10x^2$

7. $f(x) = 20x^2$ at $x = a$

8. $f(x) = 2x^3$ at $x = -3$

9. $f(x) = -3x^3$

10. $f(x) = x^4$

11. $f(x) = x^5$

12. $f(x) = 2\sqrt{x}$ at $x = 9$

13. $f(x) = 5\sqrt{2x}$ at $x = 8$

14. $f(x) = \sin x$ at $x = \dfrac{\pi}{3}$

15. $f(x) = \cos x$

16. $f(x) = x^2 + x$

17. $f(x) = x^3 + 3x + 2$

18. $f(x) = \dfrac{1}{x}$

19. $f(x) = ax^2 + bx + c$

20. $f(x) = \dfrac{1}{x^2}$

4

BASIC DIFFERENTIATION

In calculus, you're asked to do two very basic things: differentiate and integrate. In this section, you're going to learn differentiation. Integration will come later, in the second half of this book. Before we get about the business of learning how to take derivatives, however, here's a brief note about notation. Read this!

NOTATION

There are several different notations for derivatives in calculus. We'll use two different types interchangeably throughout this book, so get used to them now.

We'll refer to functions three different ways: $f(x)$, u, and y. For example, we might write: $f(x) = x^3$, $g(x) = x^4$, $h(x) = x^5$. Or we might use: $y = \sqrt{x}$. We'll also use notation like: $u = \sin x$ and $v = \cos x$. Usually, we pick the notation that causes the least confusion.

The derivatives of the functions will use notation that depends on the function. In other words:

Function	First Derivative	Second Derivative
$f(x)$	$f'(x)$	$f''(x)$
$g(x)$	$g'(x)$	$g''(x)$
y	y' or $\dfrac{dy}{dx}$	y'' or $\dfrac{d^2y}{dx^2}$

In addition, if we take the derivative of a function in general (for example, $ax^2 + bx + c$), we might enclose the expression in parentheses and use either of the following notations:

$$\left(ax^2 + bx + c\right)', \text{ or } \frac{d}{dx}\left(ax^2 + bx + c\right)$$

Sometimes math books refer to a derivative using either D_x or f_x. We're not going to use either of them.

THE POWER RULE

In the last chapter, you learned how to find a derivative using the definition of the derivative, a process that is very time-consuming and sometimes involves a lot of complex algebra. Fortunately, there's a shortcut to taking derivatives, so you'll never have to use the definition again—except when it's a question on an exam!

The basic technique for taking a derivative is called **the Power Rule**.

> Rule No. 1: If $y = x^n$, then $\dfrac{dy}{dx} = nx^{n-1}$.

That's it. Wasn't that simple? Of course, this and all of the following rules can be derived easily from the definition of the derivative. We're not going to spend your valuable time deriving any of them. (You're welcome.) We just want you to be able to use them.

Look at these next few examples of the Power Rule in action:

Example 1: If $y = x^5$, then $\dfrac{dy}{dx} = 5x^4$.

Example 2: If $y = x^{20}$, then $\dfrac{dy}{dx} = 20x^{19}$.

Notice that when the power of the function is negative, the power of the derivative is more negative. Don't go in the wrong direction.

Example 3: If $f(x) = x^{-5}$, then $f'(x) = -5x^{-6}$.

When the power is a fraction, you should be careful to get the subtraction right (you'll see the powers $\frac{1}{2}$, $\frac{1}{3}$, $\frac{3}{2}$, $-\frac{1}{2}$, and $-\frac{1}{3}$ often, so be comfortable with subtracting 1 from them).

Example 4: If $u = x^{\frac{1}{2}}$, then $\frac{du}{dx} = \frac{1}{2}x^{-\frac{1}{2}}$.

When the power is 1, the derivative is just a constant.

Example 5: If $y = x^1$, then $\frac{dy}{dx} = 1x^0 = 1$. (Because x^0 is 1 !)

When the power is 0 (that is, there is no power of x), the derivative is 0.

Example 6: If $y = x^0$, then $\frac{dy}{dx} = 0$. (Because multiplying by zero gives you zero!)

This leads to the next three rules:

> Rule No. 2: If $y = x$, then $\frac{dy}{dx} = 1$
>
> Rule No. 3: If $y = \frac{dy}{dx} = k$ (where k is a constant)
>
> Rule No. 4: If $y = k$, then $\frac{dy}{dx} = 0$ (where k is a constant)

Note: For future reference, a, b, c, n, and k always stand for constants.

These four rules are the bread and butter of derivatives. You must be able to do these in your sleep! (Impress your friends at parties!)

Example 7: If $y = 8x^4$, then $\frac{dy}{dx} = 32x^3$.

Example 8: If $y = 5x^{100}$, then $y' = 500x^{99}$.

Example 9: If $y = -3x^{-5}$, then $\dfrac{dy}{dx} = 15x^{-6}$.

Example 10: If $f(x) = 7x^{\frac{1}{2}}$, then $f'(x) = \dfrac{7}{2}x^{-\frac{1}{2}}$.

Example 11: If $y = x\sqrt{15}$, then $\dfrac{dy}{dx} = \sqrt{15}$.

Example 12: If $y = 12$, then $\dfrac{dy}{dx} = 0$.

If you have any questions about any of these 12 examples (especially the last two), review the rules. Now for one last rule.

THE ADDITION RULE

If $y = ax^n + bx^m$, where a and b are constants, then $\dfrac{dy}{dx} = a\left(nx^{n-1}\right) + b\left(mx^{m-1}\right)$

This works for subtraction, too.

Example 13: If $y = 3x^4 + 8x^{10}$, then $\dfrac{dy}{dx} = 12x^3 + 80x^9$.

Example 14: If $y = 7x^{-4} + 5x^{-\frac{1}{2}}$, then $\dfrac{dy}{dx} = -28x^{-5} - \dfrac{5}{2}x^{-\frac{3}{2}}$.

Example 15: If $y = 5x^4(2 - x^3)$, then $\dfrac{dy}{dx} = 40x^3 - 35x^6$.

Example 16: If $y = (3x^2 + 5)(x - 1)$, then $y = 3x^3 - 3x^2 + 5x - 5$ and $\dfrac{dy}{dx} = 9x^2 - 6x + 5$.

Example 17: If $y = ax^3 + bx^2 + cx + d$, then $\dfrac{dy}{dx} = 3ax^2 + 2bx + c$.

After you've worked through all 17 of these examples, you should be able to take the derivative of any polynomial with ease.

SIMPLIFY YOUR RESULTS!

As you should have noticed from the examples above, in calculus, you are often asked to convert from fractions and radicals to negative powers and fractional powers. In addition, don't freak out if your answer doesn't match any of the answer choices. Since answers to problems are often presented in simplified form, your answer may not be simplified enough.

There are two basic expressions that you'll often be asked to differentiate. You can make your life easier by memorizing the following derivatives:

$$\text{If } y = \frac{k}{x} \text{ then } \frac{dy}{dx} = -\frac{k}{x^2}$$

$$\text{If } y = k\sqrt{x} \text{ then } \frac{dy}{dx} = \frac{k}{2\sqrt{x}}$$

HIGHER ORDER DERIVATIVES

This may sound like a big deal, but it isn't. This term only refers to taking the derivative of a function more than once. You don't have to stop at the first derivative of a function; you can keep taking derivatives. The derivative of a first derivative is called the second derivative. The derivative of the second derivative is called the third derivative.

And so on.

Generally, you'll only have to take first and second derivatives, although occasionally (usually only for BC Calculus students), you'll find a use for the higher derivatives.

Function	First Derivative	Second Derivative
x^6	$6x^5$	$30x^4$
$8\sqrt{x}$	$\dfrac{4}{\sqrt{x}}$	$-2x^{\frac{3}{2}}$

Notice how we reduced the derivative in the latter example? You should be able to do this mentally.

Here are some sample problems involving the rules we discussed above. As you work, cover the answers with an index card, and then check your work after you're done. By the time you finish them, you should know the rules by heart.

PROBLEM 1. If $y = 50x^5 + \dfrac{3}{x} - 7x^{-\frac{5}{3}}$, then $\dfrac{dy}{dx} =$

Answer: $\dfrac{dy}{dx} = 50(5x^4) + \left(-\dfrac{3}{x^2}\right) - 7\left(-\dfrac{5}{3}\right)x^{-\frac{8}{3}} = 250x^4 - \dfrac{3}{x^2} + \dfrac{35}{3}x^{-\frac{8}{3}}$

Problem 2. If $y = 9x^4 + 6x^2 - 7x + 11$, then $\dfrac{dy}{dx} =$

Answer: $\dfrac{dy}{dx} = 9(4x^3) + 6(2x) - 7(1) + 0 = 36x^3 + 12x - 7$

Problem 3. If $f(x) = 6x^{\frac{3}{2}} - 12\sqrt{x} - \dfrac{8}{\sqrt{x}} + 24x^{-\frac{3}{2}}$, then $f'(x) =$

Answer: $f'(x) = 6\left(\dfrac{3}{2}x^{\frac{1}{2}}\right) - \left(\dfrac{12}{2\sqrt{x}}\right) - 8\left(-\dfrac{1}{2}x^{-\frac{3}{2}}\right) + 24\left(-\dfrac{3}{2}x^{-\frac{5}{2}}\right) = 9\sqrt{x} - \dfrac{6}{\sqrt{x}} + 4x^{-\frac{3}{2}} - 36x^{-\frac{5}{2}}$

How'd you do? Did you notice the changes in notation? How about the fractional powers, radical signs, and x's in denominators? You should be able to switch back and forth between notations, between fractional powers and radical signs, and between negative powers in a numerator and positive powers in a denominator.

PRACTICE PROBLEMS

Find the derivative of each expression and simplify. The answers are in chapter 21.

1. $(4x^2 + 1)^2$

2. $(x^5 + 3x)^2$

3. $11x^7$

4. $8x^{10}$

5. $18x^3 + 12x + 11$

6. $\dfrac{1}{2}(x^{12} + 17)$

7. $-\dfrac{1}{3}(x^9 + 2x^3 - 9)$

8. π^5

9. $\dfrac{1}{a}\left(\dfrac{1}{b}x^2 - \dfrac{2}{a}x - \dfrac{d}{x}\right)$

10. $-8x^{-8} + 12\sqrt{x}$

11. $6x^{-7} - 4\sqrt{x}$

12. $x^{-5} + \dfrac{1}{x^8}$

13. $\sqrt{x} + \dfrac{1}{x^3}$

14. $(6x^2 + 3)(12x - 4)$

15. $(3 - x - 2x^3)(6 + x^4)$

16. $e^{10} + \pi^3 - 7$

17. $\left(\dfrac{1}{x} + \dfrac{1}{x^2} \right)\left(\dfrac{4}{x^3} - \dfrac{6}{x^4} \right)$

18. $\sqrt{x} + \dfrac{1}{\sqrt{3}}$

19. $(x^2 + 8x - 4)(2x^{-2} + x^{-4})$

20. 0

21. $(x + 1)^3$

22. $\sqrt{x} + \sqrt[3]{x} + \sqrt[3]{x^2}$

23. $x(2x + 7)(x - 2)$

24. $\sqrt{x}\left(\sqrt[3]{x} + \sqrt[5]{x} \right)$

25. $ax^5 + bx^4 + cx^3 + dx^2 + ex + f$

THE PRODUCT RULE

Now that you know how to find derivatives of simple polynomials, it's time to get more complicated. (You knew this had to happen, right?) What if you had to find the derivative of this?

$$f(x) = (x^3 + 5x^2 - 4x + 1)(x^5 - 7x^4 + x)$$

You could multiply out the expression and taking the derivative of each term, like this:

$$f(x) = x^8 - 2x^7 - 39x^6 + 29x^5 - 6x^4 + 5x^3 - 4x^2 + x$$

And the derivative is:

$$f'(x) = 8x^7 - 14x^6 - 234x^5 + 145x^4 - 24x^3 + 15x^2 - 8x + 1$$

Needless to say, this process takes forever. Naturally, there's an easier way. When a function involves two terms multiplied by each other, we use **the Product Rule**.

THE PRODUCT RULE

$$\text{If } f(x) = uv, \text{ then } f'(x) = u\frac{dv}{dx} + v\frac{du}{dx}$$

To find the derivative of two things multiplied by each other, you multiply the first function by the derivative of the second, and add that to the second function multiplied by the derivative of the first. With the product rule, the order of these two operations doesn't matter. It does matter with other rules, though, so it helps to use the same order each time, just to be sure.

Let's use the Product Rule to find the derivative of our example.

$$f'(x) = (x^3 + 5x^2 - 4x + 1)(5x^4 - 28x^3 + 1) + (x^5 - 7x^4 + x)(3x^2 + 10x - 4)$$

If we were to simplify this, we'd get the same answer as before. But, here's the best part: we're not going to simplify it. One of the great things about the AP exam is that when it's difficult to simplify an expression, you almost never have to. Nonetheless, you'll often need to simplify expressions when you're taking second derivatives, or when you use the derivative in some other equation. Practice simplifying whenever possible.

Example 1: $f(x) = (9x^2 + 4x)(x^3 - 5x^2)$

$$f'(x) = (9x^2 + 4x)(3x^2 - 10x) + (x^3 - 5x^2)(18x + 4)$$

Example 2: $y = \left(\sqrt{x} + 4\sqrt[3]{x}\right)\left(x^5 - 11x^8\right)$

$$y' = \left(\sqrt{x} + 4\sqrt[3]{x}\right)\left(5x^4 - 88x^7\right) + \left(x^5 - 11x^8\right)\left(\frac{1}{2\sqrt{x}} + \frac{4}{3\sqrt[3]{x^2}}\right)$$

Example 3: $y = \left(\frac{1}{x} + \frac{1}{x^2} - \frac{1}{x^3}\right)\left(\frac{1}{x} - \frac{1}{x^3} + \frac{1}{x^5}\right)$

$$y' = \left(\frac{1}{x} + \frac{1}{x^2} - \frac{1}{x^3}\right)\left(-\frac{1}{x^2} + \frac{3}{x^4} - \frac{5}{x^6}\right) + \left(\frac{1}{x} - \frac{1}{x^3} + \frac{1}{x^5}\right)\left(-\frac{1}{x^2} - \frac{2}{x^3} + \frac{3}{x^4}\right)$$

Notice that we're not simplifying.

THE QUOTIENT RULE

What happens when you have to take the derivative of a function that is the quotient of two other functions? You guessed it: use **the Quotient Rule**.

$$\text{The Quotient Rule: If } f(x) = \frac{u}{v} \text{ then } f'(x) = \frac{v\dfrac{du}{dx} - u\dfrac{dv}{dx}}{v^2}$$

In this rule, as opposed to the Product Rule, the order in which you take the derivatives is very important, because you're subtracting instead of adding. It's always the bottom function times the derivative of the top minus the top function times the derivative of the bottom. Then divide the whole thing by the bottom function squared. A good way to remember this is to say the following:

$$\frac{"LoDeHi - HiDeLo"}{(Lo)^2}$$

You could also write:

$$f(x) = \frac{u}{v} \text{ as } f(x) = u\frac{1}{v}.$$

Then you could derive the Quotient Rule using the Product Rule:

$$f'(x) = u\left(-\frac{1}{v^2}\frac{dv}{dx}\right) + \frac{du}{dx}\frac{1}{v} = \frac{v\dfrac{du}{dx} - u\dfrac{dv}{dx}}{v^2}.$$

Here are some more examples:

Example 4: $f(x) = \dfrac{\left(x^5 - 3x^4\right)}{\left(x^2 + 7x\right)}$

$$f'(x) = \frac{\left(x^2 + 7x\right)\left(5x^4 - 12x^3\right) - \left(x^5 - 3x^4\right)(2x + 7)}{\left(x^2 + 7x\right)^2}$$

Example 5: $y = \dfrac{\left(x^{-3} - x^{-8}\right)}{\left(x^{-2} + x^{-6}\right)}$

$$\frac{dv}{dx} = \frac{\left(x^{-2} + x^{-6}\right)\left(-3x^{-4} + 8x^{-9}\right) - \left(x^{-3} - x^{-8}\right)\left(-2x^{-3} - 6x^{-7}\right)}{\left(x^{-2} + x^{-6}\right)^2}$$

We're not going to simplify these, although the Quotient Rule often produces expressions that simplify more readily than those involving the Product Rule. Sometimes it's helpful to simplify, but avoid it if at all possible. When you have to find a second derivative, however, you do have to simplify the quotient. If this is the case, the AP exam usually will give you a simple expression to deal with, such as in the example below.

Example 6: $y = \dfrac{3x+5}{5x-3}$

$$\frac{dy}{dx} = \frac{(5x-3)(3)-(3x+5)(5)}{(5x-3)^2} = \frac{(15x-9)-(15x+25)}{(5x-3)^2} = \frac{-34}{(5x-3)^2}$$

In order to take the second derivative of this, you have to use the Chain Rule.

THE CHAIN RULE

The most important rule in this chapter (and sometimes the most difficult one) is called **the Chain Rule**. It's used when you're given composite functions--that is, a function inside of another function. You'll always see one of these on the exam, so it's important to know the Chain Rule cold.

A composite function is usually written as: $f(g(x))$.

For example:

$$\text{If } f(x) = \frac{1}{x} \text{ and } g(x) = \sqrt{3x}, \text{ then } f(g(x)) = \frac{1}{\sqrt{3x}}$$

We could also find:

$$g(f(x)) = \sqrt{\frac{3}{x}}.$$

When finding the derivative of a composite function, we take the derivative of the "outside" function, with the inside function g considered as the variable, leaving the "inside" function alone. Then, we multiply this by the derivative of the "inside" function, with respect to its variable x.

Another way to write the Chain Rule is like this:

$$\text{If } y = f(g(x)) = \text{ then } y' = \left(\frac{df(g)}{dg}\right)\left(\frac{dg}{dx}\right)$$

This rule is tricky, so here are several examples. The last couple incorporate the Product Rule and the Quotient Rule as well.

Example 7: If $y = (5x^3 + 3x)^5$

then $\dfrac{dy}{dx} = 5(5x^3 + 3x)^4(15x^2 + 3)$

We just dealt with the derivative of something to the fifth power, like this:

$$y = (g)^5 \qquad \dfrac{dy}{dg} = 5(g)^4, \text{ where } g = 5x^3 + 3x.$$

Then we multiplied by the derivative of g: $(15x^2 + 3)$.

Always do it this way. The process has several successive steps, like peeling away the layers of an onion until you reach the center.

Example 8: If $y = \sqrt{x^3 - 4x}$

$$\text{then } \dfrac{dy}{dx} = \dfrac{1}{2}\left(x^3 - 4x\right)^{-\frac{1}{2}}\left(3x^2 - 4\right)$$

Again, we took the derivative of the outside function, leaving the inside alone. Then we multiplied by the derivative of the inside.

Example 9: If $y = \sqrt{\left(x^5 - 8x^3\right)\left(x^2 + 6x\right)}$, then

$$\dfrac{dy}{dx} = \dfrac{1}{2}\left[\left(x^5 - 8x^3\right)\left(x^2 + 6x\right)\right]^{-\frac{1}{2}}\left[\left(x^5 - 8x^3\right)(2x + 6) + \left(x^2 + 6x\right)\left(5x^4 - 24x^2\right)\right]$$

Messy, isn't it? That's because we used the Chain Rule and the Product Rule. Now for one with the Chain Rule and the Quotient Rule.

Example 10: If $y = \left(\dfrac{2x + 8}{x^2 - 10x}\right)^5$, then

$$\dfrac{dy}{dx} = 5\left[\dfrac{2x + 8}{x^2 - 10x}\right]^4\left[\dfrac{\left(x^2 - 10x\right)(2) - (2x + 8)(2x - 10)}{\left(x^2 - 10x\right)^2}\right]$$

As you can see, these grow quite complex, so we only simplify these as a last resort. If you must simplify, the AP exam will only have a very simple Chain Rule problem, like so:

Example 11: If $y = \sqrt{5x^3 + x}$ then $\dfrac{dy}{dx} = \dfrac{1}{2}\left(5x^3 + x\right)^{-\frac{1}{2}}\left(15x^2 + 1\right)$

Now we use the Product Rule and the Chain Rule to find the second derivative:

$$\frac{d^2y}{dx^2} = \frac{1}{2}\left(5x^3 + x\right)^{-\frac{1}{2}}(30x) + \left(15x^2 + 1\right)\left[-\frac{1}{4}\left(5x^3 + x\right)^{-\frac{3}{2}}\left(15x^2 + 1\right)\right]$$

You can also simplify this further, if necessary.

There's another representation of the Chain Rule that you need to learn.

If $y = y(v)$ and $v = v(x)$ then $\dfrac{dy}{dx} = \dfrac{dy}{dv}\dfrac{dv}{dx}$

Example 12: $y = 8v^2 - 6v$ and $v = 5x^3 - 11x$, then

$$\frac{dy}{dx} = (16v - 6)(15x^2 - 11).$$

Then substitute for v:

$$\frac{dy}{dx} = \left(16\left(5x^3 - 11x\right) - 6\right)\left(15x^2 - 11\right) = \left(80x^3 - 176x - 6\right)\left(15x^2 - 11\right)$$

Here are some solved problems. Cover the answers first, then check your work.

PROBLEM 1. Find $\dfrac{dy}{dx}$ if $y = \left(5x^4 + 3x^7\right)\left(x^{10} - 8x\right)$.

Answer: $\dfrac{dy}{dx} = \left(5x^4 + 3x^7\right)\left(10x^9 - 8\right) + \left(x^{10} - 8x\right)\left(20x^3 + 21x^6\right)$

PROBLEM 2. Find $\dfrac{dy}{dx}$ if $y = \left(x^3 + 3x^2 + 3x + 1\right)\left(x^2 + 2x + 1\right)$.

Answer: $\dfrac{dy}{dx} = \left(x^3 + 3x^2 + 3x + 1\right)(2x + 2) + \left(x^2 + 2x + 1\right)\left(3x^2 + 6x + 3\right)$

PROBLEM 3. Find $\dfrac{dy}{dx}$ if $y = \left(\sqrt{x} + \dfrac{1}{x}\right)\left(\sqrt[3]{x^2} - \dfrac{1}{x^3}\right)$.

Answer: $\dfrac{dy}{dx} = \left(\sqrt{x} + \dfrac{1}{x}\right)\left(\dfrac{2}{3}x^{-\frac{1}{3}} + \dfrac{3}{x^4}\right) + \left(\sqrt[3]{x^2} - \dfrac{1}{x^3}\right)\left(\dfrac{1}{2\sqrt{x}} - \dfrac{1}{x^2}\right)$

PROBLEM 4. Find $\dfrac{dy}{dx}$ if $y = \left(x^3 + 1\right)\left(x^2 + 5x - \dfrac{1}{x^5}\right)$.

Answer: $\dfrac{dy}{dx} = \left(x^3 + 1\right)\left(2x + 5 + \dfrac{5}{x^6}\right) + \left(x^2 + 5x - \dfrac{1}{x^5}\right)\left(3x^2\right)$

PROBLEM 5. Find $\dfrac{dy}{dx}$ if $y = \dfrac{2x - 4}{x^2 - 6}$.

Answer: $\dfrac{dy}{dx} = \dfrac{\left(x^2 - 6\right)(2) - (2x - 4)(2x)}{\left(x^2 - 6\right)^2} = \dfrac{-2x^2 + 8x - 12}{\left(x^2 - 6\right)^2}$

PROBLEM 6. Find $\dfrac{dy}{dx}$ if $y = \dfrac{x^2 + 1}{x^2 + x + 4}$.

Answer: $\dfrac{dy}{dx} = \dfrac{\left(x^2 + x + 4\right)(2x) - \left(x^2 + 1\right)(2x + 1)}{\left(x^2 + x + 4\right)^2} = \dfrac{x^2 + 6x - 1}{\left(x^2 + x + 4\right)^2}$

PROBLEM 7. Find $\dfrac{dy}{dx}$ if $y = \dfrac{x + 5}{x - 5}$.

Answer: $\dfrac{dy}{dx} = \dfrac{(x - 5)(1) - (x + 5)(1)}{(x - 5)^2} = \dfrac{-10}{(x - 5)^2}$

PROBLEM 8. Find $\dfrac{dy}{dx}$ at $x = 1$ if $y = \dfrac{(x + 3)\left(x^3 - 9\right)}{\left(x^2 + 4\right)(x - 8)}$.

Answer: You might be tempted to use the Product Rule to find the derivatives of the top and bottom, but this problem is much easier if you expand the numerator and denominator first:

$$y = \dfrac{x^4 + 3x^3 - 9x - 27}{x^3 - 8x^2 + 4x - 32}$$

$$\dfrac{dy}{dx} = \dfrac{\left(x^3 - 8x^2 + 4x - 32\right)\left(4x^3 + 9x^2 - 9\right) - \left(x^4 + 3x^3 - 9x - 27\right)\left(3x^3 - 16x + 4\right)}{\left(x^3 - 8x^2 + 4x - 32\right)^2}$$

You'd have to be deranged to attempt to simplify this expression. Luckily, you don't have to. Just plug in $x = 1$ and simplify.

$$\left.\dfrac{dy}{dx}\right|_{x=1} = \dfrac{(-35)(4) - (-32)(-9)}{\left[(-35)\right]^2} = -\dfrac{428}{1225}$$

PROBLEM 9. Find $\dfrac{dy}{dx}$ if $y = (x^4 + x)^2$.

Answer: $\dfrac{dy}{dx} = 2(x^4 + x)(4x^3 + 1)$

PROBLEM 10. Find $\dfrac{dy}{dx}$ if $y = \left(\dfrac{x+3}{x-3}\right)^3$.

Answer: $\dfrac{dy}{dx} = 3\left(\dfrac{x+3}{x-3}\right)^2\left(\dfrac{(x-3)(1)-(x+3)(1)}{(x-3)^2}\right) = -18\left(\dfrac{(x+3)^2}{(x-3)^4}\right)$

PROBLEM 11. Find $\dfrac{dy}{dx}$ if $y = \left(\dfrac{x}{4}+\dfrac{4}{x}\right)^4$.

Answer: $\dfrac{dy}{dx} = 4\left(\dfrac{x}{4}+\dfrac{4}{x}\right)^3\left(\dfrac{1}{4}-\dfrac{4}{x^2}\right)$

PROBLEM 12. Find $\dfrac{dy}{dx}$ at $x = 1$ if $y = \left[(x^3 + x)(x^4 - x^2)\right]^2$.

Answer: $\dfrac{dy}{dx} = 2\left[(x^3 + x)(x^4 - x^2)\right]\left[(x^3 + x)(4x^3 - 2x)+(x^4 - x^2)(3x^2 + 1)\right]$

Once again, plug in right away. Never simplify until after you've substituted.

at $x=1$, $\dfrac{dy}{dx} = 0$.

PRACTICE PROBLEMS

Simplify when possible. The answers are in chapter 21.

1. Find $f'(x)$ if $f(x) = \left(\dfrac{4x^3 - 3x^2}{5x^7 + 1}\right)$.

2. Find $f'(x)$ if $f(x) = (x^2 - 4x + 3)(x + 1)$.

3. Find $f'(x)$ if $f(x) = (x + 1)^{10}$.

4. Find $f'(x)$ if $f(x) = 8\sqrt{(x^4 - 4x^2)}$.

5. Find $f'(x)$ if $f(x) = \left(\dfrac{x}{x^2 + 1}\right)^3$.

6. Find $f'(x)$ if $f(x) = \sqrt[4]{\left(\dfrac{2x - 5}{5x + 2}\right)}$.

7. Find $f'(x)$ if $f(x) = \dfrac{4x^8 - \sqrt{x}}{8x^4}$.

8. Find $f'(x)$ if $f(x) = \left(x + \dfrac{1}{x}\right)\left(x^2 - \dfrac{1}{x^2}\right)$.

9. Find $f'(x)$ if $f(x) = \left(\dfrac{x}{x + 1}\right)^4$.

10. Find $f'(x)$ if $f(x) = \left(x^2 + x\right)^{100}$.

11. Find $f'(x)$ if $f(x) = \sqrt{\dfrac{x^2 + 1}{x^2 - 1}}$.

12. Find $f'(x)$ at $x = 2$ if $f(x) = \dfrac{(x + 4)(x - 8)}{(x + 6)(x - 6)}$.

13. Find $f'(x)$ at $x = 1$ if $f(x) = \dfrac{x^6 + 4x^3 + 6}{\left(x^4 - 2\right)^2}$.

14. Find $f'(x)$ at $x = 1$ if $f(x) = \left[\dfrac{x - \sqrt{x}}{x + \sqrt{x}}\right]^2$.

15. Find $f'(x)$ if $f(x) = \dfrac{x^2 - 3}{(x - 3)}$.

16. Find $f'(x)$ at $x = 1$ if $f(x) = \left(x^4 - x^2\right)\left(2x^3 + x\right)$.

17. Find $f'(x)$ at $x = 2$ if $f(x) = \dfrac{x^2 + 2x}{x^4 - x^3}$.

18. Find $f'(x)$ if $f(x) = \sqrt{x^4 + x^2}$.

19. Find $f'(x)$ at $x = 1$ if $f(x) = \dfrac{x}{\left(1+x^2\right)^2}$.

20. Find $\dfrac{dy}{dx}$ if $y = u^2 - 1$ and $u = \dfrac{1}{x-1}$.

21. Find $\dfrac{dy}{dx}$ at $x = 1$ if $y = \dfrac{t^2+2}{t^2-2}$ and $t = x^3$.

22. Find $\dfrac{dy}{dt}$ if $y = \left(x^6 - 6x^5\right)\left(5x^2 + x\right)$ and $x = \sqrt{t}$.

23. Find $\dfrac{du}{dv}$ at $v = 2$ if $u = \sqrt{x^3 + x^2}$ and $x = \dfrac{1}{v}$.

24. Find $\dfrac{dy}{dx}$ at $x = 1$ if $y = \dfrac{1+u}{1+u^2}$ and $u = x^2 - 1$.

25. Find $\dfrac{du}{dv}$ if $u = y^3$ and $y = \dfrac{x}{x+8}$ and $x = v^2$.

DERIVATIVES OF TRIG FUNCTIONS

There are a lot of trigonometry problems in calculus and on the AP calculus exams. If you're not sure of your trig, you should definitely go back and review the unit on Prerequisite Math. You'll need to remember your trig formulas, the values of the special angles, and the trig ratios, among other stuff.

In addition, angles are always referred to in radians. You can forget all about using degrees. Degrees aren't real numbers; if you use them in these problems, you'll get messed up (and your calculator will stage a revolt).

You should know the derivatives of all six trig functions. The good news is that the derivatives are pretty easy, and all you have to do is memorize them. Because the AP exam might ask you about this, though, let's use the definition of the derivative to figure out the derivative of sin x.

If $f(x) = \sin x$, then $f(x + h) = \sin(x + h)$.

Substitute this into the definition of the derivative:

$$\lim_{h \to 0} \frac{f(x+h) - f(x)}{h} = \lim_{h \to 0} \frac{\sin(x+h) - \sin x}{h}.$$

Remember that $\sin(x + h) = \sin x \cos h + \cos x \sin h$! Now simplify it:

$$\lim_{h \to 0} \frac{\sin x \cos h + \cos x \sin h - \sin x}{h}.$$

Next, rewrite this as:

$$\lim_{h \to 0} \frac{\sin x(\cos h - 1) + \cos x \sin h}{h} = \lim_{h \to 0} \frac{\sin x(\cos h - 1)}{h} + \lim_{h \to 0} \frac{\cos x \sin h}{h}.$$

Next, use some of the trigonometric limits that you memorized back in chapter 1. Specifically,

$$\lim_{h \to 0} \frac{(\cos h - 1)}{h} = 0 \text{ and } \lim_{h \to 0} \frac{\sin h}{h} = 1.$$

This gives you:

$$\lim_{h \to 0} \frac{\sin x(\cos h - 1)}{h} + \lim_{h \to 0} \frac{\cos x \sin h}{h} = \sin x(0) + \cos x(1) = \cos x.$$

The derivative of sin x is cos x. (Yes, we did this in chapter 3 as well. We're just making sure that you're paying attention.) Now that you've got that straightened out, let's try an example.

Example 1: Find the derivative of $\sin\left(\dfrac{\pi}{2} - x\right)$.

$$\frac{d}{dx} \sin\left(\frac{\pi}{2} - x\right) = \cos\left(\frac{\pi}{2} - x\right)(-1) = -\cos\left(\frac{\pi}{2} - x\right).$$

Use some of the rules of trigonometry you remember from last year (naturally!). Because

$$\sin\left(\frac{\pi}{2} - x\right) = \cos x \text{ and } \cos\left(\frac{\pi}{2} - x\right) = \sin x,$$

you can substitute into the above expression and get:

$$\frac{d}{dx} \cos x = -\sin x.$$

Now, let's derive the derivatives of the other four trigonometric functions.

Example 2: Find the derivative of $\dfrac{\sin x}{\cos x}$

You've got a quotient here, so use the Quotient Rule:

$$\frac{d}{dx} \frac{\sin x}{\cos x} = \frac{(\cos x)(\cos x) - (\sin x)(-\sin x)}{(\cos x)^2} = \frac{\cos^2 x + \sin^2 x}{\cos^2 x} = \frac{1}{\cos^2 x} = \sec^2 x.$$

Because $\dfrac{\sin x}{\cos x} = \tan x$, you should get:

$$\frac{d}{dx}\tan x = \sec^2 x.$$

Example 3: Find the derivative of $\dfrac{\cos x}{\sin x}$.

It's time for the Quotient Rule again:

$$\frac{d}{dx}\frac{\cos x}{\sin x} = \frac{(\sin x)(-\sin x)-(\cos x)(\cos x)}{(\sin x)^2} = \frac{-(\cos^2 x + \sin^2 x)}{\sin^2 x} = -\frac{1}{\sin^2 x} = -\csc^2 x.$$

Because $\dfrac{\cos x}{\sin x} = \cot x$, you get:

$$\frac{d}{dx}\cot x = -\csc^2 x.$$

Example 4: Find the derivative of $\dfrac{1}{\cos x}$.

Use the Reciprocal Rule.

$$\frac{d}{dx}\frac{1}{\cos x} = \frac{-1}{(\cos x)^2}(-\sin x) = \frac{\sin x}{\cos^2 x} = \frac{1}{\cos x}\frac{\sin x}{\cos x} = \sec x \tan x.$$

Because $\dfrac{1}{\cos x} = \sec x$, you get:

$$\frac{d}{dx}\sec x = \sec x \tan x.$$

Example 5: Find the derivative of $\dfrac{1}{\sin x}$.

You get the idea by now.

$$\frac{d}{dx}\frac{1}{\sin x} = \frac{-1}{(\sin x)^2}(\cos x) = \frac{-\cos x}{\sin^2 x} = \frac{1}{\sin x}\frac{\cos x}{\sin x} = -\csc x \cot x.$$

Because $\dfrac{1}{\sin x} = \csc x$, you get:

$$\frac{d}{dx}\csc x = -\csc x \cot x.$$

There you go. We have now found the derivatives of all six of the trigonometric functions. (A chart of them appears at the end of the book.) Now memorize them. You'll thank us later.

Let's do some more examples.

Example 6: Find the derivative of sin(5x).

$$\frac{d}{dx}\sin(5x) = \cos(5x)(5) = 5\cos(5x)$$

Example 7: Find the derivative of sec(x^2).

$$\frac{d}{dx}\sec(x^2) = \sec(x^2)\tan(x^2)(2x).$$

Example 8: Find the derivative of csc($x^3 - 5x$).

$$\frac{d}{dx}\csc(x^3 - 5x) = -\csc(x^3 - 5x)\cot(x^3 - 5x)(3x^2 - 5).$$

These derivatives are almost like formulas. You just follow the pattern and use the Chain Rule when appropriate.

Here are some solved problems. Do each problem, covering the answer first, then checking your answer.

PROBLEM 1. Find $f'(x)$ if $f(x) = \sin(2x^3)$.

Answer: Follow the rule: $f'(x) = \cos(2x^3)6x^2$.

PROBLEM 2. Find $f'(x)$ if $f(x) = \cos\left(\sqrt{3x}\right)$.

Answer: $f'(x) = -\sin\left(\sqrt{3x}\right)\left[\frac{1}{2}(3x)^{-\frac{1}{2}}(3)\right] = \frac{-3\sin\left(\sqrt{3x}\right)}{2\sqrt{3x}}$.

PROBLEM 3. Find $f'(x)$ if $f(x) = \tan\left(\frac{x}{x+1}\right)$.

Answer: $f'(x) = \sec^2\left(\frac{x}{x+1}\right)\left[\frac{(x+1)-x}{(x+1)^2}\right] = \left(\frac{1}{(x+1)^2}\right)\sec^2\left(\frac{x}{x+1}\right)$.

PROBLEM 4. Find $f'(x)$ if $f(x) = \csc(x^3 + x + 1)$.

Answer: Follow the rule: $f'(x) = -\csc(x^3 + x + 1)\cot(x^3 + x + 1)(3x^2 + 1)$.

PRACTICE PROBLEMS

Now try these problems. The answers are in chapter 21.

1. Find $\dfrac{dy}{dx}$ if $y = \sin^2 x$.

2. Find $\dfrac{dy}{dx}$ if $y = \cos x^2$.

3. Find $\dfrac{dy}{dx}$ if $y = (\tan x)(\sec x)$.

4. Find $\dfrac{dy}{dx}$ if $y = \cot 4x$.

5. Find $\dfrac{dy}{dx}$ if $y = \sqrt{\sin 3x}$.

6. Find $\dfrac{dy}{dx}$ if $y = \dfrac{1 + \sin x}{1 - \sin x}$.

7. Find $\dfrac{dy}{dx}$ if $y = \csc^2 x^2$.

8. Find $\dfrac{dy}{dx}$ if $y = 2 \sin 3x \cos 4x$.

9. Find $\dfrac{d^4 y}{dx^4}$ if $y = \sin 2x$.

10. Find $\dfrac{dy}{dx}$ if $y = \sin t - \cos t$ and $t = 1 + \cos^2 x$.

11. Find $\dfrac{dy}{dx}$ if $y = \left(\dfrac{\tan x}{1 - \tan x} \right)^2$.

12. Find $\dfrac{dr}{d\theta}$ if $r = \sec \theta \, \tan 2\theta$.

13. Find $\dfrac{dr}{d\theta}$ if $r = \cos(1 + \sin \theta)$.

14. Find $\dfrac{dr}{d\theta}$ if $r = \dfrac{\sec\theta}{1+\tan\theta}$.

15. Find $\dfrac{dy}{dx}$ if $y = \left(1+\cot\left(\dfrac{2}{x}\right)\right)^{-2}$.

16. Find $\dfrac{dy}{dx}$ if $y = \sin\left(\cos\left(\sqrt{x}\right)\right)$.

5

IMPLICIT DIFFERENTIATION

By now, it should be easy for you to take the derivative of an equation such as $y = 3x^5 - 7x$. If you're given an equation such as $y^2 = 3x^5 - 7x$, you can still figure out the derivative by taking the square root of both sides, which gives you y in terms of x. It's messy, but possible.

If you have to find the derivative of $y^2 + y = 3x^5 - 7x$, you don't have an easy way to get y in terms of x, so you can't differentiate this equation using any of the techniques we've learned so far. That's because each of those previous techniques needs to be used on an equation in which y is in terms of x. When you can't isolate y in terms of x (or if isolating y makes taking the derivative a nightmare), it's time to take the derivative **implicitly**.

Implicit differentiation is one of the easier techniques you need to learn to do in calculus, but for some reason it gives many students trouble. Suppose you have the equation $y^2 = 3x^5 - 7x$. This means that the value of y is a function of the value of x. When we take the derivative, $\dfrac{dy}{dx}$, we're looking at the rate at which y changes as x changes. Thus, when we write:

$$\frac{dy}{dx} = 2x + 1$$

we're saying that "the rate" at which y changes, with respect to how x changes, is $2x + 1$.

Now, suppose you want to find $\dfrac{dx}{dy}$. As you might imagine:

$$\frac{dx}{dy} = \frac{1}{dy \, / \, dx}.$$

So here, $\dfrac{dx}{dy} = \dfrac{1}{2x+1}$. But notice that this derivative is in terms of x, not y, and you need to find the derivative with respect to y. This derivative is an **implicit** one. When you can't isolate the variables of an equation, you often end up with a derivative that is in terms of both variables.

Another way to think of this is that there is a hidden term in the derivative $\dfrac{dx}{dx}$, and when we take the derivative, what we really get is:

$$\frac{dy}{dx} = 2x\left(\frac{dx}{dx}\right) + 1\left(\frac{dx}{dx}\right).$$

Because a fraction that has the same term in its numerator and denominator is equal to 1, we write:

$$\frac{dy}{dx} = 2x(1) + 1(1) = 2x + 1.$$

Every time we take a derivative of a term with x in it, we multiply by the term $\dfrac{dx}{dx}$, but because this is 1 in this case, we ignore it. Suppose however, that we wanted to find out how y changes with respect to t (for time). Then we would have:

$$\frac{dy}{dt} = 2x\left(\frac{dx}{dt}\right) + 1\left(\frac{dx}{dt}\right).$$

If we wanted to find out how y changes with respect to r, we would have:

$$\frac{dy}{dr} = 2x\left(\frac{dx}{dr}\right) + 1\left(\frac{dx}{dr}\right)$$

and if we wanted to find out how y changes with respect to y, we would have:

$$\frac{dy}{dy} = 2x\left(\frac{dx}{dy}\right) + 1\left(\frac{dx}{dy}\right) \text{ or } 1 = 2x\frac{dx}{dy} + \frac{dx}{dy}.$$

This is how we really do differentiation. Remember:

$$\frac{dx}{dy} = \frac{1}{\frac{dy}{dx}}.$$

When you have an equation of x in terms of y, and you want to find the derivative with respect to y, simply differentiate. But if the equation is of y in terms of x, find $\frac{dy}{dx}$ and take its reciprocal to find $\frac{dx}{dy}$.

WHEN DO YOU USE IT?

You should use implicit differentiation any time you can't write a function explicitly in terms of the variable that we want to take the derivative with respect to. Go back to our original example:

$$y^2 + y = 3x^5 - 7x.$$

To take the derivative according to the information in the last paragraph, you get:

$$2y\left(\frac{dy}{dx}\right) + 1\left(\frac{dy}{dx}\right) = 15x^4\left(\frac{dx}{dx}\right) - 7\left(\frac{dx}{dx}\right)$$

Notice how each variable is multiplied by its appropriate $\frac{d}{dx}$. Now, remembering that $\frac{dx}{dx} = 1$, rewrite the expression this way:

$$2y\left(\frac{dy}{dx}\right) + 1\left(\frac{dy}{dx}\right) = 15x^4 - 7.$$

Next, factor $\frac{dy}{dx}$ out of the left-hand side:

$$\frac{dy}{dx}(2y + 1) = 15x^4 - 7$$

Isolating $\frac{dy}{dx}$ gives you:

$$\frac{dy}{dx} = \frac{15x^4 - 7}{(2y + 1)}$$

This is the derivative you're looking for. Notice how the derivative is defined in terms of y **and** x.

Up until now, $\frac{dy}{dx}$ has been strictly in terms of x. This is why the differentiation is "implicit."

Confused? Don't sweat it. Let's do a few examples and you will get the hang of it.

Example 1: Find $\dfrac{dy}{dx}$ if $y^3 - 4y^2 = x^5 + 3x^4$.

Using implicit differentiation, you get:

$$3y^2\left(\frac{dy}{dx}\right) - 8y\left(\frac{dy}{dx}\right) = 5x^4\left(\frac{dx}{dx}\right) + 12x^3\left(\frac{dx}{dx}\right).$$

Remember that $\dfrac{dx}{dx} = 1$:

$$\frac{dy}{dx}\left(3y^2 - 8y\right) = 5x^4 + 12x^3.$$

After you factor out $\dfrac{dy}{dx}$, divide both sides by $3y^2 - 8y$:

$$\frac{dy}{dx} = \frac{5x^4 + 12x^3}{\left(3y^2 - 8y\right)}$$

Note: Now that you understand that the derivative of an x term with respect to x will always be multiplied by $\dfrac{dx}{dx}$, and that $\dfrac{dx}{dx} = 1$, we won't write $\dfrac{dx}{dx}$ anymore. You should understand that the term is implied.

Example 2: Find $\dfrac{dy}{dx}$ if $\sin y^2 - \cos x^2 = \cos y^2 + \sin x^2$.

Use implicit differentiation:

$$\cos y^2\left(2y\frac{dy}{dx}\right) + \sin x^2(2x) = -\sin y^2\left(2y\frac{dy}{dx}\right) + \cos x^2(2x).$$

Then simplify:

$$2y\cos y^2\left(\frac{dy}{dx}\right) + 2x\sin x^2 = -2y\sin y^2\left(\frac{dy}{dx}\right) + 2x\cos x^2.$$

Next, put all of the terms containing $\dfrac{dy}{dx}$ on the left and all of the other terms on the right:

$$2y\cos y^2\left(\frac{dy}{dx}\right) + 2y\sin y^2\left(\frac{dy}{dx}\right) = -2x\sin x^2 + 2x\cos x^2.$$

Next, factor out $\dfrac{dy}{dx}$:

$$\frac{dy}{dx}\left(2y\cos y^2 + 2y\sin y^2\right) = -2x\sin x^2 + 2x\cos x^2.$$

And isolate $\dfrac{dy}{dx}$:

$$\frac{dy}{dx} = \frac{-2x\sin x^2 + 2x\cos x^2}{\left(2y\cos y^2 + 2y\sin y^2\right)}.$$

This can be simplified further to:

$$\frac{dy}{dx} = \frac{-x\left(\sin x^2 - \cos x^2\right)}{y\left(\cos y^2 + \sin y^2\right)}.$$

Example 3: Find $\dfrac{dy}{dx}$ if $3x^2 + 5xy^2 - 4y^3 = 8$.

Implicit differentiation should result in:

$$6x + \left[5x\left(2y\frac{dy}{dx}\right) + (5)y^2\right] - 12y^2\left(\frac{dy}{dx}\right) = 0.$$

(Did you notice the use of the Product Rule to find the derivative of $5xy^2$? The AP exam loves to make you do this. All of the same differentiation rules that you've learned up until now still apply. We're just adding another technique.) You can simplify this to:

$$6x + 10xy\frac{dy}{dx} + 5y^2 - 12y^2\frac{dy}{dx} = 0.$$

Next, put all of the terms containing $\dfrac{dy}{dx}$ on the left and all of the other terms on the right:

$$10xy\frac{dy}{dx} - 12y^2\frac{dy}{dx} = -6x - 5y^2.$$

Next, factor out $\dfrac{dy}{dx}$:

$$\left(10xy - 12y^2\right)\frac{dy}{dx} = -6x - 5y^2.$$

Then, isolate $\dfrac{dy}{dx}$:

$$\frac{dy}{dx} = \frac{-6x - 5y^2}{\left(10xy - 12y^2\right)}.$$

Example 4: Find the derivative of $3x^2 - 4y^2 + y = 9$ at $(2,1)$.

You need to use implicit differentiation to find $\dfrac{dy}{dx}$:

$$6x - 8y\left(\frac{dy}{dx}\right) + \left(\frac{dy}{dx}\right) = 0.$$

Now, instead of rearranging to isolate $\dfrac{dy}{dx}$, plug in $(2,1)$ immediately and solve for the derivative:

$$6(2) - 8\left(\frac{dy}{dx}\right) + \left(\frac{dy}{dx}\right) = 0.$$

Simplify:

$$12 - 7\left(\frac{dy}{dx}\right) = 0 \text{, so } \frac{dy}{dx} = \frac{12}{7}.$$

Getting the hang of implicit differentiation yet? We hope so, because these next examples are slightly harder.

Example 5: Find $\dfrac{dy}{dx}$ if $\dfrac{y^2 + x^3}{y^3 - x^2} = x$.

First, cross-multiply. Do this whenever you have a quotient because Product Rule is a lot less messy than Quotient Rule for Implicit Differentiation.
We get:

$$y^2 + x^3 = x(y^3 - x^2)$$

Distribute:

$$y^2 + x^3 = xy^3 - x^3$$

Take the derivative:

$$2y\frac{dy}{dx} + 3x^2 = x\left(3y^2\frac{dy}{dx}\right) + y^3 - 3x^2$$

Simplify:

$$2y\frac{dy}{dx} + 3x^2 = 3xy^2\frac{dy}{dx} + y^3 - 3x^2$$

Put all of the terms containing $\frac{dy}{dx}$ on the left and all of the other terms on the right:

$$2y\frac{dy}{dx} - 3xy^2\frac{dy}{dx} = y^3 - 6x^2$$

Factor out $\frac{dy}{dx}$:

$$\frac{dy}{dx}\left(2y - 3xy^2\right) = y^3 - 6x^2.$$

Now you can isolate $\frac{dy}{dx}$:

$$\frac{dy}{dx} = \frac{y^3 - 6x^2}{2y - 3xy^2}.$$

Example 6: Find the derivative of $\dfrac{2x - 5y^2}{4y^3 - x^2} = -x$ at (1,1).

First, cross-multiply:

$$2x - 5y^2 = -x\left(4y^3 - x^2\right)$$

Distribute:

$$2x - 5y^2 = -4xy^3 + x^3$$

Take the derivative:

$$2 - 10y\frac{dy}{dx} = -4x\left(3y^2\frac{dy}{dx}\right) - 4y^3 + 3x^2$$

Do not simplify now. Rather, plug in (1,1) right away. This will save you from the algebra:

$$2 - 10(1)\frac{dy}{dx} = -4(1)\left(3(1)^2\frac{dy}{dx}\right) - 4(1)^3 + 3(1)^2$$

Now solve for $\dfrac{dy}{dx}$:

$$2 - 10\frac{dy}{dx} = -12\frac{dy}{dx} - 1$$

$$2\frac{dy}{dx} = -3$$

$$\frac{dy}{dx} = \frac{-3}{2}.$$

SECOND DERIVATIVES

Sometimes, you'll be asked to find a second derivative implicitly.

Example 7: Find $\dfrac{d^2y}{dx^2}$ if $y^2 + 2y = 4x^2 + 2x.$

Differentiating implicitly, you get:

$$2y\frac{dy}{dx} + 2\frac{dy}{dx} = 8x + 2.$$

Next, simplify and solve for $\dfrac{dy}{dx}$:

$$\frac{dy}{dx} = \frac{4x + 1}{y + 1}.$$

Now it's time to take the derivative again:

$$\frac{d^2y}{dx^2} = \frac{4(y+1) - (4x+1)\left(\dfrac{dy}{dx}\right)}{(y+1)^2}$$

Finally, substitute for $\dfrac{dy}{dx}$:

$$\frac{4(y+1) - (4x+1)\left(\dfrac{4x+1}{y+1}\right)}{(y+1)^2} = \frac{4(y+1)^2 - (4x+1)^2}{(y+1)^3}.$$

Try these solved problems without looking at the answers. Then check your work.

PROBLEM 1. Find $\dfrac{dy}{dx}$ if $x^2 + y^2 = 6xy$.

Answer: Differentiate with respect to x:

$$2x + 2y\frac{dy}{dx} = 6x\frac{dy}{dx} + 6y.$$

Group all of the $\dfrac{dy}{dx}$ terms on the right and the other terms on the left:

$$2y\frac{dy}{dx} - 6x\frac{dy}{dx} = 6y - 2x.$$

Now factor out $\dfrac{dy}{dx}$:

$$\frac{dy}{dx}(2y - 6x) = 6y - 2x.$$

Therefore, the first derivative is the following:

$$\frac{dy}{dx} = \frac{6y - 2x}{2y - 6x} = \frac{3y - x}{y - 3x}.$$

PROBLEM 2. Find $\dfrac{dy}{dx}$ if $x - \cos y = xy$.

Answer: Differentiate with respect to x:

$$1 + \sin y\frac{dy}{dx} = x\frac{dy}{dx} + y.$$

Grouping the terms, you get:

$$\sin y\frac{dy}{dx} - x\frac{dy}{dx} = y - 1.$$

Now factor out $\dfrac{dy}{dx}$:

$$\frac{dy}{dx}(\sin y - x) = y - 1.$$

The derivative is:

$$\frac{dy}{dx} = \frac{y-1}{\sin y - x}.$$

PROBLEM 3. Find the equation of the tangent line to $x^2 y^2 = 25$ at $(1,-5)$.

Answer: Don't worry if this idea looks strange to you. We'll discuss tangent lines in greater depth in chapter 6. Once again, differentiate with respect to x:

$$x^2 \left(2y \frac{dy}{dx} \right) + 2xy^2 = 0.$$

Now, plug in $x = 1$ and $y = -5$, and solve for $\frac{dy}{dx}$:

$$-10\frac{dy}{dx} + 50 = 0 \qquad \frac{dy}{dx} = 5.$$

Plug this into the point-slope equation of a line, and you're done:

$$y + 5 = 5(x - 1).$$

PROBLEM 4. Find the derivative of each variable with respect to t of $x^2 + y^2 = z^2$.

Answer: $2x\dfrac{dx}{dt} + 2y\dfrac{dy}{dt} = 2z\dfrac{dz}{dt}$

PROBLEM 5. Find the derivative of each variable with respect to t of $V = \dfrac{1}{3}\pi r^2 h$.

Answer: $\dfrac{dV}{dt} = \dfrac{1}{3}\pi \left(r^2 \dfrac{dh}{dt} + 2r\dfrac{dr}{dt}h \right)$

PROBLEM 6. Find $\dfrac{d^2 y}{dx^2}$ if $y^2 = x^2 - 2x$.

Answer: First, take the derivative with respect to x:

$$2y\frac{dy}{dx} = 2x - 2.$$

Then solve for $\frac{dy}{dx}$:

$$\frac{dy}{dx} = \frac{2x-2}{2y} = \frac{x-1}{y}.$$

The second derivative with respect to x becomes:

$$\frac{d^2y}{dx^2} = \frac{y(1) - (x-1)\frac{dy}{dx}}{y^2}.$$

Now substitute for $\frac{dy}{dx}$ and simplify:

$$\frac{d^2y}{dx^2} = \frac{y - (x-1)\left(\frac{x-1}{y}\right)}{y^2} = \frac{y^2 - (x-1)^2}{y^3}$$

PRACTICE PROBLEMS

Use implicit differentiation to find the following derivatives. The answers are in chapter 21.

1. Find $\frac{dy}{dx}$ if $x^3 - y^3 = y$.

2. Find $\frac{dy}{dx}$ if $x^2 - 16xy + y^2 = 1$.

3. Find $\frac{dy}{dx}$ at (2,1) if $\frac{x+y}{x-y} = 3$.

4. Find $\frac{dy}{dx}$ if $\cos y - \sin x = \sin y - \cos x$.

5. Find $\frac{dy}{dx}$ if $16x^2 - 16xy + y^2 = 1$ at (1,1).

6. Find $\dfrac{dy}{dx}$ if $x^{\frac{1}{2}} + y^{\frac{1}{2}} = 2y^2$ at $(1,1)$.

7. Find $\dfrac{dy}{dx}$ if $x\sin y + y\sin x = \dfrac{\pi}{2\sqrt{2}}$ at $\left(\dfrac{\pi}{4}, \dfrac{\pi}{4}\right)$.

8. Find $\dfrac{d^2y}{dx^2}$ if $x^2 + 4y^2 = 1$.

9. Find $\dfrac{d^2y}{dx^2}$ if $\sin x + 1 = \cos y$.

10. Find $\dfrac{d^2y}{dx^2}$ if $x^2 - 4x = 2y - 2$.

6

BASIC APPLICATIONS OF THE DERIVATIVE

EQUATIONS OF TANGENT LINES

Finding the equation of a line tangent to a certain curve at a certain point is a standard calculus problem. This is because, among other things, the derivative is the slope of a tangent line to a curve at a particular point. Thus, we can find the equation of the tangent line to a curve if we have the equation of the curve, and the point at which we want to find the tangent line. Then all we have to do is take the derivative of the equation, plug in the point to find the slope, then use the point and the slope to find the equation of the line. Let's take this one step at a time.

Suppose we have a point (x_1, y_1) and a slope m. Then the equation of the line through that point with that slope is:

$$(y - y_1) = m(x - x_1).$$

You should remember this formula from algebra. If not, memorize it!

Next, suppose that we have an equation $y = f(x)$, where (x_1, y_1) satisfies that equation. Then $f'(x_1) = m$, and we can plug all of our values into the equation for a line and get the equation of the tangent line. This is much easier to explain with a simple example.

Example 1: Find the equation of the tangent line to the curve $y = 5x^2$ at the point (3, 45).

First of all, notice that the point satisfies the equation: when $x = 3$, $y = 45$. Now take the derivative of the equation:

$$\frac{dy}{dx} = 10x.$$

Now, if you plug in $x = 3$, you'll get the slope of the curve at that point. By the way, the notation for plugging in a point is $\Big|_{x=3}$. Learn to recognize it!

$$\frac{dy}{dx}\Big|_{x=3} = 10(3) = 30.$$

Thus, we have the slope and the point, and the equation is:

$$(y - 45) = 30(x - 3).$$

It's customary to simplify the equation if it's not too onerous (look it up, vocab fans):

$$y = 30x - 45.$$

Example 2: Find the equation of the tangent line to $y = x^3 + x^2$ at (3, 36).

The derivative looks like this:

$$\frac{dy}{dx} = 3x^2 + 2x,$$

so the slope is:

$$\frac{dy}{dx}\Big|_{x=3} = 3(3)^2 + 2(3) = 33.$$

The equation looks like:

$$(y - 36) = 33(x - 3), \text{ or } y = 33x - 63.$$

Naturally, there are a couple of things that can be done to make the problems harder. First of all, you can be given only the x-coordinate. Second, the equation can be more difficult to differentiate.

In order to find the y-coordinate, all you have to do is plug the x value into the equation for the curve and solve for y. Remember this: you'll see it again!

Example 3: Find the equation of the tangent line to $y = \dfrac{2x+5}{x^2-3}$ at $x = 1$.

First, find the y-coordinate:

$$y(1) = \frac{2(1)+5}{1^2-3} = -\frac{7}{2}.$$

Second, take the derivative:

$$\frac{dy}{dx} = \frac{(x^2-3)(2)-(2x+5)(2x)}{(x^2-3)^2}.$$

You're probably dreading having to simplify this derivative. Don't waste your time! Plug in $x = 1$ right away.

$$\left.\frac{dy}{dx}\right|_{x=1} = \frac{\left(1^2-3\right)(2)-(2(1)+5)(2(1))}{\left(1^2-3\right)^2} = \frac{-4-14}{4} = -\frac{9}{2}.$$

Now, we have a slope and a point, so the equation is:

$$y + \frac{7}{2} = -\frac{9}{2}(x-1) \text{ or } 2y = -9x+2.$$

THE NORMAL LINE

Sometimes, instead of finding the equation of a tangent line, you will be asked to find the equation of a normal line. A **normal** line is simply the line perpendicular to the tangent line at the same point. You follow the same steps as with the tangent line, but you use the slope that will give you a perpendicular line. Remember what that is? It's the negative reciprocal of the slope of the tangent line.

Example 4: Find the equation of the line normal to $y = x^5 - x^4 + 1$ at $x = 2$.

First, find the y-coordinate:

$$y(2) = 2^5 - 2^4 + 1 = 17.$$

Second, take the derivative:

$$\frac{dy}{dx} = 5x^4 - 4x^3.$$

Third, find the slope at $x = 2$:

$$\frac{dy}{dx}\bigg|_{x=2} = 5(2)^4 - 4(2)^3 = 48.$$

Fourth, take the negative reciprocal of 48, which is $-\dfrac{1}{48}$.

Finally, the equation becomes:

$$y - 17 = -\frac{1}{48}(x - 2).$$

Try these solved problems. Do each problem, covering the answer first, then checking your answer.

PROBLEM 1. Find the equation of the tangent line to the graph of $y = 4 - 3x - x^2$ at the point (2,–6).

Answer: First, take the derivative of the equation:

$$\frac{dy}{dx} = -3 - 2x.$$

Now, plug in $x = 2$ to get the slope of the tangent line:

$$\frac{dy}{dx} = -3 - 2(2) = -7.$$

Third, plug the slope and the point into the equation for the line:

$$y - (-6) = -7\,(x - 2).$$

This simplifies to $y = -7x + 8$.

PROBLEM 2. Find the equation of the normal line to the graph of $y = 6 - x - x^2$ at $x = -1$.

Answer: Once again, take that derivative:

$$\frac{dy}{dx} = -1 - 2x.$$

And then plug in $x = -1$ to get the slope of the tangent:

$$\frac{dy}{dx} = -1 - 2(-1) = 1.$$

Plug $x = -1$ into the original equation to get the y-coordinate:

$$y = 6 + 1 - 1 = 6.$$

Use the negative reciprocal of the slope in the second step to get the slope of the normal line:

$$m = -1.$$

Finally, plug the slope and the point into the equation for the line:

$$y - 6 = -1\,(x + 1).$$

This simplifies to $y = -x + 5$.

PROBLEM 3. Find the equations of the tangent and normal lines to the graph of $y = \dfrac{10x}{x^2 + 1}$ at the point (2,4).

Answer: This problem will put your algebra to the test. You have to use the Quotient Rule to take the derivative of this messy thing:

$$\frac{dy}{dx} = \frac{(x^2 + 1)(10) - (10x)(2x)}{(x^2 + 1)^2}.$$

Second, plug in $x = 2$ to get the slope of the tangent:

$$\frac{dy}{dx} = \frac{(5)(10) - (20)(4)}{5^2} = -\frac{30}{25} = -\frac{6}{5}.$$

Now, plug the slope and the point into the equation for the tangent line:

$$y - 4 = -\frac{6}{5}(x - 2).$$

which simplifies to $6x + 5y = 32$. The equation of the normal line must then be:

$$y - 4 = \frac{5}{6}(x - 2),$$

which simplifies to $-5x + 6y = 14$.

PROBLEM 4. Find the equation of the tangent line to the graph of $y = \sqrt{x^2 + 5}$ at the point (–2,3).

Answer: First, take the derivative of the equation:

$$\frac{dy}{dx} = \frac{1}{2}(x^2 + 5)^{-\frac{1}{2}}(2x).$$

Now, we plug in $x = -2$ to get the slope of the tangent line:

$$\frac{dy}{dx} = \frac{1}{2}(9)^{-\frac{1}{2}}(-4) = -\frac{2}{3}.$$

Third, we plug the slope and the point into the equation for the line:

$$y - 3 = -\frac{2}{3}(x + 2),$$

which simplifies to $2x + 3y = 5$.

PROBLEM 5. The curve $y = ax^2 + bx + c$ passes through the point (2,4) and is tangent to the line $y = x + 1$ at (0,1). Find a, b, and c.

Answer: The curve passes through (2,4), so if you plug in $x = 2$, you'll get $y = 4$. Therefore,

$$4 = 4a + 2b + c.$$

Second, the curve also passes through the point (0,1), so $c = 1$.

Because the curve is tangent to the line $y = x + 1$ at (0,1), they must both have the same slope at that point. The slope of the line is 1. The slope of the curve is the first derivative:

$$\frac{dy}{dx} = 2ax + b$$

and, at (0,1) $\frac{dy}{dx} = b$. Therefore, $b = 1$.

Now that you know b and c, plug them back into the equation from the first step and solve for a:

$$4 = 4a + 2 + 1, \text{ and } a = \frac{1}{4}.$$

PROBLEM 6. Find the points on the curve $y = 2x^3 - 3x^2 - 12x + 20$ where the tangent is parallel to the x-axis.

Answer: The x-axis is a horizontal line, so it has slope zero. Therefore, you want to know where the derivative of this curve is zero. Take the derivative:

$$\frac{dy}{dx} = 6x^2 - 6x - 12.$$

And set it equal to zero and solve for x. Get accustomed to doing this: it's one of the most common questions in differential calculus.

$$\frac{dy}{dx} = 6x^2 - 6x - 12 = 0$$

$$6(x^2 - x - 2) = 0$$

$$6(x - 2)(x + 1) = 0$$

$$x = 2 \text{ or } x = -1$$

Third, find the y-coordinates of these two points.

$$y = 2(8) - 3(4) - 12(2) + 20 = 0$$

$$y = 2(-1) - 3(1) - 12(-1) + 20 = 27$$

Therefore, the points are (2,0) and (–1,27).

PRACTICE PROBLEMS

Now try these problems. The answers are in chapter 21.

1. Find the equation of the tangent to the graph of $y = 3x^2 - x$ at $x = 1$.

2. Find the equation of the tangent to the graph of $y = x^3 - 3x$ at $x = 3$.

3. Find the equation of the normal to the graph of $y = \sqrt{8x}$ at $x = 2$.

4. Find the equation of the tangent to the graph of $y = \dfrac{1}{\sqrt{x^2 + 7}}$ at $x = 3$.

5. Find the equation of the normal to the graph of $y = \dfrac{x+3}{x-3}$ at $x = 4$.

6. Find the equation of the tangent to the graph of $y = 4 - 3x - x^2$ at (0,4).

7. Find the equation of the tangent to the graph of $y = 2x^3 - 3x^2 - 12x + 20$ at $x = 2$.

8. Find the equation of the tangent to the graph of $y = \dfrac{x^2 + 4}{x - 6}$ at $x = 5$.

9. Find the equation of the tangent to the graph of $y = \sqrt{x^3 - 15}$ at (4,7).

10. Find the equation of the tangent to the graph of $y = (x^2 + 4x + 4)^2$ at $x = -2$.

11. Find the values of x where the tangent to the graph of $y = 2x^3 - 8x$ has slope equal to the slope of $y = x$.

12. Find the equation of the normal to the graph of $y = \dfrac{3x+5}{x-1}$ at $x = 3$.

13. Find the values of x where the normal to the graph of $(x - 9)^2$ is parallel to the y-axis.

14. Find the coordinates where the tangent to the graph of $y = 8 - 3x - x^2$ is parallel to the x-axis.

15. Find the values of a, b, and c where the curves $y = x^2 + ax + b$ and $y = cx + x^2$ have a common tangent line at (–1,0).

TWO NEW THEOREMS

Now it's time to discuss two more applications of the derivative that you'll hear about a lot in calculus: the Mean Value Theorem and Rolle's Theorem. Both of them are tested frequently on the AP exam; neither of them is too difficult.

The Mean Value Theorem for Derivatives

If $y = f(x)$ is continuous on the interval $[a,b]$, and is differentiable everywhere on the interval (a,b), then there is at least one number c between a and b such that:

$$\frac{f(b) - f(a)}{b - a} = f'(c).$$

In other words, there's some point in the interval where the slope of the tangent line equals the slope of the secant line that connects the endpoints of the interval. (The function has to be continuous at the endpoints of the interval, but it doesn't have to be differentiable at the endpoints. Is this important? Maybe to mathematicians, but probably not to you!) You can see this graphically in the figure below:

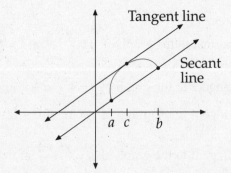

Example 1: Suppose you have the function $f(x) = x^2$, and you're looking at the interval $[1,3]$. The Mean Value Theorem for Derivatives (this is often abbreviated MVTD) states that there is some number c such that:

$$f'(c) = \frac{3^2 - 1^2}{3 - 1} = 4.$$

Because $f'(x) = 2x$, plug in c for x and solve: $2c = 4$ so $c = 2$. Notice that 2 is in the interval. This is what the MVTD predicted! If you don't get a value for c within the interval, something went wrong; either you made a mistake, or the function is not continuous and differentiable in the required interval.

Example 2: Consider the function $f(x) = x^3 - 12x$ on the interval $[-2,2]$. The MVTD states that there is a c such that

$$f'(c) = \frac{\left(2^3 - 24\right) - \left((-2)^3 + 24\right)}{2 - (-2)} = -8$$

Then $f'(c) = 3c^2 - 12 = -8$ and $c = \pm\dfrac{2}{\sqrt{3}}$ (which is approximately 1.155).

Notice that here there are two values of c that satisfy the MVTD. That's allowed. In fact, there can be infinitely many values, depending on the function.

Example 3: Consider the function $f(x) = \dfrac{1}{x}$ on the interval [–2,2].

Follow the MVTD:

$$f'(c) = \frac{\dfrac{1}{2} - \left(-\dfrac{1}{2}\right)}{2 - (-2)} = \frac{1}{4}.$$

Then:

$$f'(c) = \frac{-1}{c^2} = \frac{1}{4}.$$

There is no value of c that will satisfy this equation! We expected this. Why? Because $f(x)$ is not continuous at $x = 0$, which is in the interval. Suppose the interval had been [1,3], eliminating the discontinuity The result would have been:

$$f'(c) = \frac{\dfrac{1}{3} - (1)}{3 - 1} = -\frac{1}{3} \text{ and } f'(c) = \frac{-1}{c^2} = -\frac{1}{3}; \ c = \pm\sqrt{3}.$$

Since $c = -\sqrt{3}$ is not in the interval, but $c = \sqrt{3}$ is, the answer is $c = \sqrt{3}$.

Example 4: Consider the function $f(x) = x^2 - x - 12$ on the interval [–3,4].

Follow the MVTD:

$$f'(c) = \frac{0 - 0}{7} = 0 \text{ and } f'(c) = 2c - 1 = 0 \text{, so } c = \frac{1}{2}.$$

In this last example, you discovered where the derivative of the equation equaled zero. This is going to be the single most common problem you'll solve in differential calculus. So now, we've got an important tip for you:

> When you don't know what to do, take the derivative of the equation and set it equal to zero!!!

Remember this advice for the rest of AP calculus.

ROLLE'S THEOREM

Now let's learn Rolle's Theorem, which is a special case of the MVTD.

If $y = f(x)$ is continuous on the interval $[a,b]$, and is differentiable everywhere on the interval (a,b) and if $f(a) = f(b) = 0$, then there is at least one number c between a and b such that $f'(c) = 0$.

> Graphically, this means that a continuous, differentiable curve has a horizontal tangent between any two points where it crosses the x-axis.

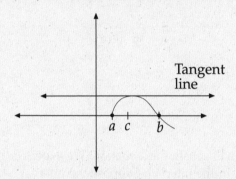

Example 4 was an example of Rolle's Theorem, but let's do another.

Example 5: Consider the function $f(x) = \dfrac{x^2}{2} - 6x$ on the interval $[0,12]$.

First, show that:

$$f(0) = \frac{0}{2} - 6(0) = 0 \text{ and } f(12) = \frac{144}{2} - 6(12) = 0.$$

Then find:

$$f'(x) = x - 6 \text{ so } f'(c) = c - 6.$$

If you set this equal to zero (remember what we told you!), you get $c = 6$. This value of c falls in the interval, so the theorem holds for this example.

As with the MVTD, you'll run into problems with the theorem when the function is not continuous and differentiable over the interval. This is where you need to look out for a trap set by ETS. Otherwise, just follow what we did here and you won't have any trouble with either Rolle's Theorem or the MVTD. Try these example problems, and cover the responses until you check your work.

PROBLEM 1. Find the values of c that satisfy the Mean Value Theorem for $f(x) = x^2 + 2x - 1$ on the interval $[0,1]$.

Answer: First, find $f(0)$ and $f(1)$:

$$f(0) = 0^2 + 2(0) - 1 = -1 \text{ and } f(1) = 1^2 + 2(1) - 1 = 2.$$

Then,

$$\frac{2-(-1)}{1-0} = \frac{3}{1} = 3 = f'(c).$$

Next, find $f'(x)$:

$$f'(x) = 2x + 2.$$

Thus, $f'(c) = 2c + 2 = 3$, and $c = \dfrac{1}{2}$.

PROBLEM 2. Find the values of c that satisfy the Mean Value Theorem for $f(x) = x^3 + 1$ on the interval $[1,2]$.

Answer: Find $f(1) = 1^3 + 1 = 2$ and $f(2) = 2^3 + 1 = 9$. Then:

$$\frac{9-2}{2-1} = 7 = f'(c).$$

Next, $f'(x) = 3x^2$, so $f'(c) = 3c^2 = 7$ and $c = \pm\sqrt{\dfrac{7}{3}}$.

Notice that there are two answers for c, but only one of them is in the interval. The answer is

$c = \sqrt{\dfrac{7}{3}}$.

PROBLEM 3. Find the values of c that satisfy the Mean Value Theorem for $f(x) = x + \dfrac{1}{x}$ on the interval $[-4,4]$.

Answer: First, because the function is not continuos on the interval, there won't be a solution for c.

Let's show that this is true. Find $f(-4) = -4 - \dfrac{1}{4} = -\dfrac{17}{4}$ and $f(4) = 4 + \dfrac{1}{4} = \dfrac{17}{4}$. Then,

$$\frac{\dfrac{17}{4} - \left(-\dfrac{17}{4}\right)}{4 - (-4)} = \frac{17}{16} = f'(c).$$

Next, $f'(x) = 1 - \dfrac{1}{x^2}$. Therefore, $f'(c) = 1 - \dfrac{1}{c^2} = \dfrac{17}{16}$.

There's no solution to this equation.

PROBLEM 4. Find the values of c that satisfy Rolle's Theorem for $f(x) = x^4 - x$ on the interval $[0,1]$.

Answer: Show that $f(0) = 0^4 - 0 = 0$ and that $f(1) = 1^4 - 1 = 0$.

Next, find $f'(x) = 4x^3 - 1$. By setting $f'(c) = 4c^3 - 1 = 0$ and solving, you'll see that $c = \sqrt[3]{\dfrac{1}{4}}$, which

is in the interval.

PRACTICE PROBLEMS

Now try these problems. The answers are in chapter 21.

1. Find the values of c that satisfy the Mean Value Theorem for $f(x) = 3x^2 + 5x - 2$ on the interval $[-1,1]$.

2. Find the values of c that satisfy the Mean Value Theorem for $f(x) = x^3 + 24x - 16$ on the interval $[0,4]$.

3. Find the values of c that satisfy the Mean Value Theorem for $f(x) = x^3 + 12x^2 + 7x$ on the interval $[-4,4]$.

4. Find the values of c that satisfy the Mean Value Theorem for $f(x) = \dfrac{6}{x} - 3$ on the interval $[1,2]$.

5. Find the values of c that satisfy the Mean Value Theorem for $f(x) = \dfrac{6}{x} - 3$ on the interval $[-1,2]$.

6. Find the values of c that satisfy Rolle's Theorem for $f(x) = x^2 - 8x + 12$ on the interval $[2,6]$.

7. Find the values of c that satisfy Rolle's Theorem for $f(x) = x^3 - x$ on the interval $[-1,1]$.

8. Find the values of c that satisfy Rolle's Theorem for $f(x) = x(1 - x)$ on the interval $[0,1]$.

9. Find the values of c that satisfy Rolle's Theorem for $f(x) = 1 - \dfrac{1}{x^2}$ on the interval $[-1,1]$.

10. Find the values of c that satisfy Rolle's Theorem for $f(x) = x^{\frac{2}{3}} - x^{\frac{1}{3}}$ on the interval $[0,1]$.

MORE APPLICATIONS

Here's another chapter of material involving more ways to apply the derivative to several other types of problems. This stuff focuses mainly on using the derivative to aid in graphing a function, etc.

MAXIMA AND MINIMA

One of the most common applications of the derivative is to find a maximum or minimum value of a function. These values can be called extreme values, optimal values, or critical points. Each of these problems involves the same, very simple principle:

> A maximum or a minimum of a function occurs at a point where the derivative of a function is zero.

At a point where the first derivative equals zero, the curve has a horizontal tangent line, at which point it could be reaching either a "peak" (maximum) or a "valley" (minimum).

There are a few exceptions to every rule. This rule is no different.

If the derivative of a function is zero at a certain point, it is usually a maximum or minimum—but not always.

There are two different kinds of maxima and minima: relative and absolute. A **relative** maximum or minimum means that the curve has a horizontal tangent line at that point, but it is not the highest or lowest value that the function attains. In the figure below, the two indicated points are relative maxima/minima.

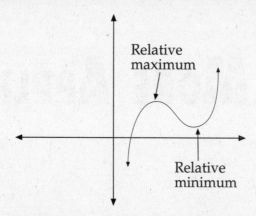

An **absolute** maximum or minimum is either a point where the curve has a horizontal tangent line or a point at an end of the domain of the function where the function attains its highest or lowest value. In the figure below, the two indicated points are absolute maxima/minima. A relative maximum can also be an absolute maximum.

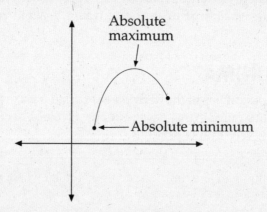

A typical word problem will ask you to find a maximum or a minimum value of a function, as it pertains to a certain situation. Sometimes you're given the equation; other times, you have to figure it out for yourself. Once you have the equation, you find its derivative and set it equal to zero. The values you get are called **critical values**. Then, test these values to determine whether each value is a maximum or a minimum. The simplest way to do this is with the **second derivative test**:

If a function has a critical value at $x = c$, then that value is a relative maximum if $f''(c) < 0$ and it is a relative minimum if $f''(c) > 0$.

If the second derivative is also zero at $x = c$, then the point is neither a maximum or a minimum but a point of inflection. More about that later.

It's time to do some examples.

Example 1: Find the minimum value on the curve $y = ax^2$, if $a > 0$.

Take the derivative and set it equal to zero:

$$\frac{dy}{dx} = 2ax = 0.$$

The first derivative is equal to zero at $x = 0$. By plugging 0 back into the original equation, we can solve for the y-coordinate of the minimum (the y-coordinate is also 0, so the point is at the origin).

In order to determine if this is a maximum or a minimum, take the second derivative:

$$\frac{d^2y}{dx^2} = 2a.$$

Because a is positive, the second derivative is positive and the critical point we obtained from the first derivative is a minimum point. Had a been negative, the second derivative would have been negative and a maximum would have occurred at the critical point.

Example 2: A manufacturing company has determined that the total cost of producing an item can be determined from the equation $C = 8x^2 - 176x + 180$, where x is the number of units that the company makes. How many units should the company manufacture in order to minimize the cost?

Once again, take the derivative of the cost equation and set it equal to zero:

$$\frac{dC}{dx} = 16x - 176 = 0.$$

$$x = 11.$$

This tells us that 11 is a critical point of the equation. Now we need to figure out if this is a maximum or a minimum using the second derivative:

$$\frac{d^2C}{dx^2} = 16.$$

Since 16 is always positive, any critical value is going to be a minimum. Therefore, the company should manufacture 11 units in order to minimize its cost.

Example 3: A rocket is fired into the air, and its height in meters at any given time t can be calculated using the formula $h(t) = 1600 + 196t - 4.9t^2$. Find the maximum height of the rocket and the time at which it occurs.

Take the derivative and set it equal to zero:

$$\frac{dh}{dt} = 196 - 9.8t$$

$$t = 20$$

Now we know that 20 is a critical point of the equation. Now use the second derivative test:

$$\frac{d^2h}{dt^2} = -9.8$$

This is always negative, so any critical value is a maximum. To determine the maximum height of the rocket, plug $t = 20$ into the equation:

$$h(20) = 1600 + 196(20) - 4.9(20^2) = 3560 \text{ meters.}$$

The technique is always the same: (a) take the derivative of the equation; (b) set it equal to zero; and (c) use the second derivative test. The hardest part of these word problems is when you have to set up the equation yourself. The following is a classic AP problem:

Example 4: Max wants to make a box with no lid from a rectangular sheet of cardboard that is 18 inches by 24 inches. The box is to be made by cutting a square of side x from each corner of the sheet and folding up the sides (see figure below). Find the value of x that maximizes the volume of the box.

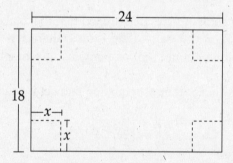

After we cut out the squares of side x and fold up the sides, the dimensions of the box will be:

width: $18 - 2x$
length: $24 - 2x$
depth: x.

Using the formula for the volume of a rectangular prism, we can get an equation for the volume in terms of x:

$$V = x(18 - 2x)(24 - 2x).$$

Multiply the terms together (and be careful with your algebra):

$$V = x(18 - 2x)(24 - 2x) = 4x^3 - 84x^2 + 432x.$$

Now take the derivative:

$$\frac{dV}{dx} = 12x^2 - 168x + 432.$$

Set the derivative equal to zero, and solve for x:

$$12x^2 - 168x + 432 = 0$$

$$x^2 - 14x + 36 = 0$$

$$x = \frac{14 \pm \sqrt{196 - 144}}{2} = 7 \pm \sqrt{13} \approx 3.4, 10.6.$$

Common sense tells us that you can't cut out two square pieces that measure 10.6 inches to a side (the sheet's only 18 inches wide!), so the maximizing value has to be 3.4 inches. Here's the second derivative test, just to be sure:

$$\frac{d^2V}{dx^2} = 24x - 168.$$

At $x = 3.4$,

$$\frac{d^2V}{dx^2} = 24(3.4) - 168 = -86.4$$

so the volume of the box will be maximized when $x = 3.4$.

Therefore, the dimensions of the box that maximize the volume are approximately: $11.2 \times 17.2 \times 3.4$.

CRITICAL POINTS WITHIN AN INTERVAL

Sometimes, particularly when the domain of a function is restricted, you have to test the endpoints of the interval as well. This is because the highest or lowest value of a function may be at an endpoint of that interval; the critical value you obtained from the derivative might be just a local maximum or minimum. For the purposes of the AP exam, however, endpoints are considered separate from critical values.

Example 5: Find the absolute maximum and minimum values of $y = x^3 - x$ on the interval $[-3,3]$.

So far, the song remains the same. Take the derivative and set it equal to zero:

$$\frac{dy}{dx} = 3x^2 - 1 = 0.$$

Solve for x:

$$x = \pm \frac{1}{\sqrt{3}}.$$

Test the critical points:

$$\frac{d^2y}{dx^2} = 6x.$$

At $x = \frac{1}{\sqrt{3}}$, we have a minimum. At $x = -\frac{1}{\sqrt{3}}$, we have a maximum.

$$\text{At } x = -\frac{1}{\sqrt{3}}, \, y = -\frac{1}{3\sqrt{3}} + \frac{1}{\sqrt{3}} = \frac{2}{3\sqrt{3}} \approx .385.$$

$$\text{At } x = \frac{1}{\sqrt{3}}, \, y = \frac{1}{3\sqrt{3}} - \frac{1}{\sqrt{3}} = -\frac{2}{3\sqrt{3}} \approx -.385.$$

Now it's time to check the endpoints of the interval:

$$\text{At } x = -3, \, y = -24.$$
$$\text{At } x = 3, \, y = 24.$$

We can see that the function actually has a *lower* value at $x = -3$ than at its "minimum" when $x = \frac{1}{\sqrt{3}}$. Similarly, the function has a *higher* value at $x = 3$ than at its "maximum" of $x = -\frac{1}{\sqrt{3}}$. This means that the function has a "local minimum" at $x = \frac{1}{\sqrt{3}}$, and an "absolute minimum" when $x = -3$. And, the function has a "local maximum" at $x = -\frac{1}{\sqrt{3}}$, and an "absolute maximum" at $x = 3$.

Here's another classic AP problem:

Example 6: A rectangle is to be inscribed in a semicircle with radius 4, with one side on the semicircle's diameter. What is the largest area this rectangle can have?

Let's look at this on the coordinate axes. The equation for a circle of radius 4, centered at the origin, is $x^2 + y^2 = 16$; a semicircle has the equation $y = \sqrt{16 - x^2}$. Our rectangle can then be expressed as a function of x, where the height is $\sqrt{16 - x^2}$ and the base is $2x$. See the figure below:

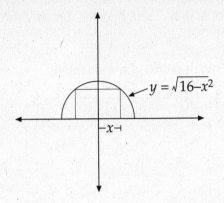

The area of the rectangle is: $A = 2x\sqrt{16 - x^2}$. Let's take the derivative of the area:

$$\frac{dA}{dx} = 2\sqrt{16 - x^2} - \frac{2x^2}{\sqrt{16 - x^2}}.$$

The derivative is not defined at $x = \pm 4$. Setting the derivative equal to zero we get:

$$2\sqrt{16 - x^2} - \frac{2x^2}{\sqrt{16 - x^2}} = 0$$

$$2\sqrt{16 - x^2} = \frac{2x^2}{\sqrt{16 - x^2}}$$

$$2(16 - x^2) = 2x^2$$

$$32 - 2x^2 = 2x^2$$

$$32 = 4x^2$$

$$x = \pm\sqrt{8}.$$

Note that the domain of this function is $-4 \le x \le 4$, so these numbers serve as endpoints of the interval. Let's compare the critical values and the endpoints:

When $x = -4$, $y = 0$ and the area is 0.
When $x = 4$, $y = 0$ and the area is 0.

When $x = \sqrt{8}$, $y = \sqrt{8}$ and the area is 16.

Thus, the maximum area occurs when $x = \sqrt{8}$ and the area equals 16.
Try some of these solved problems on your own. As always, cover the answers as you work.

PROBLEM 1. A rectangular field, bounded on one side by a building, is to be fenced in on the other three sides. If 3,000 feet of fence is to be used, find the dimensions of the largest field that can be fenced in.

Answer: First, let's make a rough sketch of the situation.

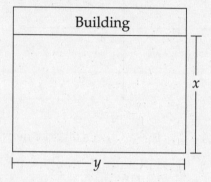

If we call the length of the field y and the width of the field x, the formula for the area of the field becomes:

$$A = xy.$$

The perimeter of the fencing is equal to the sum of two widths and the length:

$$2x + y = 3000.$$

Now, solve this second equation for y:

$$y = 3000 - 2x.$$

When you plug this expression into the formula for the area, you get a formula for A in terms of x:

$$A = x(3000 - 2x) = 3000x - 2x^2.$$

Next, take the derivative, set it equal to zero, and solve for x:

$$\frac{dA}{dx} = 3000 - 4x = 0$$

$$x = 750.$$

Let's check to make sure it's a maximum. Find the second derivative:

$$\frac{d^2A}{dx^2} = -4.$$

Since we have a negative result, $x = 750$ is a maximum. Finally, if we plug in $x = 750$ and solve for y, we find that $y = 1500$. The largest field will measure 750 feet by 1,500 feet.

PROBLEM 2. A poster is to contain 100 square inches of picture surrounded by a 4-inch margin at the top and bottom and a 2-inch margin on each side. Find the overall dimensions that will minimize the total area of the poster.

Answer: First, make a sketch.

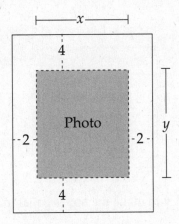

Let the area of the picture be $xy = 100$. The total area of the poster is $A = (x + 4)(y + 8)$. Then, expand the equation:

$$A = xy + 4y + 8x + 32.$$

Substitute $xy = 100$ and $y = \dfrac{100}{x}$ into the area equation, and we get:

$$A = 132 + \frac{400}{x} + 8x.$$

Now take the derivative and set it equal to zero:

$$\frac{dA}{dx} = 8 - \frac{400}{x^2} = 0.$$

Solving for x, we find that $x = \sqrt{50}$. Now solve for y by plugging $x = \sqrt{50}$ into the equation: $y = 2\sqrt{50}$. Then check that these dimensions give us a minimum:

$$\frac{d^2A}{dx^2} = \frac{800}{x^3}.$$

This is positive when x is positive, so the minimum area occurs when $x = \sqrt{50}$. Thus the overall dimensions of the poster are $4 + \sqrt{50}$ inches by $8 + 2\sqrt{50}$ inches.

PROBLEM 3. An open-top box with a square bottom and rectangular sides is to have a volume of 256 cubic inches. Find the dimensions that require the minimum amount of material.

Answer: First, make a sketch of the situation:

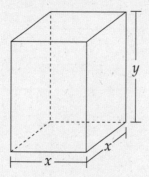

The amount of material necessary to make the box is equal to the surface area:

$$S = x^2 + 4xy.$$

The formula for the volume of the box is $x^2 y = 256$.

If we solve the latter equation for y, $y = \dfrac{256}{x^2}$, and plug it into the former equation, we get:

$$S = x^2 + 4x\frac{256}{x^2} = x^2 + \frac{1{,}024}{x}.$$

Now take the derivative and set it equal to zero:

$$\frac{dS}{dx} = 2x - \frac{1{,}024}{x^2} = 0.$$

If we solve this for x, we get $x^3 = 512$ and $x = 8$. Solving for y, we get $y = 4$.
Check that these dimensions give us a minimum:

$$\frac{d^2 A}{dx^2} = 2 + \frac{2{,}048}{x^3}$$

This is positive when x is positive, so the minimum surface area occurs when $x = 8$. The dimensions of the box should be 8 inches by 8 inches by 4 inches.

PROBLEM 4. Find the point on the curve $y = \sqrt{x}$ that is a minimum distance from the point (4,0).

Answer: First, make that sketch.

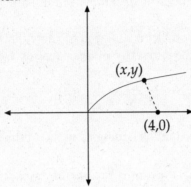

Using the distance formula:

$$D^2 = (x - 4)^2 + (y - 0)^2 = x^2 - 8x + 16 + y^2$$

Because $y = \sqrt{x}$,

$$D^2 = x^2 - 8x + 16 + x = x^2 - 7x + 16.$$

Next, let $L = D^2$. We can do this because the minimum value of D^2 will occur at the same value of x as the minimum value of D. Therefore, it's simpler to minimize D^2 rather than D (because we won't have to take a square root!):

$$L = x^2 - 7x + 16.$$

Now take the derivative and set it equal to zero:

$$\frac{dL}{dx} = 2x - 7 = 0.$$

$$x = \frac{7}{2}.$$

Solving for y, we get $y = \sqrt{\dfrac{7}{2}}$.

Finally, because $\dfrac{d^2L}{dx^2} = 2$, the point $\left(\dfrac{7}{2}, \sqrt{\dfrac{7}{2}}\right)$ is the minimum distance from the point (4,0).

PROBLEM 5. Suppose that the revenue of a company can be represented with the function $r(x) = 48x$, and the company's cost function is $c(x) = x^3 - 12x^2 + 60x$, where x represents thousands of units and revenue and cost are represented in thousands of dollars. What production level maximizes profit, and what is the maximum profit to the nearest thousand dollars?

Answer: In order to solve this problem, you need to know that profit equals revenue minus cost. Since $P = r - c$, you should take the derivative of this equation with respect to x:

$$\frac{dP}{dx} = \frac{dr}{dx} - \frac{dc}{dx}.$$

Now, take the derivative of each of the equations in the problem:

$$\frac{dr}{dx} = 48 \text{ and } \frac{dc}{dx} = 3x^2 - 24x + 60.$$

Plug these values into the previous equation to get:

$$\frac{dP}{dx} = 48 - 3x^2 + 24x - 60 = -3x^2 + 24x - 12.$$

Now set $\frac{dP}{dx}$ equal to zero and solve for x:

$$-3x^2 + 24x - 12 = 0$$

$$x^2 - 8x + 4 = 0$$

$$x = \frac{8 \pm \sqrt{48}}{2} \approx 7.46, \ 0.54.$$

At $x = 0.54$, revenue is 25.92 and cost is 29.06. At $x = 7.46$, revenue is 358.08 and cost is 194.94. The first value of x gives us a negative profit and the second value of x gives us a positive one, so the maximum profit must occur at $x \approx 7.46$ (thousand) units. The maximum profit, therefore, must be approximately $358.08 - 194.94 = 163.14$, or \$163,140.

PRACTICE PROBLEMS

Now try these problems on your own. The answers are in chapter 21.

1. A rectangle has its base on the x-axis and its two upper corners on the parabola $y = 12 - x^2$. What is the largest possible area of the rectangle?

2. An open rectangular box is to be made from a 9×12 inch piece of tin by cutting squares of side x from the corners and folding up the sides. What should x be to maximize the volume of the box?

3. A 384 square meter plot of land is to be enclosed by a fence and divided into two equal parts by another fence parallel to one pair of sides. What dimensions of the outer rectangle will minimize the amount of fence used?

4. What is the radius of a cylindrical soda can with volume of 512 cubic inches that will use the minimum material?

5. A swimmer is at a point 500 m from the closest point on a straight shoreline. She needs to reach a cottage located 1800 m down shore from the closest point. If she swims at 4 m/s and she walks at 6 m/s, how far from the cottage should she come ashore so as to arrive at the cottage in the shortest time?

6. Find the closest point on the curve $x^2 + y^2 = 1$ to the point (2,0).

7. A window consists of an open rectangle topped by a semicircle and is to have a perimeter of 288 inches. Find the radius of the semicircle that will maximize the area of the window.

8. The range of a projectile is $R = \dfrac{v_0^2 \sin 2\theta}{g}$, where v_0 is its initial velocity, g is the acceleration due to gravity, and is a constant, and θ is its firing angle. Find the angle that maximizes the projectile's range.

9. A computer company determines that its profit equation (in millions of dollars) is given by $P = x^3 - 48x^2 + 720x - 1000$, where x is the number of thousands of units of software sold and $0 \leq x \leq 40$. Optimize the manufacturer's profit.

CURVE SKETCHING

Another topic on which students spend a lot of time in calculus is curve sketching. In the old days, whole courses (called "Analytic Geometry" or something equally dull) were devoted to the subject, and students had to master a wide variety of techniques to learn how to sketch a curve accurately.

Fortunately (or unfortunately, depending on your point of view), students no longer need to be as good at analytic geometry. There are two reasons for this: (1) the AP exam only tests a few types of curves; and (2) you can use a graphing calculator. Because of the calculator, you can get an idea of the shape of the curve, and all you need to do is find important points to label the graph. We use calculus to find some of these points.

When it's time to sketch a curve, we'll show you a four-part analysis that'll give you all the information you need.

TEST THE FUNCTION

Find where $f(x) = 0$. This tells you the function's x-intercepts (or roots). By setting $x = 0$, we can determine the y-intercepts. Then, find any horizontal and/or vertical asymptotes.

TEST THE FIRST DERIVATIVE

Find where $f'(x) = 0$. This tells you the critical points. We can determine whether the curve is rising or falling, as well as where the maxima and minima are. It's also possible to determine if the curve has any points where it's nondifferentiable.

TEST THE SECOND DERIVATIVE

Find where $f''(x) = 0$. This shows you where any points of inflection are. (These are points where the graph of a function changes concavity.) Then we can determine where the graph curves upward and where it curves downward.

TEST END BEHAVIOR

Look at what the general shape of the graph will be, based on the values of y for very large values of $\pm x$. Using this analysis, we can always come up with a sketch of a curve.

And now, the rules:

(1) When $f'(x) > 0$, the curve is rising; when $f'(x) < 0$, the curve is falling; when $f'(x) = 0$, the curve is at a critical point.

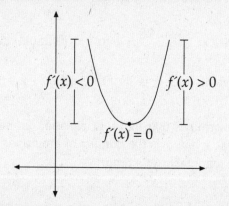

(2) When $f''(x) > 0$, the curve is "concave up"; when $f''(x) < 0$, the curve is "concave down"; when $f''(x) = 0$, the curve is at a point of inflection.

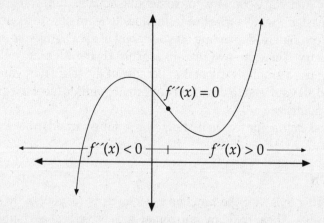

(3) The y-coordinates of each critical point are found by plugging the x-value into the original equation.

As always, this stuff will sink in better if we try a few examples.

Example 1: Sketch the equation $y = x^3 - 12x$.

Step 1: Find the x-intercepts:

$$x^3 - 12x = 0$$

$$x(x^2 - 12) = 0$$

$$x\left(x - \sqrt{12}\right)\left(x + \sqrt{12}\right) = 0$$

$$x = 0, \pm\sqrt{12}$$

The curve has x-intercepts at $\left(\sqrt{12}, 0\right), \left(-\sqrt{12}, 0\right)$, and $(0,0)$.

Next, find the y-intercepts:

$$y = (0)^3 - 12(0) = 0.$$

The curve has a y-intercept at $(0,0)$.

There are no asymptotes, because there's no place where the curve is undefined (you won't have asymptotes for curves that are simple polynomials).

Step 2: Take the derivative of the function to find the critical points:

$$\frac{dy}{dx} = 3x^2 - 12.$$

Set the derivative equal to zero and solve for x:

$$3x^2 - 12 = 0$$
$$3(x^2 - 4) = 0$$
$$3(x - 2)(x + 2) = 0$$
$$\text{so } x = 2, -2.$$

Next, plug $x = 2, -2$ into the original equation to find the y-coordinates of the critical points:

$$y = (2)^3 - 12(2) = -16$$
$$y = (-2)^3 - 12(-2) = 16.$$

Thus, we have critical points at $(2, -16)$ and $(-2, 16)$.

Step 3: Now, take the second derivative to find any points of inflection:

$$\frac{d^2y}{dx^2} = 6x.$$

This equals zero at $x = 0$. We already know that when $x = 0$, $y = 0$, so the curve has a point of inflection at $(0, 0)$.

Now, plug the critical values into the second derivative to determine whether each is a maximum Thus, there is an inflection point at $x=0$. Now, plug in the critical points. $f''(2) = 6(2) = 12$, this is positive, so the curve has a minimum at $(2, -16)$ and the curve is concave up at that point. However, $6(-2) = -12$. This value is negative, so the curve has a maximum at $(-2,16)$ and the curve is concave down there.

Armed with this information, we can now plot the graph:

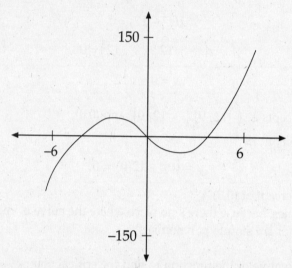

If you've got a TI-82 calculator (or some other gizmo that graphs functions), you can check your work to get an idea of what the graph looks like. Using a TI-82, push the **GRAPH** button, and set $y_1 = x^3 - 12x$. This is a cubic equation, so you'll need to adjust the window a little bit. Push the **WINDOW** button and set **xmin** to -6, **xmax** to 6, **ymin** to -150, and **ymax** to 150, **step** 30.

Now, check out your graph. Does it match?

You might be wondering where we came up with those window settings. Think about the critical points. You know that the critical values are -2, 0, and 2, so you don't need to make the x-axis very wide. Look at the critical y-coordinates, though: at $x = 6$, the y-coordinate is 144, and at $x = -6$, the y-coordinate is -144. From this information, you'll have to stretch the y-axis.

It's time to try another. Try to contain your excitement.

Example 2: Sketch the graph of $y = x^4 + 2x^3 - 2x^2 + 1$.

Follow the same routine as the previous example.
Step 1: First let's find the x-intercepts:

$$x^4 + 2x^3 - 2x^2 + 1 = 0.$$

Hmm. This is a stumper. If the equation doesn't factor easily, it's best not to bother to find the function's roots. Convenient, huh? The good news is that if the roots aren't easy to find, ETS won't ask you to find them.

Next, let's find the y-intercepts.

$$y = (0)^4 + 2(0)^3 - 2(0)^2 + 1.$$

The curve has a y-intercept at $(0,1)$.
There are no asymptotes because there is no place where the curve is undefined.

Step 2: Now we take the derivative to find the critical points:

$$\frac{dy}{dx} = 4x^3 + 6x^2 - 4x.$$

Set the derivative equal to zero:

$$4x^3 + 6x^2 - 4x = 0$$
$$2x(2x^2 + 3x - 2) = 0$$
$$2x(2x - 1)(x + 2) = 0$$

$$x = 0, \frac{1}{2}, -2.$$

Next, plug these three values into the original equation to find the y-coordinates of the critical points. We already know that when $x = 0$, $y = 1$.

When $x = \frac{1}{2}$, $y = \left(\frac{1}{2}\right)^4 + 2\left(\frac{1}{2}\right)^3 - 2\left(\frac{1}{2}\right)^2 + 1 = \frac{13}{16}$.

When $x = -2$, $y = (-2)^4 + 2(-2)^3 - 2(-2)^2 + 1 = -7$.

Thus, we have critical points at $(0,1)$, $\left(\frac{1}{2}, \frac{13}{16}\right)$ and $(-2,-7)$.

Step 3: Take the second derivative to find any points of inflection:

$$\frac{d^2y}{dx^2} = 12x^2 + 12x - 4.$$

Set this equal to zero:

$$12x^2 + 12x - 4 = 0$$
$$3x^2 + 3x - 1 = 0$$

$$x = \frac{-3 \pm \sqrt{21}}{6} \approx .26, -1.26.$$

Therefore, the curve has points of inflection at $x = \frac{-3 \pm \sqrt{21}}{6}$.

Now, solve for the y-coordinates:

$$(.26, .90) \text{ and } (-1.26, -3.66).$$

We can now plug the critical values into the second derivative to determine whether each is a maximum or a minimum.

$$12(0)^2 + 12(0) - 4 = -4.$$

This is negative, so the curve has a maximum at $(0,1)$; the curve is concave down there.

$$12\left(\frac{1}{2}\right)^2 + 12\left(\frac{1}{2}\right) - 4 = 5.$$

This is positive, so the curve has a minimum at $\left(\dfrac{1}{2}, \dfrac{13}{16}\right)$; the curve is concave up there.

$$12(-2)^2 + 12(-2) - 4 = 20 .$$

This is positive, so the curve has a minimum at $(-2, -7)$ and the curve is also concave up there. We can now plot the graph:

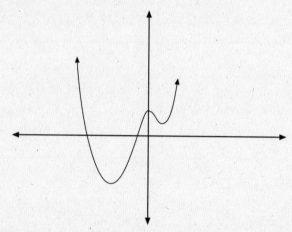

Finally, use your calculator to check the graph. You can also use the calculator to find the two roots of this graph. (Remember the ones we couldn't figure out before?) Using the **CALC** menu, item 2 (**ROOT**), you can get the two values; they're approximately at $(-.59, 0)$ and $(-2.69, 0)$.

FINDING A CUSP

If the derivative of a function approaches ∞ from one side of a point and $-\infty$ from the other, and if the function is continuous at that point, then the curve has a "cusp" at that point. In order to find a cusp, you need to add a little to your graphing routine. You need to look at points where the first derivative is undefined, as well as where it's zero.

Another example beckons.

Example 3: Sketch the graph of $y = 2 - x^{\frac{2}{3}}$.

Step 1: Find the x-intercepts:

$$2 - x^{\frac{2}{3}} = 0$$

$$x^{\frac{2}{3}} = 2 \qquad x = 2^{\frac{3}{2}} = \pm 2\sqrt{2} .$$

The x-intercepts are at $\left(\pm 2\sqrt{2},\, 0\right)$.

Next, find the y-intercepts:

$$y = 2 - (0)^{\frac{2}{3}} = 2 .$$

The curve has a y-intercept at $(0,2)$.
There are no asymptotes because there is no place where the curve is undefined.

Step 2: Now take the derivative to find the critical points:

$$\frac{dy}{dx} = -\frac{2}{3}x^{-\frac{1}{3}}.$$

What's next? You guessed it! Set the derivative equal to zero:

$$-\frac{2}{3}x^{-\frac{1}{3}} = 0.$$

There are no values of x for which the equation is zero. But here's the new stuff to deal with: at $x = 0$, the derivative is undefined. If you look at the limit as x approaches 0 from both sides, we can determine whether the graph has a cusp.

$$\lim_{x \to 0^+} -\frac{2}{3}x^{-\frac{1}{3}} = -\infty \text{ and } \lim_{x \to 0^-} -\frac{2}{3}x^{-\frac{1}{3}} = \infty$$

Therefore, the curve has a cusp at $(0, 2)$.
There aren't any other critical points. But, we can see that when $x < 0$, the derivative is positive (which means that the curve is rising to the left of zero); and when $x > 0$ the derivative is negative (which means that the curve is falling to the right of zero).

Step 3: Now, we take the second derivative to find any points of inflection.

$$\frac{d^2y}{dx^2} = \frac{2}{9}x^{-\frac{4}{3}}.$$

Again, there's no way for this term to equal zero. In fact, the second derivative is positive at all values of x except 0. Therefore, the graph is concave up everywhere.
Now it's time to graph this:

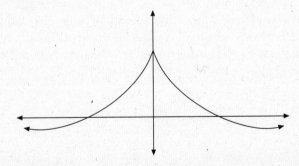

RATIONAL FUNCTIONS

There's one other type of graph you should know about: a rational function. In order to graph a rational function, you need to know how to find that function's asymptotes.

HOW TO FIND ASYMPTOTES

A line $y = c$ is a horizontal asymptote of the graph of $y = f(x)$ if:

$$\lim_{x \to \infty} f(x) = c \text{ or if } \lim_{x \to -\infty} f(x) = c.$$

A line $x = k$ is a vertical asymptote of the graph of $y = f(x)$ if:

$$\lim_{x \to k^+} f(x) = \pm\infty \text{ or if } \lim_{x \to k^-} f(x) = \pm\infty .$$

Example 4: Sketch the graph of $y = \dfrac{3x}{x+2}$.

Step 1: Find the x-intercepts. A fraction can be equal to zero only when it's numerator is equal to zero (provided that the denominator is not also zero there). All we have to do is set $3x = 0$, and you get $x = 0$. Thus, the graph has an x-intercept at (0,0). Note: this is also the y-intercept.

Next, look for asymptotes. Guess what? We've got one! The denominator is undefined at $x = -2$, and if we take the left- and right-hand limits of the function:

$$\lim_{x \to -2^+} \frac{3x}{x+2} = -\infty \text{ and } \lim_{x \to -2^-} \frac{3x}{x+2} = \infty.$$

The curve has a vertical asymptote at $x = -2$.

If we take $\lim\limits_{x \to \infty} \dfrac{3x}{x+2} = 3$ and $\lim\limits_{x \to -\infty} \dfrac{3x}{x+2} = 3$, so the curve has a horizontal asymptote at $y = 3$.

Step 2: Now take the derivative to figure out the critical points:

$$\frac{dy}{dx} = \frac{(x+2)(3) - (3x)(1)}{(x+2)^2} = \frac{6}{(x+2)^2}.$$

There are no values of x which make the derivative equal to zero. Since the numerator is 6 and the denominator is a square, the derivative will always be positive (the curve is always rising). You should note that the derivative is undefined at $x = -2$, but you already know that there's an asymptote at $x = -2$, so you don't need to examine this point further.

Step 3: Now it's time for the second derivative:

$$\frac{d^2y}{dx^2} = \frac{-12}{(x+2)^3}.$$

Again, there's no way to set this equal to zero. The expression is positive when $x < -2$, so the graph is concave up when $x < -2$. The second derivative is negative when $x > -2$, so it's concave down when $x > -2$.

Now plot the graph:

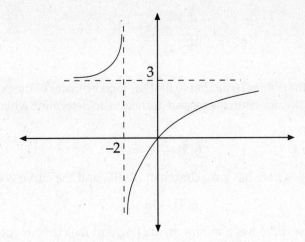

We could go on with examples like this until you're blue in the face. However, the sketches we've shown you so far are the types you're most likely to see on the AP exam. So, now it's time to practice some problems. Do each problem, covering the answer first, then checking your answer.

PROBLEM 1. Sketch the graph of $y = x^3 - 9x^2 + 24x - 10$. Plot all extrema, points of inflection, and asymptotes.

Answer: Follow the three steps.
First, see if the x-intercepts are easy to find. This is a cubic equation that isn't easily factored. So skip this step.
Next, find the y-intercepts by setting $x = 0$.

$$y = (0)^3 - 9(0)^2 + 24(0) - 10 = -10.$$

The curve has a y-intercept at $(0, -10)$.
There are no asymptotes, because the curve is a simple polynomial.
Next, find the critical points using the first derivative:

$$\frac{dy}{dx} = 3x^2 - 18x + 24.$$

Set the derivative equal to zero and solve for x:

$$3x^2 - 18x + 24 = 0$$
$$3(x^2 - 6x + 8) = 0$$
$$3(x - 4)(x - 2) = 0$$
$$x = 2, 4.$$

Plug $x = 2$ and $x = 4$ into the original equation to find the y-coordinates of the critical points.
When $x = 2$, $y = 10$. When $x = 4$, $y = 6$.
Thus, we have critical points at $(2, 10)$ and $(4, 6)$.

In our third step, the second derivative indicates any points of inflection:

$$\frac{d^2y}{dx^2} = 6x - 18.$$

This equals zero at $x = 3$.

Next, plug $x = 3$ into the original equation to find the y-coordinates of the point of inflection, which is at $(3, 8)$. Plug the critical values into the second derivative to determine whether each is a maximum or a minimum:

$$6(2) - 18 = -6.$$

This is negative, so the curve has a maximum at $(2, 10)$, and the curve is concave down there.

$$6(4) - 18 = 6.$$

This is positive, so the curve has a minimum at $(4,6)$, and the curve is concave up there.
It's graph-plotting time.

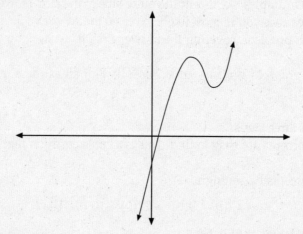

Problem 2. Sketch the graph of $y = 8x^2 - 16x^4$. Plot all extrema, points of inflection, and asymptotes.

Answer:
Factor the polynomial:

$$8x^2(1 - 2x^2) = 0.$$

Solving for x, we get $x = 0$ (a double root), $x = \dfrac{1}{\sqrt{2}}$, and $x = -\dfrac{1}{\sqrt{2}}$.

Find the y-intercepts: when $x = 0$, $y = 0$.
There are no asymptotes, because the curve is a simple polynomial.
Find the critical points using the first derivative:

$$\frac{dy}{dx} = 16x - 64x^3.$$

Set the derivative equal to zero and solve for x. You get $x = 0$, $x = \dfrac{1}{2}$, and $x = -\dfrac{1}{2}$.

Next, plug $x = 0$, $x = \dfrac{1}{2}$, and $x = -\dfrac{1}{2}$ into the original equation to find the y-coordinates of the critical points:

When $x = 0$, $y = 0$. When $x = \dfrac{1}{2}$, $y = 1$. When $x = -\dfrac{1}{2}$, $y = 1$.

Thus, there are critical points at $(0, 0)$, $\left(\dfrac{1}{2}, 1\right)$ and $\left(-\dfrac{1}{2}, 1\right)$.

Take the second derivative to find any points of inflection:

$$\frac{d^2y}{dx^2} = 16 - 192x^2.$$

This equals zero at $x = \dfrac{1}{\sqrt{12}}$ and $x = -\dfrac{1}{\sqrt{12}}$.

Next, plug $x = \dfrac{1}{\sqrt{12}}$ and $x = -\dfrac{1}{\sqrt{12}}$ into the original equation to find the y-coordinates of the

points of inflection, which are at $\left(\dfrac{1}{\sqrt{12}}, \dfrac{5}{9}\right)$ and $\left(-\dfrac{1}{\sqrt{12}}, \dfrac{5}{9}\right)$.

Now determine whether the points are maxima or minima.
At $x = 0$, we have a minimum; the curve is concave up there.

At $x = \dfrac{1}{2}$, it's a maximum, and the curve is concave down.

At $x = -\dfrac{1}{2}$, it's also a maximum (still concave down).
Now plot:

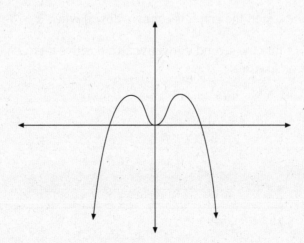

PROBLEM 3. Sketch the graph of $y = \left(\dfrac{x-4}{x+3}\right)^2$. Plot all extrema, points of inflection, and asymptotes.

Answer: This should seem rather routine by now.
Find the x-intercepts, by setting the numerator equal to zero; $x = 4$. The graph has an x-intercept at $(4,0)$. (It's a double root).

Next, find the y-intercept by plugging in $x = 0$: $y = \dfrac{16}{9}$.

The denominator is undefined at $x = -3$, so there's a vertical asymptote at that point.
Look at the limits:

$$\lim_{x \to \infty} \left(\frac{x-4}{x+3}\right)^2 = 1 \text{ and } \lim_{x \to -\infty} \left(\frac{x-4}{x+3}\right)^2 = 1.$$

The curve has a horizontal asymptote at $y = 1$.
It's time for the first derivative:

$$\frac{dy}{dx} = 2\left(\frac{x-4}{x+3}\right)\frac{(x+3)(1)-(x-4)(1)}{(x+3)^2} = \frac{14x-56}{(x+3)^3}.$$

The derivative is zero when $x = 4$, and the derivative is undefined at $x = -3$. (There's an asymptote there, so we can ignore the point. If the curve were *defined* at $x = -3$, then it would be a critical point, as you'll see in the next example.)
Now for the second derivative:

$$\frac{d^2y}{dx^2} = \frac{(x+3)^3(14)-(14x-56)3(x+3)^2}{(x+3)^6} = \frac{-28x+210}{(x+3)^4}.$$

This is zero when $x = \dfrac{15}{2}$. The second derivative is positive (and the graph is concave up) when

$x < \dfrac{15}{2}$, and it's negative (and the graph is concave down) when $x > \dfrac{15}{2}$.

We can now plug $x = 4$ into the second derivative. It's positive there, so $(4,0)$ is a minimum.
Your graph should look like this:

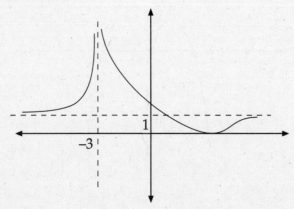

PROBLEM 4. Sketch the graph of $y = (x-4)^{\frac{2}{3}}$. Plot all extrema, points of inflection, and asymptotes.

Answer: By inspection, the x-intercept is at $x = 4$.

Next, find the y-intercepts. When $x = 0$, $y = \sqrt[3]{16} \approx 2.52$.

No asymptotes exist because there's no place where the curve is undefined.

The first derivative is:

$$\frac{dy}{dx} = \frac{2}{3}(x-4)^{-\frac{1}{3}}.$$

Set it equal to zero:

$$\frac{2}{3}(x-4)^{-\frac{1}{3}} = 0.$$

This can never equal zero. But, at $x = 4$ the derivative is undefined, so this is a critical point. If you look at the limit as x approaches 4 from both sides, you can see if there's a cusp:

$$\lim_{x \to 4^+} \frac{2}{3}(x-4)^{-\frac{1}{3}} = \infty \quad \text{and} \quad \lim_{x \to 4^-} \frac{2}{3}(x-4)^{-\frac{1}{3}} = -\infty.$$

The curve has a cusp at $(4,0)$.

There were no other critical points. But, we can see that when $x > 4$, the derivative is positive and the curve is rising; when $x < 4$ the derivative is negative, and the curve is falling.

The second derivative is:

$$\frac{d^2y}{dx^2} = -\frac{2}{9}(x-4)^{-\frac{4}{3}}.$$

No value of x can set this equal to zero. In fact, the second derivative is negative at all values of x except 4. Therefore, the graph is concave down everywhere.

Your graph should look like this:

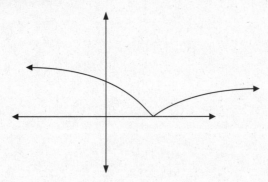

PRACTICE PROBLEMS

It's time for you to try some of these on your own. Sketch each of the graphs below and check the answers in chapter 21.

1. $y = x^3 - 9x - 6$

2. $y = -x^3 - 6x^2 - 9x - 4$

3. $y = \left(x^2 - 4\right)\left(9 - x^2\right)$

4. $y = \dfrac{x^4}{4} - 2x^2$

5. $y = \dfrac{x - 3}{x + 8}$

6. $y = \dfrac{x^2 - 4}{x - 3}$

7. $y = \dfrac{2x}{x^2 - 25}$

8. $y = 3 + x^{\frac{2}{3}}$

9. $y = x^{\frac{2}{3}}\left(3 - 2x^{\frac{1}{3}}\right)$

10. $y = \dfrac{3x^2}{x^2 - 4}$

8

MOTION

Are you having fun so far? This chapter deals with two different types of word problems that involve motion: related rates and the relationship between velocity and acceleration of a particle. The subject matter might seem arcane, but once you get the hang of them, you'll see that these aren't so hard, either. Besides, the AP only tests a few basic problem types.

RELATED RATES

The idea behind these problems is very simple. In a typical problem, you'll be given an equation relating two or more variables. These variables will change with respect to time, and you'll use derivatives to determine how the rates of change are related. (Hence the name: related rates.) Sounds easy, doesn't it?

Example 1: A circular pool of water is expanding at the rate of $16\pi \frac{in^2}{\text{sec}}$. At what rate is the radius expanding when the radius is 4 inches?

 Note: the pool is expanding in square inches per second. We've been given the rate that the area is changing, and we need to find the rate of change of the radius. What equation relates the area of a circle to its radius? $A = \pi r^2$.

Step 1: Set up the equation and take the derivative of this equation with respect to t (time), and you'll get this:

$$\frac{dA}{dt} = 2\pi r \frac{dr}{dt}.$$

In this equation, dA/dt represents the rate at which the area is changing, and $\frac{dr}{dt}$ is the rate that the radius is changing. The simplest way to explain this is that whenever you have a variable in an equation (r, for example), the derivative with respect to time $\left(\frac{dr}{dt}\right)$ represents the rate at which that variable is increasing or decreasing.

Step 2: Now we can plug in the values for the rate of change of the area and for the radius. (Never plug in the values until after you have taken the derivative or you will get nonsense!)

$$16\pi = 2\pi(4)\frac{dr}{dt}.$$

Solving for $\frac{dr}{dt}$, we get:

$$16\pi = 8\pi\frac{dr}{dt} \text{ and } \frac{dr}{dt} = 2.$$

The radius is changing at a rate of $2\frac{in}{sec}$. It's important to note that this is the rate only when the radius is 4 inches. As the circle gets bigger and bigger, the radius will expand slower and slower. (If you don't believe us, pour a can of chocolate syrup on your kitchen floor and watch the circular puddle expand. Tell your folks it's an experiment.)

Example 2: A 25-foot long ladder is leaning against a wall and sliding toward the floor. If the foot of the ladder is sliding away from the base of the wall at a rate of $15\frac{feet}{sec}$, how fast is the top of the ladder sliding down the wall when the top of the ladder is 7 feet from the ground?

Here's another classic related rates problem. As always, a picture is worth 1,000 words:

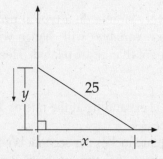

You can see that the ladder forms a right triangle with the wall. Let x stand for the distance from the foot of the ladder to the base of the wall, and let y represent the distance from the top of the ladder to the ground. What's our favorite theorem that deals with right triangles? It's time for our good buddy Pythagoras. The Pythagorean Theorem tells us here that $x^2 + y^2 = 25^2$. Now we have an equation that relates the variables to each other.

Now take the derivative of the equation with respect to t:

$$2x\frac{dx}{dt} + 2y\frac{dy}{dt} = 0.$$

It's downhill from here. Just plug in what you know and solve. Since we're looking for the rate that the vertical distance is changing, we're going to solve for $\frac{dy}{dt}$.

Let's see what we know. We're given the rate that the ladder is sliding away from the wall: $\frac{dx}{dt} = 15$. The distance from the ladder to the top of the wall is 7 feet ($y = 7$). To find x, use The Pythagorean Theorem. If we plug in $y = 7$ to the equation $x^2 + y^2 = 25^2$, $x = 24$.
Now plug all this information into the derivative equation:

$$2(24)(15) + 2(7)\frac{dy}{dt} = 0.$$

$$\frac{dy}{dt} = \frac{-360}{7}\frac{\text{feet}}{\text{sec}}$$

Example 3: A spherical balloon is expanding at a rate of $60\pi\,\frac{\text{in}^3}{\text{sec}}$. How fast is the surface area of the balloon expanding when the radius of the balloon is 4 in?

Step 1: You're given the rate at which the volume's expanding, and you know the equation that relates volume to radius. But you have to relate radius to surface area as well, because you've got to find the surface area's rate of change. This means that you'll need the equations for volume of a sphere:

$$V = \frac{4}{3}\pi r^3$$

and surface area of a sphere:

$$A = 4\pi r^2.$$

You're trying to find $\frac{dA}{dt}$, but A is given in terms of r, so you have to get $\frac{dr}{dt}$ first. Because we know the volume, if we work with the equation that gives us volume in terms of radius, we can find $\frac{dr}{dt}$. From there, work with the other equation to find $\frac{dA}{dt}$. If we take the derivative of the equation with respect to t we get: $\frac{dV}{dt} = 4\pi r^2 \frac{dr}{dt}$. Plugging in for $\frac{dV}{dt}$ and for r, we get: $60\pi = 4\pi(4)^2 \frac{dr}{dt}$.

Solving for $\frac{dr}{dt}$ we get:

$$\frac{dr}{dt} = \frac{15}{16}\frac{\text{in}}{\text{sec}}.$$

Step 2: Now, we take the derivative of the other equation with respect to t:

$$\frac{dA}{dt} = 8\pi r \frac{dr}{dt}.$$

We can plug in for r and $\frac{dr}{dt}$ from the previous step and we get:

$$\frac{dA}{dt} = 8\pi(4)\frac{15}{16} = \frac{480\pi}{16}\frac{\text{in}^2}{\text{sec}} = 30\pi\frac{\text{in}^2}{\text{sec}}.$$

One final example.

Example 4: An underground conical tank, standing on its vertex, is being filled with water at the rate of $18\pi\dfrac{\text{ft}^3}{\text{min}}$. If the tank has a height of 30 feet and a radius of 15 feet, how fast is the water level rising when the water is 12 feet deep?

This "cone" problem is also typical. The key point to getting these right is knowing that the ratio of the height of a right circular cone to its radius is constant. By telling us that the height of the cone is 30 and the radius is 15, we know that at any level, the height of the water will be twice its radius, or $h = 2r$.

The volume of a cone is $V = \dfrac{1}{3}\pi r^2 h$. (You'll learn to derive this formula through integration in chapter 15.)

You must find the rate at which the water is rising (the height is changing), or $\dfrac{dh}{dt}$. Therefore, you want to eliminate the radius from the volume. By substituting $\dfrac{h}{2} = r$ into the equation for volume, we get:

$$V = \frac{1}{3}\pi\left(\frac{h}{2}\right)^2 h = \frac{\pi h^3}{12}.$$

Differentiate both sides with respect to t:

$$\frac{dV}{dt} = \frac{\pi}{12}3h^2\frac{dh}{dt}.$$

Now we can plug in and solve for $\dfrac{dh}{dt}$:

$$18\pi = \frac{\pi}{12}3(12)^2\frac{dh}{dt}$$

$$\frac{dh}{dt} = \frac{1}{2}\frac{\text{feet}}{\text{min}}.$$

In order to solve related rates problems, you have to be good at determining relationships between variables. Once you figure that out, the rest is a piece of cake. Many of these problems involve geometric relationships, so review the formulas for the volumes and areas of cones, spheres, boxes, etc. Once you get the hang of setting up the problems, you'll see that these problems follow the same predictable patterns. Then they won't seem so bad.

Especially after you look through these sample problems. The answers are right there below each question, so whip out your index card and get to work.

PROBLEM 1. A circle is increasing in area at the rate of 16π in² / s. How fast is the radius increasing when the radius is 2 in?

Answer: Use the expression that relates the area of a circle to its radius: $A = \pi R^2$
Next, take the derivative of the expression with respect to t:

$$\frac{dA}{dt} = 2\pi R \frac{dR}{dt}.$$

Now, plug in $\frac{dA}{dt} = 16\pi$ and $R = 2$

$$16\pi = 2\pi(2)\frac{dR}{dt}.$$

When you solve for $\frac{dR}{dt}$, you'll get $\frac{dR}{dt} = 4$ in/sec.

PROBLEM 2. A rocket is rising vertically at a rate of 5,400 miles per hour. An observer on the ground is standing 20 miles from the rocket's launch point. How fast (in radians per second) is the angle of elevation between the ground and the observer's line of sight of the rocket increasing when the rocket is at an elevation of 40 miles? (Notice that velocity is given in miles per hour and the answer asks for radians per second. In situations like this one, you have to be sure to convert the units properly, or you'll get nailed.)

Answer: First, draw that picture.

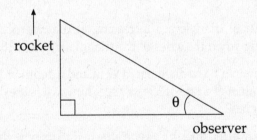

Now create the equation that relates the angle of elevation to the rocket's altitude:

$$\tan\theta = \frac{H}{20}.$$

If we take the derivative of both sides of this expression with respect to t, we get:

$$\sec^2\theta\,\frac{d\theta}{dt} = \frac{1}{20}\frac{dH}{dt}.$$

We know that $\dfrac{dH}{dt} = 5400$ miles per hour, but the problem asks for a time unit in seconds; we need to convert this number. There are 3,600 seconds in an hour, so $\dfrac{dH}{dt} = \dfrac{3}{2}$ miles per second. Next, we know that $\tan\theta = \dfrac{H}{20}$, so when $H = 40$, $\tan\theta = 2$. Because $1 + \tan^2\theta = \sec^2\theta$, we get: $\sec^2\theta = 5$. Plug this information in:

$$5\frac{d\theta}{dt} = \frac{1}{20}\left(\frac{3}{2}\right) \text{ and } \frac{d\theta}{dt} = \frac{3}{200} \text{ radians per second.}$$

PRACTICE PROBLEMS

The answers are in chapter 21.

1. Oil spilled from a tanker spreads in a circle whose circumference increases at a rate of 40 ft / sec. How fast is the area of the spill increasing when the circumference of the circle is 100π ft?

2. A spherical balloon is inflating at a rate of 27π in³/sec. How fast is the radius of the balloon increasing when the radius is 3 in?

3. Cars A and B leave a town at the same time. Car A heads due south at a rate of 80 km / hr and car B heads due west at a rate of 60 km / hr. How fast is the distance between the cars increasing after three hours?

4. A cylindrical tank with a radius of 6 meters is filling with fluid at a rate of 108π m³/sec. How fast is the height increasing?

5. The sides of an equilateral triangle are increasing at the rate of 27 in/sec. How fast is the triangle's area increasing when the sides of the triangle are each 18 inches long?

6. An inverted conical container has a diameter of 42 in and a depth of 15 in. If water is flowing out of the vertex of the container at a rate of 35π in³/sec , how fast is the depth of the water dropping when the height is 5 inches?

7. A boat is being pulled toward a dock by a rope attached to its bow through a pulley on the dock 7 feet above the bow. If the rope is hauled in at a rate of 4 ft/sec, how fast is the boat approaching the dock when 25 feet of rope is out?

8. A 6-foot-tall woman is walking at the rate of 4 ft/sec away from a street lamp that is 24 feet tall. How fast is the length of her shadow changing?

9. The voltage, V, in an electrical circuit is related to the current, I, and the resistance, R, by the equation $V = IR$. The current is decreasing at –4 amps/sec as the resistance increases at 20 ohms/sec. How fast is the voltage changing when the voltage is 100 volts and the current is 20 amps?

10. The minute hand of a clock is 6 inches long. Starting from noon, how fast is the area of the sector swept out by the minute hand increasing in in^2/min at any instant?

POSITION, VELOCITY, AND ACCELERATION

Every AP exam has a question on position, velocity, or acceleration (all of these fall under the category of rectilinear motion). It's one of the traditional areas of physics where calculus comes in handy. Some of these problems require the use of integral calculus, which we don't talk about until the second half of this book. So, this unit is divided in half; you'll see the other half later.

If you have a function that gives you the position of an object (usually called a "particle") at a specified time, then the derivative of that function with respect to time is the velocity of the object, and the second derivative is the acceleration. These are usually represented by the following:

Position: $x(t)$ or sometimes $s(t)$.

Velocity: $v(t)$, which is found by using $x'(t)$.

Acceleration: $a(t)$, which is found by using $x''(t)$ or $v'(t)$.

By the way, speed is the absolute value of velocity.

Example 1: If the position of a particle at a time t is given by the equation $x(t) = t^3 - 11t^2 + 24t$, find the velocity and the acceleration of the particle at time $t = 5$.

First, take the derivative of $x(t)$:

$$x'(t) = 3t^2 - 22t + 24 = v(t).$$

Second, plug in $t = 5$ to find the velocity at that time:

$$v(5) = 3(5^2) - 22(5) + 24 = -11.$$

Third, take the derivative of $v(t)$ to find $a(t)$:

$$v'(t) = 6t - 22 = a(t).$$

Finally, plug in $t = 5$ to find the acceleration at that time:

$$a(5) = 6(5) - 22 = 8.$$

See the negative velocity? The sign of the velocity is important, because it indicates the direction of the speed. Make sure that you know the following:

When the velocity is negative the particle is moving to the left.

When the velocity is positive the particle is moving to the right.

When the velocity and acceleration of the particle have the same signs, the particle's speed is increasing.

When the velocity and acceleration of the particle have the opposite signs, the particle's speed is decreasing (or slowing down).

When the velocity is zero and the acceleration is not zero, the particle is momentarily stopped and changing direction.

These equations are usually functions of time (t). Typically, t is greater than zero, but it doesn't have to be.

Example 2: If the position of a particle is given by $x(t) = t^3 - 12t^2 + 36t + 18$, where $t > 0$, find the point at which the particle changes direction.

The derivative is:
$$x'(t) = v(t) = 3t^2 - 24t + 36.$$

Set it equal to zero and solve for t:
$$x'(t) = 3t^2 - 24t + 36 = 0$$
$$t^2 - 8t + 12 = 0$$
$$(t - 2)(t - 6) = 0$$
$$t = 2 \text{ or } t = 6.$$

You need to check that the acceleration is not 0. $x''(t) = 6t - 24$. This equals 0 at $t = 4$. Therefore, the particle is changing direction at $t = 2$ or $t = 6$.

Example 3: Given the same position function as in Example 2, find the interval of time during which the particle is slowing down.

When $0 < t < 2$ and $t > 6$, the particle's velocity is positive; when $2 < t < 6$, the particle's velocity is negative. You can verify this by graphing the function and seeing when it's above or below the x-axis. Or, try some points in the regions between the roots and outside the roots. Now, we need to determine the same information about the acceleration:
$$a(t) = v'(t) = 6t - 24.$$

So the acceleration will be negative when $t < 4$, and positive when $t > 4$.

So we have:

Time	Velocity	Acceleration
$0 < t < 2$	Positive	Negative
$2 < t < 4$	Negative	Negative
$4 < t < 6$	Negative	Positive
$t > 6$	Positive	Positive

Whenever the velocity and acceleration have opposite signs, the particle is slowing down, so the particle is slowing down during the first two seconds ($0 < t < 2$) and between the fourth and sixth seconds ($4 < t < 6$).

Another typical question you'll be asked is to find the distance a particle has traveled from one time to another. This is the distance that the particle has covered without regard to the sign, not just the displacement. In other words, if the particle had an odometer on it, what would it read? Usually, all you have to do is plug the two times into the position function and find the difference.

Example 4: How far does a particle travel between the eighth and tenth seconds if its position function is $x(t) = t^2 - 6t$?

Find $x(10) - x(8) = (100 - 60) - (64 - 48) = 24$.

Be careful about one very important thing: if the velocity changes sign during the problem's time interval, you'll get the wrong answer if you simply follow the method in the paragraph above. For example, suppose we had the same position function as above but we wanted to find the distance that the particle travels from $t = 2$ to $t = 4$:

$$x(4) - x(2) = (-8) - (-8) = 0.$$

This is wrong. The particle travels from –8 back to –8, but it hasn't stood still. To fix this problem, divide the time interval into the time when the velocity is negative and the time when the velocity is positive, and add the absolute values of each distance. Here, the velocity is $v(t) = 2t - 6$. The velocity is negative when $t < 3$ and positive when $t > 3$. So we find the absolute value of the distance traveled from $t = 2$ to $t = 3$ and add to that the absolute value of the distance traveled from $t = 3$ to $t = 4$. Because $x(t) = t^2 - 6t$,

$$\left|x(3) - x(2)\right| + \left|x(4) - x(3)\right| = \left|-9 + 8\right| + \left|-8 + 9\right| = 2.$$

This is the distance that the particle traveled.

Example 5: Given the position function $x(t) = t^4 - 8t^2$, find the distance that the particle travels from $t = 0$ to $t = 4$.

First, find the first derivative ($v(t) = 4t^3 - 16t$) and set it equal to zero:

$$4t^3 - 16t = 0 \quad 4t(t^2 - 4) = 0 \quad t = 0, 2, -2.$$

So we need to divide the time interval into $t = 0$ to $t = 2$ and $t = 2$ to $t = 4$:

$$\left|x(2) - x(0)\right| + \left|x(4) - x(2)\right| = 16 + 144 = 160.$$

Here are some solved problems. Do each problem, covering the answer first, then checking your answer.

PROBLEM 1. Find the velocity and acceleration of a particle whose position function is $x(t) = 2t^3 - 21t^2 + 60t + 3$, for $t > 0$.

Answer: Time to take some derivatives:

$$v(t) = 6t^2 - 42t + 60$$

$$a(t) = 12t - 42.$$

PROBLEM 2. Given the position function in problem 1, find when the particle's speed is increasing.

Answer: First, set $v(t) = 0$:

$$6t^2 - 42t + 60 = 0$$
$$t^2 - 7t + 10 = 0$$
$$(t - 2)(t - 5) = 0$$
$$t = 2, t = 5.$$

You should be able to determine that the velocity is positive from $0 < t < 2$, negative from $2 < t < 5$, and positive again from $t > 5$.

Now, set $a(t) = 0$:

$$12t - 42 = 0$$
$$t = \frac{7}{2}.$$

You should be able to determine that the acceleration is negative from $0 < t < \frac{7}{2}$ and positive from $t > \frac{7}{2}$.

The intervals where the velocity and the acceleration have the same sign are $2 < t < \frac{7}{2}$ and $t > 5$.

PROBLEM 3. Given the position of a particle is found by $x(t) = t^3 - 6t^2 + 1$, $t > 0$, find the distance that the particle travels from $t = 2$ to $t = 5$.

Answer: First, find $v(t)$.

$$v(t) = 3t^2 - 12t.$$

Second, set $v(t) = 0$ and find the critical values:

$$3t^2 - 12t = 0 \quad 3t(t - 4) = 0 \quad t = \{0, 4\}.$$

Since the particle changes direction after four seconds, you have to figure out two time intervals separately (from $t = 2$ to $t = 4$ and from $t = 4$ to $t = 5$) and add the absolute values of the distances:

$$|x(4) - x(2)| + |x(5) - x(4)| = |(-31) - (-15)| = |(-24) - (-31)| = 23.$$

PROBLEM 4. If $x(t) = t^3 - 6t^2 + 9t + 1$, $t > 0$, find when the particle is changing direction.

Answer: First, find $v(t)$.:

$$v(t) = 3t^2 - 12t + 9.$$

Second, set $v(t) = 0$:

$$3t^2 - 12t + 9 = 0$$

$$t^2 - 4t + 3 = 0$$

$$t = 1, 3.$$

Third, find $a(t)$:

$$a(t) = 6t - 12.$$

Check to see if the acceleration is zero at either $t = 1$ or $t = 3$: $a(1) = -6$, $a(3) = 6$. Because the acceleration is not zero at these times, and the velocity is zero, the particle changes direction at $t = 1$ and $t = 3$.

PRACTICE PROBLEMS

Now try these problems. The answers are in chapter 21.

1. Find the velocity and acceleration of a particle whose position function is $x(t) = t^3 - 9t^2 + 24t$, $t > 0$.

2. Find the velocity and acceleration of a particle whose position function is $x(t) = \sin(2t) + \cos(t)$.

3. If the position function of a particle is $x(t) = \dfrac{t}{t^2 + 9}$, $t > 0$, find when the particle is changing direction.

4. If the position function of a particle is $x(t) = \sin\left(\dfrac{t}{2}\right)$, $0 < t < 4\pi$, find when the particle is changing direction.

5. If the position function of a particle is $x(t) = 3t^2 + 2t + 4$, $t > 0$, find the distance that the particle travels from $t = 2$ to $t = 5$.

6. If the position function of a particle is $x(t) = t^2 + 8t$, $t > 0$, find the distance that the particle travels from $t = 0$ to $t = 4$.

7. If the position function of a particle is $x(t) = 2\sin^2 t + 2\cos^2 t$, $t > 0$ find the velocity and acceleration of the particle.

8. If the position function of a particle is $x(t) = t^3 + 8t^2 - 2t + 4$, $t > 0$, find when the particle is changing direction.

9. If the position function of a particle is $x(t) = 2t^3 - 6t^2 + 12t - 18$, $t > 0$, find when the particle is changing direction.

10. If the position function of a particle is $x(t) = \sin^2 2t$, $t > 0$, find the distance that the particle travels from $t = 0$ to $t = 2$.

Exponential and Logarithmic Functions, Part One

As with trigonometric functions, you'll be expected to remember all of the logarithmic and exponential functions you've studied in the past. If you're not sure about any of this stuff, review the unit on Prerequisite Mathematics. Also, this is only Part One of our treatment of exponents and logs. Much of what you need to know about these functions requires knowledge of integrals (the second half of the book), so we'll discuss them again later.

THE DERIVATIVE OF ln x

When you studied logs in the past, you probably concentrated on common logs (that is, those with a base of 10), and avoided natural logarithms (base e) as much as possible. Well, we have bad news for you; most of what you'll see from now on involves natural logs. In fact, common logs almost never show up in calculus. But that's okay. All you have to do is memorize a bunch of rules, and you'll be fine.

Rule No. 1: If $y = \ln x$, then $\dfrac{dy}{dx} = \dfrac{1}{x}$.

This rule has a corollary that incorporates the Chain Rule and is actually a better rule to memorize:

Rule No. 2: If $y = \ln u$, then $\dfrac{dy}{dx} = \dfrac{1}{u}\dfrac{du}{dx}$.

Remember: u is a function of x, and $\dfrac{du}{dx}$ is its derivative.

You'll see how simple this rule is after we try a few examples.

Example 1: Find the derivative of $f(x) = \ln(x^3)$.

$$f'(x) = \frac{3x^2}{x^3} = \frac{3}{x}.$$

You could have done this another way. If you recall your rules of logarithms:

$$\ln(x^3) = 3\ln x.$$

Therefore, $f'(x) = 3\left(\dfrac{1}{x}\right) = \dfrac{3}{x}$.

Example 2: Find the derivative of $f(x) = \ln(5x - 3x^6)$.

$$f'(x) = \frac{\left(5 - 18x^5\right)}{\left(5x - 3x^6\right)}.$$

Example 3: Find the derivative of $f(x) = \ln(\cos x)$.

$$f'(x) = \frac{-\sin x}{\cos x} = -\tan x.$$

Finding the derivative of a natural logarithm is just a matter of following a simple formula.

THE DERIVATIVE OF e^x

As you'll see in Rule No. 3, the derivative of e^x is probably the easiest thing that you'll ever have to do in calculus.

> Rule No. 3: If $y = e^x$ then $\dfrac{dy}{dx} = e^x$.

That's not a typo. The derivative is the same thing as itself!

Incorporating the chain rule, we get a good formula for finding the derivative:

> Rule No. 4: If $y = e^u$ then $\dfrac{dy}{dx} = e^u \dfrac{du}{dx}$.

And you were worried that all of this logarithm and exponential stuff was going to be hard!

Example 4: Find the derivative of $f(x) = e^{3x}$.

$$f(x) = e^{3x}(3) = 3e^{3x}$$

Example 5: Find the derivative of $f(x) = e^{x^3}$.

$$f'(x) = e^{x^3}\left(3x^2\right) = 3x^2 e^{x^3}$$

Example 6: Find the derivative of $f(x) = e^{\tan x}$.

$$f(x) = \left(\sec^2 x\right)e^{\tan x}$$

Example 7: Find the second derivative of $f(x) = e^{x^2}$.

$$f'(x) = 2xe^{x^2}$$
$$f''(x) = 2e^{x^2} + 4x^2 e^{x^2}$$

Once again, it's just a matter of following a formula.

THE DERIVATIVE OF $\log_a x$

This derivative is actually a little trickier than the derivative of a natural log. First, if you remember your logarithm rules about change of base, we can rewrite $\log_a x$ this way:

$$\log_a x = \frac{\ln x}{\ln a}.$$

Review the unit on Prerequisite Mathematics if this leaves you scratching your head. Anyway, because $\ln a$ is a constant, we can take the derivative and we get:

$$\frac{1}{\ln a}\frac{1}{x}.$$

This leads us to our next rule:

Rule No. 5: If $y = \log_a x$ then $\dfrac{dy}{dx} = \dfrac{1}{x \ln a}$.

Once again, incorporating the chain rule gives us a better formula:

Rule No. 6: If $y = \log_a u$ then $\dfrac{dy}{dx} = \dfrac{1}{u \ln a}\dfrac{du}{dx}$.

Example 8: Find the derivative of $f(x) = \log_{10} x$.

$$f'(x) = \frac{1}{x \ln 10}$$

Note: we refer to the $\log_{10} x$ as $\log x$.

Example 9: Find the derivative of $f(x) = \log_8\left(x^2 + x\right)$.

$$f'(x) = \frac{2x+1}{\left(x^2 + x\right)\ln 8}$$

Example 10: Find the derivative of $f(x) = \log_e x$.

$$f'(x) = \frac{1}{x \ln e} = \frac{1}{x}$$

You can expect this result from Rules 1 and 2 involving natural logs.

Actually, this rule isn't so hard either, but it's beginning to become a lot to memorize. And there's more where that came from.

THE DERIVATIVE OF a^x

You should recall from your precalculus days that we can rewrite a^x as $e^{x \ln a}$. Keep in mind that $\ln a$ is just a constant, which gives us the next rule:

Rule No. 7: If $y = a^x$ then $\dfrac{dy}{dx} = \left(e^{x\ln a}\right)\ln a = a^x(\ln a)$.

Given the pattern of this chapter, you can guess what's coming: another rule that incorporates the Chain Rule.

Rule No. 8: If $y = a^u$ then $\dfrac{dy}{dx} = a^u(\ln a)\dfrac{du}{dx}$.

And now, some examples:

Example 11: Find the derivative of $f(x) = 3^x$.

$$f'(x) = 3^x \ln 3$$

Example 12: Find the derivative of $f'(x) = 8^{4x^5}$

$$f'(x) = 8^{4x^5}\left(20x^4\right)\ln 8$$

Example 13: Find the derivative of $f(x) = \pi^{\sin x}$.

$$f'(x) = \pi^{\sin x}(\cos x)\ln \pi$$

Finally, here's every nasty teacher's favorite exponential derivative:

Example 14: Find the derivative of $f(x) = x^x$.

First, rewrite this as: $x = e^{x\ln x}$. Then take the derivative:

$$f'(x) = e^{x\ln x}\left(\ln x + \frac{x}{x}\right) = e^{x\ln x}(\ln x + 1) = x^x(\ln x + 1)$$

Would you have thought of that? Remember this trick. It might come in handy!

Okay. Ready for some practice? Here are some more solved problems. Cover the solutions and get cracking.

PROBLEM 1. Find the derivative of $y = 3\ln(5x^2 + 4x)$.

Answer: Use Rule No. 2.

$$\frac{dy}{dx} = 3\frac{10x + 4}{5x^2 + 4x} = \frac{30x + 12}{5x^2 + 4x}$$

PROBLEM 2. Find the derivative of $f(x) = \ln\left(\sin\left(x^5\right)\right)$.

Answer:

$$f'(x) = \frac{5x^4 \cos\left(x^5\right)}{\sin\left(x^5\right)} = 5x^4 \cot\left(x^5\right)$$

PROBLEM 3. Find the derivative of $f(x) = e^{3x^7 - 4x^2}$.

Answer: Use Rule No. 4.

$$f'(x) = \left(21x^6 - 8x\right)e^{3x^7 - 4x^2}$$

PROBLEM 4. Here's a doozy: find the derivative of $y = e^{\frac{\cos 3x}{\sin 2x}}$.

Answer:

$$\frac{dy}{dx} = e^{\frac{\cos 3x}{\sin 2x}}\left[\frac{\sin 2x(-3\sin 3x) - \cos 3x(2\cos 2x)}{(\sin 2x)^2}\right] = e^{\frac{\cos 3x}{\sin 2x}}\left[\frac{-3\sin 2x \sin 3x - 2\cos 3x \cos 2x}{\sin^2 2x}\right]$$

PROBLEM 5. Find the derivative of $f(x) = \log_4(\tan x)$.

Answer: Use Rule No. 6.

$$f'(x) = \frac{1}{\ln 4}\frac{\sec^2 x}{\tan x}$$

PROBLEM 6. Find the derivative of $y = \log_8 \sqrt{\frac{x^3}{1+x^2}}$.

Answer: First use the rules of logarithms to rewrite the equation:

$$y = \frac{1}{2}\left[3\log_8 x - \log_8\left(1+x^2\right)\right]$$

Now it's much easier to find the derivative:

$$\frac{dy}{dx} = \frac{1}{2}\left[3\frac{1}{x\ln 8}\right] - \frac{1}{2}\left[\frac{2x}{\ln 8\left(1+x^2\right)}\right] = \frac{1}{2\ln 8}\left[\frac{3}{x} - \frac{2x}{\left(1+x^2\right)}\right].$$

PROBLEM 7. Find the derivative of $y = 5^{\sqrt{x}}$.

Answer: Use Rule No. 8.

$$\frac{dy}{dx} = 5^{\sqrt{x}} \frac{1}{2\sqrt{x}} \ln 5 = \frac{5^{\sqrt{x}} \ln 5}{2\sqrt{x}}.$$

PROBLEM 8. Find the derivative of

Answer:

$$\frac{dy}{dx} = e^{3\sec x + 2\ln x}\left(3\sec x \tan x + \frac{2}{x}\right)$$

PROBLEM 9. Find the derivative of $y = \dfrac{e^{x^3}}{5^{\cos x}}$.

Answer: Here, you need to use the Quotient Rule and Rules Nos. 4 and 8.

$$\frac{dy}{dx} = \frac{5^{\cos x}\left(3x^2 e^{x^3}\right) - e^{x^3}\left(5^{\cos x} \ln 5(-\sin x)\right)}{\left(5^{\cos x}\right)^2} = 5^{\cos x} e^{x^3} \frac{\left(3x^2\right) + (\sin x \ln 5)}{5^{2\cos x}} = e^{x^3} \frac{3x^2 + \sin x \ln 5}{5^{\cos x}}$$

PROBLEM 10. Find the derivative of $y = \sqrt[4]{\log_5 e^{x^3}}$.

Answer: First use the rules of logarithms to rewrite the equation:

$$y = \sqrt[4]{\log_5 e^{x^3}} = \sqrt[4]{\frac{\ln e^{x^3}}{\ln 5}} = \sqrt[4]{\frac{x^3}{\ln 5}} = \frac{x^{\frac{3}{4}}}{(\ln 5)^{\frac{1}{4}}}$$

Now, the derivative's easy:

$$\frac{dy}{dx} = \frac{\frac{3}{4}}{(x \ln 5)^{\frac{1}{4}}} = \frac{3}{4(x \ln 5)^{\frac{1}{4}}}$$

PRACTICE PROBLEMS

Now find the derivative of each of the following functions. The answers are in chapter 21.

1. $f(x) = \ln(x^4 + 8)$

2. $f(x) = \ln(3x\sqrt{3+x})$

3. $f(x) = \ln(\cot x - \csc x)$

4. $f(x) = x \ln \cos 3x - x^3$

5. $f(x) = \ln\left(\dfrac{5x^2}{\sqrt{5+x^2}}\right)$

6. $f(x) = e^{x \cos x}$

7. $f(x) = e^{-3x} \sin 5x$

8. $f(x) = e^{\ln \frac{1}{x}}$

9. $f(x) = \dfrac{e^{\tan 4x}}{4x}$

10. $f(x) = e^{\pi x} - \ln e^{\pi x}$

11. $f(x) = \log_{12}(x^3)$

12. $f(x) = \log_6(3x \tan x)$

13. $f(x) = \dfrac{\log_4 x}{e^{4x}}$

14. $f(x) = \log \sqrt{10^{3x}}$

15. $f(x) = \ln x \log x$

16. $f(x) = e^{3x} - 3^{ex}$

17. $f(x) = 10^{\sin x}$

18. $f(x) = (\sin x)^{\tan x}$

19. $f(x) = \ln(10^x)$

20. $f(x) = x^5 5^x$

10

OTHER STUFF ON THE
AP EXAMS

This chapter is devoted to other topics involving differential calculus that are seen on the AP exam every once in a while.

THE DERIVATIVE OF AN INVERSE FUNCTION

ETS occasionally asks a question about finding the derivative of an inverse function. To do this, you only need to learn this simple formula.

Suppose we have a function $x = f(y)$ that is defined and differentiable at $y = a$. Suppose we also know that the $f^{-1}(x)$ exists at $x = c$. Thus, $f(a) = c$ and $f^{-1}(c) = a$. Then, because $\dfrac{dy}{dx} = \dfrac{1}{\dfrac{dx}{dy}}$

$$\frac{d}{dx} f^{-1}(x) \bigg|_{x=c} = \frac{1}{\left[\dfrac{d}{dy} f(y) \right]_{y=a}}$$

The short translation of this is: we can find the derivative of a function's inverse at a particular point by taking the reciprocal of the derivative at that point's corresponding y-value. These examples should help clear up any confusion.

Example 1: If $f(x) = x^2$, find a derivative of $f^{-1}(x)$ at $x = 9$.

First, notice that $f(3) = 9$. One of the most confusing parts of finding the derivative of an inverse function is that when you're asked to find the derivative at a value of x, they're really asking you for the derivative of the inverse of the function at the value that corresponds to $f(x) = 9$. This is because x-values of the inverse correspond to $f(x)$ values of the original function.

The rule is very simple: when you're asked to find the derivative of $f^{-1}(x)$ at $x = c$, you take the reciprocal of the derivative of $f(x)$ at $x=a$, where $f(a)=c$.

We know that $\dfrac{d}{dx} f(x) = 2x$. This means that we're going to plug $x = 3$ into the formula (because $f(3) = 9$). This gives us:

$$\left.\frac{1}{2x}\right|_{x=3} = \frac{1}{6}.$$

We can verify this by finding the inverse of the function first and then taking the derivative. The inverse of the function $f(x) = x^2$ is the function $f^{-1}(x) = \sqrt{x}$. Now we find the derivative and evaluate it at $f(3) = 9$:

$$\frac{d}{dx}\sqrt{x} = \left.\frac{1}{2\sqrt{x}}\right|_{x=9} = \frac{1}{6}.$$

Remember the rule: find the value, a, of $f(x)$ that gives you the value of x that the problem asks for. Then plug that value, a, into the derivative of the *inverse* function.

Example 2: Find a derivative of the inverse of $y = x^3 - 1$ when $y = 7$.

First, we need to find the x-value that corresponds to $y = 7$. A little algebra tells us that this is $x = 2$. Then:

$$\frac{dy}{dx} = 3x^2 \text{ and } \frac{1}{\dfrac{dy}{dx}} = \frac{1}{3x^2}.$$

Therefore, the derivative of the inverse is:

$$\left.\frac{1}{3x^2}\right|_{x=2} = \frac{1}{12}.$$

Verify it: the inverse of the function $y = x^3 - 1$ is the function $y = \sqrt[3]{x+1}$. The derivative of this latter function is:

$$\left.\frac{1}{3\sqrt[3]{(x+1)^2}}\right|_{x=7} = \frac{1}{12}.$$

Let's do one more.

Example 3: Find a derivative of the inverse of $y = x^2 + 4$ when $y = 29$.

At $y = 29$, $x = 5$. The derivative of the function is:

$$\frac{dy}{dx} = 2x,$$

so the derivative of the inverse is:

$$\frac{1}{2x}\bigg|_{x=5} = \frac{1}{10}.$$

It's not that hard, once you get the hang of it. This is all you'll be required to know involving derivatives of inverses. Naturally, there are ways to create harder problems, but the AP stays away from them and sticks to simpler stuff.

Here are some solved problems. Do each problem, cover the answer first, then check your answer.

PROBLEM 1. Find a derivative of the inverse of $y = f(x) = 2x^3 + 5x + 1$ at $y = 8$.

Answer: First, we take the derivative of $f(x)$:

$$f'(x) = 6x^2 + 5.$$

A possible value of x is $x = 1$.

Then, we use the formula to find the derivative of the inverse:

$$\frac{1}{f'(1)} = \frac{1}{11}.$$

PROBLEM 2. Find a derivative of the inverse of $y = f(x) = 3x^3 - x + 7$ at $y = 9$.

Answer: First, take the derivative of $f(x)$:

$$f'(x) = 9x^2 - 1.$$

A possible value of x is $x = 1$.

Then, use the formula to find the derivative of the inverse:

$$\frac{1}{f'(1)} = \frac{1}{8}.$$

PROBLEM 3. Find a derivative of the inverse of $y = \dfrac{8}{x^3}$ at $y = 1$.

Answer: Take the derivative of $f(x)$:

$$f'(x) = -\frac{24}{x^4}.$$

Find the value of x where $y = 1$:

$$1 = \frac{8}{x^3}$$
$$x = 2.$$

Use the formula:

$$\frac{1}{\dfrac{dy}{dx}\Big|_{x=2}} = \frac{1}{\left(-\dfrac{24}{x^4}\right)\Big|_{x=2}} = -\frac{2}{3}.$$

Voila! Here's one more.

PROBLEM 4. Find a derivative of the inverse of $y = 2x - x^3$ at $y = 1$.

Answer: The derivative of the function is:

$$\frac{dy}{dx} = 2 - 3x^2.$$

Next, find the value of x where $y = 1$. By inspection, $y = 1$ when $x = 1$. Then, we use the formula to find the derivative of the inverse:

$$\frac{1}{\dfrac{dy}{dx}\Big|_{x=1}} = \frac{1}{\left(2 - 3x^2\right)\big|_{x=1}} = -1.$$

PRACTICE PROBLEMS

Find a derivative of the inverse of each of the following functions. The answers are in chapter 21.

1. $y = x + \dfrac{1}{x}$ at $y = \dfrac{17}{4}$.

2. $y = 3x - 5x^3$ at $y = 2$.

3. $y = e^x$ at $y = e$.

4. $f(x) = x^7 - 2x^5 + 2x^3$ at $y = 1$.

5. $y = x + x^3$ at $y = -2$.

6. $y = 4x - x^3$ at $y = 3$.

7. $y = \ln x$ at $y = 0$.

8. $y = x^{\frac{1}{3}} + x^{\frac{1}{5}}$ at $y = 2$.

NEWTON'S METHOD

Newton's Method is a simple technique to learn, and it's gotten much easier now that you can use a calculator during the exam. The method, which is properly called the Newton-Raphson Method, is designed to help you get a very accurate approximation of the roots of a function that you can't factor.

Words of wisdom: Before you use the method, you have to make an educated guess of the root's approximate value. Therefore, you should always plot the function before you use Newton's Method, so you can make a good first guess.

The method works like this: Call the first approximation of the root x_1. You can find a second approximation of the root, x_2, using the formula:

$$x_{n+1} = x_n - \frac{f(x_n)}{f'(x_n)}.$$

Example 1: Find the positive root of $x^2 - 5 = 0$ using Newton's Method.

First, you need a good guess for the answer. We know that the exact answer is $x = \sqrt{5}$, so a good first guess is $x_1 = 2$ (you don't have to be that close, so don't complicate your life by making your first guess too difficult).

Next, find the derivative of $x^2 - 5$, which is $2x$. Then your formula is:

$$x_{n+1} = x_n - \frac{f(x_n)}{f'(x_n)} = x_n - \frac{x_n^2 - 5}{2x_n}.$$

Now, plug in $x_1 = 2$:

$$x_2 = x_1 - \frac{x_1^2 - 5}{2x_1} = 2 - \frac{4 - 5}{4} = 2.25.$$

Use this approximation to come up with a better approximation:

$$x_3 = x_2 - \frac{x_2^2 - 5}{2x_2} = 2.25 - \frac{(2.25)^2 - 5}{2(2.25)} = 2.23611$$

We can keep going until we get the accuracy that we desire.

$$x_4 = x_3 - \frac{x_3^2 - 5}{2x_3} = 2.23611 - \frac{(2.23611)^2 - 5}{2(2.23611)} = 2.23607$$

Since $\sqrt{5} \approx 2.236067977$, we're pretty accurate at this point. The AP usually asks for accuracy to two or three decimal places, so we could have stopped after x_3.

Example 2: Find the real root of $x^3 + 3x + 1 = 0$ to three decimal places.

What's a good guess for this one? Notice that all of the signs of the function are positive; therefore, it cannot have a positive root. If we plug in 0, we get 1. If we plug in –1, we get –3. So, a good guess is somewhere between 0 and –1. Try –0.5.

Next, we need the derivative:

$$f'(x) = 3x^2 + 3.$$

Use the formula:

$$x_2 = x_1 - \frac{x_1^3 + 3x_1 + 1}{3x_1^2 + 3} = (-0.5) - \frac{(-0.5)^3 + 3(-0.5) + 1}{3(-0.5)^2 + 3} = -.3333$$

$$x_3 = (-0.3333) - \frac{(-0.3333)^3 + 3(-0.3333) + 1}{3(-0.3333)^2 + 3} = -.3222$$

$$x_4 = (-0.3222) - \frac{(-0.3222)^3 + 3(-0.3222) + 1}{3(-0.3222)^2 + 3} = -.3222$$

Therefore, the answer is –.322.

USING YOUR CALCULATOR

One of the great things about using your calculator for the exam is that you can program it to perform Newton's Method. In fact, ETS actually suggests putting the following program in your calculator! Using the TI-81, TI-82, or TI-83, input the following:

Input X
Lbl 1
X – Y₁/NDeriv(Y₁,.001) → X
Disp X
Pause
Goto 1

Name the program NEWTON. You're supposed to put the function in the graph section under Y_1, and you end the program by pressing the ON key and then QUIT.

Let's practice this with another example:

Example 3: Use Newton's Method, with an initial guess of $x_1 = 2$, to find the positive root of $f(x) = -x^2 + 2x + 1$ to three decimal places.

We know that we have a positive root, because at $x = 0$, $f(0) = 1$, but $f(x)$ is negative when x is large. In the TI calculator, enter the equation $-x^2 + 2x + 1$ in Y_1. Then run the program. When the program prompts you, input 2. You'll see:

$$2.5$$
$$2.4167$$
$$2.4142$$
$$2.4142$$

At this point, you can stop. The answer is 2.414.

Wasn't that easy? Newton's Method shouldn't give you much trouble. If a question shows up on the AP, either the arithmetic will be very simple (if it's not in the calculator section of the exam), or you'll be able to use the program (if it is in the calculator portion of the exam).

Try these other examples, to make sure you have the technique down pat. Then check your work.

PROBLEM 1. Use Newton's Method, with an initial guess of $x_1 = 2$, to find x_3, the third approximation of the positive root, of $f(x) = x^4 - 3x^2 - 8$.

Answer: We already have a first approximation of $x_1 = 2$, so next we need the derivative of $x^4 - 3x^2 - 8$, which is $4x^3 - 6x$.

Remember, the formula is:

$$x_{n+1} = x_n - \frac{f(x_n)}{f'(x_n)}.$$

Plug in $x_1 + 2$:

$$x_2 = x_1 - \frac{x_1^4 - 3x_1^2 - 8}{4x_1^3 - 6x_1} = 2 - \frac{(2)^4 - 3(2)^2 - 8}{4(2)^3 - 6(2)} = 2.2$$

Now use this approximation for the next one:

$$x_3 = x_2 - \frac{x_2^4 - 3x_2^2 - 8}{4x_2^3 - 6x_2} = 2.2 - \frac{(2.2)^4 - 3(2.2)^2 - 8}{4(2.2)^3 - 6(2.2)} = 2.17$$

PROBLEM 2. Use Newton's Method to find the positive root of $f(x) = x^3 + x - 3$ to three decimal places.

Answer: The first guess can be the hardest part of the problem. If you plug in 0, you get -3. If you plug in 1, the result is -1. If $x = 2$, we get 7. Since plugging in 1 gives you a negative result and plugging in 2 gives you a positive one, a good guess is somewhere between 1 and 2. Try $x_1 = 1.5$.

Plug $x_1 = 1.5$ into the derivative of the function, which is $3x^2 + 1$, and you get:

$$x_2 = x_1 - \frac{x_1^3 + x_1 - 3}{3x_1^2 + 1} = 1.5 - \frac{(1.5)^3 + (1.5) - 3}{3(1.5)^2 + 1} = 1.2581.$$

Now use this approximation to come up with a better approximation:

$$x_3 = x_2 - \frac{x_2{}^3 + x_2 - 3}{3x_2{}^2 + 1} = 1.2581 - \frac{(1.2581)^3 + (1.2581) - 3}{3(1.2581)^2 + 1} = 1.2147$$

Continue this process until you're accurate to three decimal places:

$$x_4 = x_3 - \frac{x_3{}^3 + x_3 - 3}{3x_3{}^2 + 1} = 1.2147 - \frac{(1.2147)^3 + (1.2147) - 3}{3(1.2147)^2 + 1} = 1.2134$$

$$x_5 = x_4 - \frac{x_4{}^3 + x_4 - 3}{3x_4{}^2 + 1} = 1.2134 - \frac{(1.2134)^3 + (1.2134) - 3}{3(1.2134)^2 + 1} = 1.2134$$

Bingo. The answer is 1.213

PROBLEM 3. Find the real root of $-x^3 + 2x^2 + 1 = 0$ to three decimal places.

Answer: Based on the strategy we discussed earlier, a good first guess is $x_1 = 2.5$. (If you plug in 2, the result is 1. If you plug in 3, you get –8.)

The derivative of $-x^3 + 2x^2 + 1$ is $-3x^2 + 4x$. Plug in $x_1 = 2.5$, and you get:

$$x_2 = x_1 - \frac{-x_1{}^3 + 2x_1{}^2 + 1}{-3x_1{}^2 + 4x_1} = 2.5 - \frac{-(2.5)^3 + 2(2.5)^2 + 1}{-3(2.5)^2 - 4(2.5)} = 2.2571.$$

And so it goes . . .

$$x_3 = x_2 - \frac{-x_2{}^3 + 2x_2{}^2 + 1}{-3x_2{}^2 + 4x_2} = 2.2571 - \frac{-(2.2571)^3 + 2(2.2571)^2 + 1}{-3(2.2571)^2 - 4(2.2571)} = 2.2076$$

Continue this process until you're accurate to three decimal places:

$$x_4 = x_3 - \frac{-x_3{}^3 + 2x_3{}^2 + 1}{-3x_3{}^2 + 4x_3} = 2.2076 - \frac{-(2.2076)^3 + 2(2.2076)^2 + 1}{-3(2.2076)^2 - 4(2.2076)} = 2.2056$$

$$x_5 = x_4 - \frac{-x_4{}^3 + 2x_4{}^2 + 1}{-3x_4{}^2 + 4x_4} = 2.2056 - \frac{-(2.2056)^3 + 2(2.2056)^2 + 1}{-3(2.2056)^2 - 4(2.2056)} = 2.2056$$

You're done. The answer is 2.206.

This method isn't too difficult, but it can be tedious and time-consuming. We recommend that you use a calculator to speed up the process.

PRACTICE PROBLEMS

Use Newton's Method to solve the following problems, and solve for all roots to three decimal places. The answers are in chapter 21.

1. With an initial guess of $x_1 = 1$, find x_3, the third approximation of the positive root, of
   $$f(x) = x^5 + x - 3.$$

2. With an initial guess of $x_1 = -2$, find x_3, the third approximation of the negative root, of
   $$f(x) = -x^4 + 4x^2 - x.$$

3. Find the real root of $f(x) = x^3 - 4x^2 + 3x - 1$.

4. Find the real roots of $f(x) = x^{10} - 8x^5 + 1$.

5. Find the real root of $f(x) = -x^5 + x^4 - 10$.

6. Find the real roots of $f(x) = x^3 - 5x^2 - 8$.

7. Find the real roots of $f(x) = x^5 + x^4 - x^2 - 3$.

8. Find the positive real root of $f(x) = \sin x - \dfrac{x}{4}$.

   Attention AB students: That's it for Unit One. The rest of the stuff in this chapter pertains only to those poor saps taking BC Calc. You AB folks can move on to integration in Unit Two.

DERIVATIVES OF PARAMETRIC FUNCTIONS

Although these can seem very difficult, the questions about parametric equations on the AP exam tend to be very straightforward. As we keep pointing out, don't be intimidated by the difficult topics; ETS tends to keep the questions simple. By contrast, ETS's questions on simpler topics tend to be trickier.

What is a Parametric Function?

Let's use an analogy. Suppose you're driving a car and you want to determine a function that describes your position on the road. There are two ways that you could arrive at your position. You could figure it out based on how far you've traveled, or you could determine it based on how long you've been traveling. If you let y represent your position and x the distance, you could find your position by $y = f(x)$.

If, on the other hand, you wanted to use the time that you've traveled, you can use two functions: $x = g(t)$, to determine the distance you've traveled, and $y = h(t)$, to determine your position. These latter equations are called "parametric equations." They enable you to define x and y in terms of another variable (usually t), rather than in terms of each other. Parametric equations also follow all of the standard derivative rules.

To find $\dfrac{dy}{dx}$ using parametric equations, use the rule:

$$\frac{dy/dt}{dx/dt} = \frac{dy}{dx}.$$

For example, suppose that $x = t^2$ and $y = t^4$. Then:

$$\frac{dx}{dt} = 2t \text{ and } \frac{dy}{dt} = 4t^3, \text{ so}$$

$$\frac{dy}{dx} = \frac{4t^3}{2t} = 2t^2.$$

Now, because $x = t^2$, $\dfrac{dy}{dx} = 2x$.

We can verify this by first solving for y in terms of x and then differentiating. The result is:

$$y = x^2 \text{ and } \frac{dy}{dx} = 2x.$$

You might also be asked to turn an equation in parametric form into an equation in Cartesian form. The simplest thing to do is to solve the equation for t and substitute that equation into the other one. Sometimes the relationship is obvious, as in this example.

Example 1: What curve is represented by $x = \cos t$ and $y = \sin t$, where $0 \le t \le 2\pi$?

We know from trigonometry that $\sin^2 t + \cos^2 t = 1$, so we can substitute y for $\sin t$ and x for $\cos t$. If we square both and add them, we get $x^2 + y^2 = 1$. This is the equation of a circle, centered at the origin, with radius 1.

You can test this by picking values of t and finding the coordinates by using the two equations. For example:

$$\text{At } t = \frac{\pi}{2}, \ x = 0 \text{ and } y = 1; \text{ or}$$

$$\text{At } t = \frac{\pi}{6}, \ x = \frac{\sqrt{3}}{2} \text{ and } y = \frac{1}{2}.$$

Example 2: What curve is represented by $x = t$ and $y = t^2$?

If you substitute x for t, you'll find that $y = x^2$. Thus, this curve is a parabola with vertex at the origin.

Example 3: What curve is represented by $x = a \cos t$ and $y = b \sin t$, where $0 \le t \le 2\pi$?

First, rewrite the two equations:

$$\frac{x}{a} = \cos t \text{ and } \frac{y}{b} = \sin t.$$

Now, because we know that $\sin^2 t + \cos^2 t = 1$, we know that:

$$\left(\frac{x}{a}\right)^2 + \left(\frac{y}{b}\right)^2 = 1.$$

If $a \ne b$, this figure is an ellipse, centered at the origin, with axes a and b. If $a = b$, it's a circle, centered at the origin, with radius a.

Now you know how to convert an equation from parametric form to Cartesian form. You'll also need to know how to work with the parametric equations, even if you can't figure out how to convert them into Cartesian form (sometimes you don't have to). These frequently will be equations of motion.

Example 4: A particle's position in the xy-plane at any time t is given by $x = 2t^2 + 3$ and $y = t^4$. Find:

(a) the x-component of the particle's velocity at time $t = 5$; (b) $\dfrac{dy}{dx}$; and (c) the times at which the x- and y-components of the velocity are the same.

(a) All you do is take the derivative with respect to t:

$$\frac{dx}{dt} = 4t.$$

At time $t = 5$, this is 20.

(b) You know $\dfrac{dx}{dt}$ from part (a), so now compute $\dfrac{dy}{dt}$. This is $4t^3$. Using the rule, $\dfrac{dy}{dx} = \dfrac{dy/dt}{dx/dt}$:

$$\frac{4t^3}{2t} = 2t^2.$$

To express the answer in terms of x instead of the parameter, solve for t^2 in terms of x:

$$\frac{x-3}{2} = t^2.$$

Then substitute:

$$\frac{dy}{dx} = 2\left(\frac{x-3}{2}\right) = x - 3.$$

(c) The components of the velocity are the same when $4t^3 = 4t$.

$$4t^3 - 4t = 0$$
$$4t(t^2 - 1) = 0$$
$$t = 0, \pm 1$$

Now you know all you need to about the basics of parametrized curves. There will be other types of problems involving parametric equations, but you'll see them later in the book.

Try these sample problems, working with an index card, as usual.

PROBLEM 1. Find the Cartesian equation represented by the parametric equations $x = 4 \cos t$ and $y = 4 \sin t$, $0 \le t \le 2\pi$.

Answer: From Example 1 above, you know that $\sin^2 t + \cos^2 t = 1$. If you take each equation and rearrange the terms:

$$\frac{x}{4} = \cos t \text{ and } \frac{y}{4} = \sin t.$$

Next, substitute into $\sin^2 t + \cos^2 t = 1$:

$$\frac{x^2}{16} + \frac{y^2}{16} = 1, \text{ or}$$

$$x^2 + y^2 = 16.$$

This is a circle, centered at the origin, with radius 4.

PROBLEM 2. Find the Cartesian equation represented by the parametric equations $x = \sqrt{t}$ and $y = t$, $t \ge 0$.

Answer: Here, you need to notice that if you square the equation for x, you get $x^2 = t$. Therefore, $y = x^2$. This is a parabola with a vertex at the origin.

PROBLEM 3. Find an equation of the line tangent to the curve $x = 2 \cos t$ and $y = 3 \sin t$ at $t = \frac{\pi}{4}$.

Answer: You can, of course, eliminate t, as before, obtaining:

$$\left(\frac{x}{2}\right)^2 + \left(\frac{y}{3}\right)^2 = 1,$$

and proceed as in example 4 in chapter 5, but it's easier to retain the parameter t here and reach the answer. First, find the slope of the tangent line $\frac{dy}{dx}$ using the formula $\frac{dy}{dx} = \frac{dy/dt}{dx/dt}$:

$$\frac{dy}{dt} = 3 \cos t \text{ and } \frac{dx}{dt} = -2 \sin t.$$

Therefore, $\dfrac{dy}{dx} = \dfrac{3\cos t}{-2\sin t}$.

If we evaluate this at $t = \dfrac{\pi}{4}$:

$$\frac{dy}{dx} = \frac{3\left(\dfrac{1}{\sqrt{2}}\right)}{-2\left(\dfrac{1}{\sqrt{2}}\right)} = -\frac{3}{2}.$$

At $t = \dfrac{\pi}{4}$, $x = \dfrac{2}{\sqrt{2}}$ and $y = \dfrac{3}{\sqrt{2}}$. Now you can find the equation of the tangent line:

$$\left(y - \frac{3}{\sqrt{2}}\right) = -\frac{3}{2}\left(x - \frac{2}{\sqrt{2}}\right).$$

This can be rewritten as $3x + 2y - 6\sqrt{2} = 0$.

PROBLEM 4. A particle's position at time t is determined by the equations $x = 3 + 2t^2$ and $y = 4t^4$, $t \geq 0$. Find the x- and y-components of the particle's velocity and the times when these components are equal.

Answer: First, figure out the x- and y-components of the velocity:

$$\frac{dx}{dt} = 4t \text{ and } \frac{dy}{dt} = 16t^3.$$

These are equal when $16t^3 = 4t$. Solving for t:

$$t = 0, \pm\frac{1}{2}.$$

Throw out the negative value of t, the answer is $t = 0, \dfrac{1}{2}$.

PRACTICE PROBLEMS

Now try these problems. The answers are in chapter 21.

1. Find the Cartesian equation of the curve represented by $x = \sec^2 t - 1$ and $y = \tan t$, $-\dfrac{\pi}{2} < t < \dfrac{\pi}{2}$.

2. Find the Cartesian equation of the curve represented by $x = t$ and $y = \sqrt{1 - t^2}$, $-1 < t < 1$.

3. Find the Cartesian equation of the curve represented by $x = 4t + 3$ and $y = 16t^2 - 9$, $-\infty < t < \infty$.

4. Find the equation of the tangent line to $x = t^2 + 4$ and $y = 8t$ at $t = 6$.

5. Find the equation of the tangent line to $x = \sec t$ and $y = \tan t$ at $t = \dfrac{\pi}{4}$.

6. The motion of a particle is given by $x = -2t^2$ and $y = t^3 - 3t + 9$, $t \geq 0$. Find the coordinates of the particle when its instantaneous direction of motion is horizontal.

7. The motion of a particle is given by $x = \ln t$ and $y = t^2 - 4t$. Find the coordinates of the particle when its instantaneous direction of motion is horizontal.

8. The motion of a particle is given by $x = 2\sin t - 1$ and $y = \sin t - \dfrac{t}{2}$, $0 \leq t < 2\pi$. Find the times when the horizontal and vertical components of the particle's velocity are the same.

L'HOPITAL'S RULE

L'Hopital's Rule is a way to find the limit of certain kinds of expressions that are indeterminate forms.

If the limit of an expression results in $\dfrac{0}{0}$ or $\dfrac{\infty}{\infty}$, the limit is called "indeterminate" and you can use L'Hopital's Rule to evaluate these expressions:

If $f(c) = g(c) = 0$, and if $f'(c)$ and $g'(c)$ exist, and if $g'(c) \neq 0$, then:

$$\lim_{x \to c} \frac{f(x)}{g(x)} = \frac{f'(c)}{g'(c)}$$

Similarly,

If $f(c) = g(c) = \infty$, and if $f'(c)$ and $g'(c)$ exist, and if $g'(c) \neq 0$, then:

$$\lim_{x \to c} \frac{f(x)}{g(x)} = \frac{f'(c)}{g'(c)}$$

In other words, if the limit of the function gives us an undefined expression, like $\dfrac{0}{0}$ or $\dfrac{\infty}{\infty}$,

L'Hopital's Rule says you can take the derivative of the top and the derivative of the bottom and see if we get a determinate expression. If not, you can keep trying.

Example 1: Find $\lim\limits_{x \to 0} \dfrac{\sin x}{x}$.

First, notice that plugging in 0 results in $\dfrac{0}{0}$, which is indeterminate. Take the derivative of the top and of the bottom:

$$\lim_{x \to 0} \frac{\cos x}{1}.$$

The limit equals 1. A snap, right?

Example 2: Find $\lim\limits_{x \to 0} \dfrac{2x - \sin x}{x}$.

If you plug in 0, you get $\dfrac{0-0}{0}$, which is indeterminate. Now, take the derivative of the top and of the bottom:

$$\lim_{x \to 0} \frac{2 - \cos x}{1}.$$

This limit also equals 1.

Example 3: Find $\lim\limits_{x \to 0} \dfrac{\sqrt{4+x} - 2}{2x}$.

Again, plugging in 0 is no help; you get $\dfrac{0}{0}$. Take the derivative of the top and bottom:

$$\text{Find } \lim_{x \to 0} \frac{\dfrac{1}{2\sqrt{4+x}}}{2} = \frac{1}{8}.$$

Example 4: Find $\lim\limits_{x \to 0} \dfrac{\sqrt{4+x} - 2 - \dfrac{x}{4}}{2x^2}$.

Take the derivative of the top and bottom:

$$\lim_{x \to 0} \frac{\dfrac{1}{2\sqrt{4+x}} - \dfrac{1}{4}}{4x}.$$

Now if you take the limit, we still get $\dfrac{0}{0}$. So keep going!

$$\lim_{x \to 0} \frac{-\dfrac{1}{4}(4+x)^{-\frac{3}{2}}}{4} = -\frac{1}{128}.$$

Now let's try a couple of $\frac{\infty}{\infty}$ forms.

Example 5: Find $\lim\limits_{x\to\infty}\dfrac{5x-8}{3x+1}$.

Now, the limit is $\frac{\infty}{\infty}$. The derivative of the top and bottom is:

$$\lim_{x\to\infty}\frac{5}{3}=\frac{5}{3}.$$

Don't you wish that you had learned this back when you first did limits?

Example 6: Find $\lim\limits_{x\to\frac{\pi}{2}}\dfrac{\sec x}{1+\sec x}$.

The derivative of the top and bottom is:

$$\lim_{x\to\frac{\pi}{2}}\frac{\sec x\tan x}{\sec x\tan x}=1.$$

Example 7: Find $\lim\limits_{x\to0^+}x\cot x$.

Taking this limit results in $(0)(\infty)$, which is also indeterminate. (But you can't use the rule yet!) If you rewrite this expression as $\lim\limits_{x\to0^+}\dfrac{x}{\tan x}$, it's of the form $\dfrac{0}{0}$ and we can use L'Hopital's Rule:

$$\lim_{x\to0^+}\frac{1}{\sec^2 x}=1.$$

That's all that you need to know about L'Hopital's Rule. Just check to see if the limit results in an indeterminate form. If it does, use the rule until you get a determinate form.

Here are some more examples, in case you've still got any lingering doubts:

PROBLEM 1. Find $\lim\limits_{x\to0}\dfrac{\sin 8x}{x}$.

Answer: First, notice that plugging in 0 gives us an indeterminate result: $\dfrac{0}{0}$. Now, take the derivative of the top and of the bottom:

$$\lim_{x\to0}\frac{8\cos 8x}{1}=8.$$

PROBLEM 2. Find $\lim\limits_{\theta \to 0} \dfrac{\tan\theta}{\theta}$.

Answer: You still get $\dfrac{0}{0}$ if you plug in 0. Take the derivative of the top and of the bottom:

$$\lim_{x \to 0} \frac{\sec^2\theta}{1} = 1.$$

PROBLEM 3. Find $\lim\limits_{x \to \infty} xe^{-2x}$.

Answer: First, rewrite this expression as $\dfrac{x}{e^{2x}}$. Notice that when x nears infinity, the expression becomes $\dfrac{\infty}{\infty}$, which is indeterminate. Take the derivative of the top and bottom:

$$\lim_{x \to \infty} \frac{1}{2e^{2x}} = 0.$$

PROBLEM 4. Find $\lim\limits_{x \to \infty} \dfrac{5x^3 - 4x^2 + 1}{7x^3 + 2x - 6}$.

Answer: We learned this in chapter 1, remember? Now we'll use L'Hopital's Rule. At first glance, the limit is indeterminate: $\dfrac{\infty}{\infty}$. Let's take some derivatives:

$$\frac{15x^2 - 8x}{21x^2 + 2}.$$

This is still indeterminate, so it's time to take the derivative of the top and bottom again:

$$\frac{30x - 8}{42x}.$$

It's still indeterminate! If you try it one more time, you'll get a fraction with no variables: $\dfrac{30}{42}$, which can be simplified to $\dfrac{5}{7}$ (as we expected).

PROBLEM 5. Find $\lim\limits_{x \to \frac{\pi}{2}} \dfrac{x - \dfrac{\pi}{2}}{\cos x}$.

Answer: Plugging in $\dfrac{\pi}{2}$ gives you the indeterminate response of $\dfrac{0}{0}$. The derivative is:

$$-\frac{1}{\sin x}.$$

When we take the limit of this expression, we get -1.

PRACTICE PROBLEMS

Now find these limits using L'Hopital's Rule. The answers are in chapter 21.

1. Find $\lim\limits_{x \to 0} \dfrac{\sin 3x}{\sin 4x}$.

2. Find $\lim\limits_{x \to \pi} \dfrac{x - \pi}{\sin x}$.

3. Find $\lim\limits_{x \to 0} \dfrac{x - \sin x}{x^3}$.

4. Find $\lim\limits_{x \to 0} \dfrac{e^{3x} - e^{5x}}{x}$.

5. Find $\lim\limits_{x \to 0} \dfrac{\tan x - x}{\sin x - x}$.

6. Find $\lim\limits_{x \to \infty} \dfrac{x^5}{e^{5x}}$.

7. Find $\lim\limits_{x \to \infty} \dfrac{x^5 + 4x^3 - 8}{7x^5 - 3x^2 - 1}$.

8. Find $\lim\limits_{x \to 0^+} \dfrac{\ln(\sin x)}{\ln(\tan x)}$.

9. Find $\lim\limits_{x \to 0^+} \dfrac{\cot 2x}{\cot x}$.

10. Find $\lim\limits_{x \to 0^+} \dfrac{x}{\ln(x + 1)}$.

DIFFERENTIALS

Another small part of the BC curriculum is a topic called **differentials.** Sometimes this is called a "linearization." A differential is a very small quantity that corresponds to a change in a number. We use the symbol Δx to denote a differential. What are differentials used for? The AP exam mostly wants you to use them to approximate the value of a function or find the error of an approximation.

Recall the formula for the definition of the derivative:

$$f'(x) = \lim_{h \to 0} \frac{f(x + h) - f(x)}{h}.$$

Replace h with Δx, which also stands for a very small increment of x, and get rid of the limit:

$$f'(x) \approx \frac{f(x + \Delta x) - f(x)}{\Delta x}.$$

Notice that this is no longer equal to the derivative, but an approximation of it. If Δx is kept small, the approximation remains somewhat accurate. Next, rearrange the equation as follows:

$$f(x + \Delta x) \approx f(x) + f'(x)\Delta x$$

This is our formula for differentials. It says that "the value of a function, at x plus a little bit, equals the value of the function at x, plus the product of the derivative of the function at x with the little bit." Sounds like it's time to see this formula in action.

Example 1: Use differentials to approximate the $\sqrt{9.01}$.

You can start by letting $x = 9$, $\Delta x = +.01$, $f(x) = \sqrt{x}$. Next we need to find $f'(x)$:

$$f'(x) = \frac{1}{2\sqrt{x}}.$$

Now plug into the formula:

$$f(x + \Delta x) \approx f(x) + f'(x)\Delta x$$
$$\sqrt{x + \Delta x} \approx \sqrt{x} + \frac{1}{2\sqrt{x}}\Delta x$$

Now if we plug in $x = 9$ and $\Delta x = +.01$:

$$\sqrt{9.01} \approx \sqrt{9} + \frac{1}{2\sqrt{9}}(.01) \approx 3.001666666.$$

If you enter $\sqrt{9.01}$ into your calculator, you get: 3.001666204. As you can see, our answer is a pretty good approximation. It's not so good, however, when Δx is too big. How big is too big? Good question.

Example 2: Use differentials to approximate the $\sqrt{9.5}$.

Let $x = 9$, $\Delta x = +.5$, $f(x) = \sqrt{x}$ and plug into what you found in Example 1:

$$\sqrt{9.5} \approx \sqrt{9} + \frac{1}{2\sqrt{9}}(.5) \approx 3.083333333.$$

However, $\sqrt{9.5}$ equals 3.082207001 on a calculator. This is only good to two decimal places. As the ratio of $\dfrac{\Delta x}{x}$ grows larger, the approximation gets less accurate and we start to get away from the actual value.

There's another approximation formula that you'll need to know for the AP exam. This formula is used to estimate the error in a measurement, or to find the effect on a formula when a small change in measurement is made. The formula is:

$$dy = f'(x)dx$$

This notation may look a little confusing. It says that the change in a measurement dy, due to a differential dx, is found by multiplying the derivative of the equation for y by the differential. Let's do an example.

Example 3: The radius of a circle is increased from 3 to 3.04. Estimate the change in area.

Let $A = \pi r^2$. Then our formula says that $dA = A'dr$, where A' is the derivative of the area with respect to r, and $dr = .04$ (the change). First, find the derivative of the area: $A = 2\pi r$. Now plug into the formula:

$$dA = 2\pi r dr = 2\pi(3)(.04) = .754.$$

The actual change in the area is from 9π to 9.2416π, which is approximately .759. As you can see, this approximation formula is pretty accurate.

Here are some sample problems involving this differential formula. Try them out, then check your work against the answers directly beneath.

PROBLEM 1. Use differentials to approximate $(3.98)^4$.

Answer: Let $f(x) = x^4$, $x = 4$, and $\Delta x = -.02$. Next, find $f'(x)$, which is: $f'(x) = 4x^3$. Now plug into the formula:

$$f(x + \Delta x) \approx f(x) + f'(x)\Delta x$$
$$\sqrt{x + \Delta x} \approx x^4 + 4x^3 \Delta x$$

If you plug in $x = 4$ and $\Delta x = -.02$, you get:

$$(3.98)^4 \approx 4^4 + 4(4)^3(-.02) \approx 250.88.$$

Check $(3.98)^4$ by using your calculator; you should get 250.9182722. Not a bad approximation.

You're probably asking yourself, why can't I just use my calculator every time? Because most math teachers are dedicated to teaching you several complicated ways to calculate things without your calculator.

PROBLEM 2. Use differentials to approximate sin 46°.

Answer: This is a tricky question. The formula doesn't work if you use degrees. Here's why: let $f(x) = \sin x$, $x = 45°$, and $x = 1°$. The derivative is $f'(x) = \cos x$.

If you plug this information into the formula, you get: $\sin 46° \quad \sin 45° + \cos 45°(1°) = \sqrt{2}$. You should recognize that this is nonsense for two reasons: (1) the sine of any angle is between –1 and 1; and (2) the answer should be close to $\sin 45° = \dfrac{1}{\sqrt{2}}$.

What went wrong? You have to use radians! As we mentioned before, angles in calculus problems are measured in radians, not degrees.

Let $f(x) = \sin x$, $x = \dfrac{\pi}{4}$, and $\Delta x = \dfrac{\pi}{180}$. Now plug into the formula:

$$\sin\left(\frac{46\pi}{180}\right) \approx \sin\frac{\pi}{4} + \left(\cos\frac{\pi}{4}\right)\left(\frac{\pi}{180}\right) = 0.7194.$$

PROBLEM 3. The radius of a sphere is measured to be 4 cm with an error of ±.01cm. Use differentials to approximate the error in the surface area.

Answer: Now it's time for the other differential formula. The formula for the surface area of a sphere is:

$$S = 4\pi r^2$$

The formula says that $dS = S'dr$, so first, we find the derivative of the surface area ($S' = 8 \ r$) and plug away:

$$dS = 8\pi r dr = 8\pi(4)(\pm.01) = \pm1.0053$$

This looks like a big error, but given that the surface area of a sphere with radius 4 is approximately 201cm², the error is quite small.

PRACTICE PROBLEMS

Use the differential formulas in this chapter to solve these problems. The answers are in chapter 21.

1. Approximate $\sqrt{25.02}$.

2. Approximate $\sqrt[3]{63.97}$.

3. Approximate tan 61°.

4. Approximate $(9.99)^3$.

5. The side of a cube is measured to be 6 in. with an error of ±0.02 in. Estimate the error in the volume of the cube.

6. When a spherical ball bearing is heated, its radius increases by 0.01 mm. Estimate the change in volume of the ball bearing when the radius is 5 mm.

7. A side of an equilateral triangle is measured to be 10 cm. Estimate the change in the area of the triangle when the side shrinks to 9.8 cm.

8. A cylindrical tank is constructed to have a diameter of 5 meters and a height of 20 meters. Find the error in the volume if:

   (a) the diameter is exact, but the height is 20.1 meters; and

   (b) the height is exact, but the diameter is 5.1 meters.

LOGARITHMIC DIFFERENTIATION

There's one last topic in differential calculus that you BC students need to know: logarithmic differentiation. It's a very simple and handy technique used to find the derivatives of expressions that involve a lot of algebra. By employing the rules of logarithms, we can find the derivatives of expressions that would otherwise require a messy combination of the Chain Rule, the Product Rule, and the Quotient Rule.

First, let's review a couple of rules of logarithms (remember, when we refer to a logarithm in calculus, we mean the natural log (base e), not the common log):

$$\ln A + \ln B = \ln(AB)$$

$$\ln A - \ln B = \ln\left(\frac{A}{B}\right)$$

$$\ln A^B = B \ln A$$

For example, we can rewrite $\ln(x + 3)^5$ as $5\ln(x +3)$.
As a quick review exercise, how could you rewrite

$$\ln \frac{x^2}{(x+1)^2}?$$

Answer: $2\ln x - 2\ln(x + 1)$.

The other important thing to remember from chapter 9 is:

$$\frac{d}{dx}\ln u = \frac{1}{u}\frac{du}{dx}.$$

Bearing these in mind, we can use these rules to find the derivative of a complicated expression without using all that mind-bending algebra. Instead, take the log of both sides of an expression and then use the rules of logarithms to simplify it before finding the derivative.

Example 1: Find the derivative of $y = \dfrac{x+3}{2x-5}$.

The Quotient Rule works here just fine, but check this out instead. Take the log of both sides:

$$\ln y = \ln \frac{x+3}{2x-5}.$$

Using log rules, you can rewrite the expression this way:

$$\ln y = \ln(x+3) - \ln(2x-5).$$

Now take the derivative of both sides:

$$\frac{1}{y}\frac{dy}{dx} = \frac{1}{x+3} - \frac{2}{2x-5}.$$

Multiply both sides by y, and you get the following:

$$\frac{dy}{dx} = y\left(\frac{1}{x+3} - \frac{2}{2x-5}\right).$$

Finally, substitute for y:

$$\frac{dy}{dx} = \left(\frac{x+3}{2x-5}\right)\left(\frac{1}{x+3} - \frac{2}{2x-5}\right).$$

You might be thinking, Why go through the hassle? After all, you could have done this with the Quotient Rule and it would have involved less work. But, as you'll see, this technique comes in very handy when the algebra is especially rotten. Also, we don't usually substitute back for the y term. Instead, just leave it in the derivative (unless it needs to be defined only in terms of x).

Example 2: Find the derivative of y, where $y^2 = (x + 4)^3 (x - 2)^5$.

Take the log of both sides:

$$\ln y^2 = \ln\left[(x+4)^3(x-2)^5\right].$$

Next, use those log rules to get:

$$2 \ln y = 3 \ln(x + 4) + 5 \ln(x{-}2).$$

Take the derivative of both sides:

$$\frac{2}{y}\frac{dy}{dx} = \frac{3}{x+4} + \frac{5}{x-2}.$$

Finally, multiply both sides by $\frac{y}{2}$:

$$\frac{dy}{dx} = \frac{y}{2}\left[\frac{3}{x+4} + \frac{5}{x-2}\right].$$

That's it. Now, wasn't that easier? Substitute back for y only if you have to.

Example 3: Find the derivative of $y = \sqrt{\frac{x+3}{x-5}}\sqrt[3]{\frac{2x+7}{5x-1}}$.

Holy cripes! What a problem! Watch how using logs will save you. Take the log of both sides:

$$\ln y = \ln\left(\sqrt{\frac{x+3}{x-5}}\sqrt[3]{\frac{2x+7}{5x-1}}\right).$$

Now you can simplify using log rules:

$$\ln y = \frac{1}{2}\left[\ln(x+3) - \ln(x-5)\right] + \frac{1}{3}\left[\ln(2x+7) - \ln(5x-1)\right].$$

Take the derivative:

$$\frac{1}{y}\frac{dy}{dx} = \frac{1}{2}\left[\frac{1}{x+3} - \frac{1}{x-5}\right] + \frac{1}{3}\left[\frac{1}{2x+7} - \frac{1}{5x-1}\right]$$

And then multiply by y:

$$\frac{dy}{dx} = \frac{y}{2}\left[\frac{1}{x+3} - \frac{1}{x-5}\right] + \frac{y}{3}\left[\frac{1}{2x+7} - \frac{1}{5x-1}\right].$$

Example 4: Find the derivative of $y = \sqrt{\frac{x+3}{5-x}}$.

Take the log of both sides:

$$\ln y = \ln\left(\sqrt{\frac{x+3}{5-x}}\right).$$

Simplify using the log rules:

$$\ln y = \frac{1}{2}\left[\ln(x+3) - \ln(5-x)\right].$$

Take the derivative:

$$\frac{1}{y}\frac{dy}{dx} = \frac{1}{2}\left[\frac{1}{x+3}+\frac{1}{5-x}\right].$$

Multiply by y:

$$\frac{dy}{dx} = \frac{y}{2}\left[\frac{1}{x+3}+\frac{1}{5-x}\right].$$

And you're done.

That's all there is to logarithmic differentiation. It's a really helpful tool for simplifying complicated derivatives. There's generally one question on the AP exam that uses logarithmic differentiation, and it's usually not that complicated.

Try these solved problems on your own. Cover the answers first, then check your work.

PROBLEM 1. Use logarithmic differentiation to find $\dfrac{dy}{dx}$ if $y = \dfrac{x^2\sqrt{1-x^3}}{\left(x^2+1\right)^2}$.

Answer: Take the log of both sides:

$$\ln y = \ln\left[\frac{x^2\sqrt{1-x^3}}{\left(x^2+1\right)^2}\right].$$

Next, use the log rules to simplify the expression:

$$\ln y = \ln x^2 + \ln\sqrt{1-x^3} - \ln\left(x^2+1\right)^2$$

$$\ln y = 2\ln x + \frac{1}{2}\ln\left(1-x^3\right) - 2\ln\left(x^2+1\right).$$

Taking the derivative of both sides, you should get:

$$\frac{1}{y}\frac{dy}{dx} = \frac{2}{x} - \frac{3x^2}{2\left(1-x^3\right)} - \frac{4x}{x^2+1}.$$

Then get the derivative by itself by multiplying both sides by y:

$$\frac{dy}{dx} = y\left[\frac{2}{x} - \frac{3x^2}{2\left(1-x^3\right)} - \frac{4x}{x^2+1}\right].$$

PROBLEM 2. Use logarithmic differentiation to find $\dfrac{dy}{dx}$ if $y = \dfrac{\left(x^2 - 5x\right)\cos^2 x}{\left(x^3 + 1\right)^5}$.

Answer: The four-step process should seem second nature by now. Take the log of both sides:

$$\ln y = \ln\left[\frac{\left(x^2 - 5x\right)\cos^2 x}{\left(x^3 + 1\right)^5}\right].$$

Simplify the expression using log rules:

$$\ln\, y = \ln\left(x^2 - 5x\right) + \ln\left(\cos^2 x\right) - \ln\left(x^3 + 1\right)$$

$$\ln\, y = \ln\left(x^2 - 5x\right) + 2\ln(\cos x) - 5\ln\left(x^3 + 1\right).$$

Take the derivative of both sides:

$$\frac{1}{y}\frac{dy}{dx} = \frac{2x - 5}{x^2 - 5} - \frac{2\sin x}{\cos x} - \frac{15x^2}{x^3 + 1}$$

$$\frac{1}{y}\frac{dy}{dx} = \frac{2x - 5}{x^2 - 5x} - 2\tan x - \frac{15x^2}{x^3 + 1}.$$

And multiply by y:

$$\frac{dy}{dx} = y\left[\frac{2x - 5}{x^2 - 5x} - 2\tan x - \frac{15x^2}{x^3 + 1}\right].$$

All right. One more time. That's all. We promise.

PROBLEM 3. Use logarithmic differentiation to find $\dfrac{dy}{dx}$ if $y = \sqrt{\dfrac{\sec x}{\left(4 - x^3\right)^5}}$.

Answer: Take the log of both sides:

$$\ln y = \ln\sqrt{\frac{\sec x}{\left(4 - x^3\right)^5}}.$$

Simplify:

$$\ln y = \frac{1}{2}\left[\ln(\sec x) - 5\ln\left(4 - x^3\right)\right]$$

Take the derivative:

$$\frac{1}{y}\frac{dy}{dx} = \frac{1}{2}\left[\frac{\sec x \tan x}{\sec x} + \frac{15x^2}{4-x^3}\right]$$

$$\frac{1}{y}\frac{dy}{dx} = \frac{1}{2}\left[\tan x + \frac{15x^2}{4-x^3}\right].$$

Multiply both sides by y:

$$\frac{dy}{dx} = \frac{y}{2}\left[\tan x + \frac{15x^2}{4-x^3}\right].$$

PRACTICE PROBLEMS

Use logarithmic differentiation to find the derivative of each of the following problems. The answers are in chapter 21.

1. $y = x\left(\sqrt[4]{1-x^3}\right).$

2. $y = \sqrt{\dfrac{1+x}{1-x}}.$

3. $y = \dfrac{\left(x^3+5\right)^{\frac{3}{2}} \sqrt[3]{4-x^2}}{x^4-x^2+6}.$

4. $y = \dfrac{\sin x \cos x}{\sqrt{x^3-4}}.$

5. $y = \dfrac{\left(4x^2-8x\right)^3\left(5-3x^4+7x\right)^4}{\left(x^2+x\right)^3}.$

6. $y = \dfrac{x-1}{x \tan x}.$

7. $y = \left(x-x^2\right)^2\left(x^3+x^4\right)^3\left(x^6-x^5\right)^4.$

8. $y = \sqrt[4]{\dfrac{x(1-x)(1+x)}{\left(x^2-1\right)(5-x)}}.$

11

THE INTEGRAL

Welcome to the other half of calculus! This, unfortunately, is the more difficult half, but don't worry: we'll get you through it. In differential calculus, you learned all of the fun things that you can do with the derivative. Now you'll learn to do the reverse: how to take an integral. As you might imagine, there's a bunch of new fun things that you can do with integrals, too. (Trust us.)

It's also time for a new symbol $\int$, which stands for integration. An integral actually serves several different purposes, but the first, and most basic, one is that of the antiderivative.

THE ANTIDERIVATIVE

An antiderivative is a derivative in reverse. Therefore, we're going to reverse some of the rules we learned with derivatives and apply them to integrals. For example, we know that the derivative of x^2 is $2x$. We use antidifferentiation if we're given the derivative of a function and we have to figure out the original function. Thus, the antiderivative of $2x$ is x^2. (Actually, the answer is slightly more complicated than that, but we'll get into that in a few moments.)

Let's learn our first rule, which is a variation of the Power Rule (sort of):

If $f(x) = x^n$, then the antiderivative of $f(x)$ is:

$$\frac{x^{n+1}}{n+1} \text{ (except when } n = -1\text{).}$$

IMPORTANT STUFF TO KNOW

Now we need to add some info here to make sure that you get this absolutely correct. First, as far as notation goes, it is traditional to write the antiderivative of a function using its upper case letter , so the antiderivative of $f(x)$ is $F(x)$, the antiderivative of $g(x)$ is $G(x)$, and so on.

The second idea is very important: each function has more than one antiderivative. In fact, there are an infinite number of antiderivatives of a function. Let's go back to our example to help illustrate this.

Remember that the antiderivative of $2x$ is x^2? Well, consider: if you take the derivative of $x^2 + 1$, you get $2x$. The same is true for $x^2 + 2$, $x^2 - 1$, and so on. In fact, if *any* constant is added to x^2, the derivative is still $2x$ because the derivative of a constant is zero.

Because of this, we write the antiderivative of $2x$ as $x^2 + C$; where C stands for any constant.

Finally, whenever you take the integral (or antiderivative) of a function of x, you always add the term dx (or dy if it's a function of y, etc.) to the integrand (the thing inside the integral). You'll learn why later.

For now, just remember that you must always use the dx symbol, and teachers love to take points off for forgetting the dx. Don't ask why, but they do!

So now, let's write the Power Rule correctly:

If $f(x) = x^n$ then $\int f(x)dx = \dfrac{x^{n+1}}{n+1} + C$ (except when $n = -1$).

Example 1: Find $\int x^3 dx$.

Using the Power Rule, we get:

$$\int x^3 dx = \frac{x^4}{4} + C.$$

Don't forget the constant C, or your teachers will take points off for that, too!

Example 2: Find $\int x^{-3} dx$.

The Power Rule works with negative exponents, too:

$$\int x^{-3}dx = \frac{x^{-2}}{-2} + C.$$

Not terribly hard, is it? Now it's time for a few more rules that look remarkably similar to the rules for derivatives that we saw in chapter 4:

$$\int kf(x)dx = k\int f(x)dx$$

$$\int [f(x) + g(x)]dx = \int f(x)dx + \int g(x)dx$$

$$\int kdx = kx + C$$

Here are a few more examples to make you an expert.

Example 3: $\int 5dx = 5x + C.$

Example 4: $\int 7x^3 dx = \frac{7x^4}{4} + C.$

Example 5: $\int (3x^2 + 2x)dx = x^3 + x^2 + C.$

Example 6: $\int \sqrt{x}\, dx = \frac{x^{\frac{3}{2}}}{\frac{3}{2}} + C = \frac{2x^{\frac{3}{2}}}{3} + C.$

INTEGRALS OF TRIG FUNCTIONS

The integrals of some trigonometric functions follow directly from the derivative formulas in chapter 4:

$$\int \sin ax\, dx = -\frac{\cos ax}{a} + C$$

$$\int \csc^2 ax\, dx = -\frac{\cot ax}{a} + C$$

$$\int \cos ax \, dx = \frac{\sin ax}{a} + C$$

$$\int \sec ax \tan ax \, dx = \frac{\sec ax}{a} + C$$

$$\int \sec^2 ax \, dx = \frac{\tan ax}{a} + C$$

$$\int \csc ax \cot ax \, dx = -\frac{\csc ax}{a} + C$$

We didn't mention the integrals of tangent, cotangent, secant, and cosecant, because you need to know some rules about logarithms to figure them out. We'll get to them in a few chapters. Notice also that each of the answers is divided by a constant. This is to account for the Chain Rule. Let's do some examples.

Example 7: Check the integral $\int \sin 5x \, dx = -\frac{\cos 5x}{5} + C$ by differentiating the answer.

$$\frac{d}{dx}\left[-\frac{\cos 5x}{5} + C\right] = -\frac{1}{5}(-\sin 5x)(5) = \sin 5x.$$

Notice how the constant is accounted for in the answer?

Example 8: $\int \sec^2 3x \, dx = \frac{\tan 3x}{3} + C$

Example 9: $\int \cos \pi x \, dx = \frac{\sin \pi x}{\pi} + C$

Example 10: $\int \sec\left(\frac{x}{2}\right)\tan\left(\frac{x}{2}\right)dx = 2\sec\left(\frac{x}{2}\right) + C$

If you're not sure if you have the correct answer when you take an integral, you can always check by differentiating the answer and seeing if you get what you started with. Try to get in the habit of doing that at the beginning, because it'll help you build confidence in your ability to find integrals properly. You'll see that, although you can differentiate just about any expression that you'll normally encounter, you won't be able to integrate many of the functions you see.

ADDITION AND SUBTRACTION

By using the rules for addition and subtraction, we can integrate most polynomials.

Example 11: Find $\int \left(x^3 + x^2 - x\right) dx$.

We can break this into separate integrals, which gives us:

$$\int x^3 dx + \int x^2 dx - \int x\, dx.$$

Now you can integrate each of these individually:

$$\frac{x^4}{4} + C + \frac{x^3}{3} + C - \frac{x^2}{2} + C.$$

You can combine the constants into one constant (it doesn't matter how many C's we use, because their sum is one collective constant whose derivative is zero):

$$\frac{x^4}{4} + \frac{x^3}{3} - \frac{x^2}{2} + C.$$

Sometimes you'll be given information about the function you're seeking that will enable you to solve for the constant. Often, this is an "initial value", which is the value of the function when the variable is zero. As we've seen, normally there are an infinite number of solutions for an integral, but when we solve for the constant, there's only one.

Example 12: Find the equation of y where $\dfrac{dy}{dx} = 3x + 5$ and $y = 6$ when $x = 0$.

Let's put this in integral form:

$$y = \int (3x + 5)\, dx$$

Integrating, we get:

$$y = \frac{3x^2}{2} + 5x + C.$$

Now we can solve for the constant because we know that $y = 6$ when $x = 0$:

$$6 = \frac{3(0)^2}{2} + 5(0) + C.$$

Therefore, $C = 6$ and the equation is:

$$y = \frac{3x^2}{2} + 5x + 6.$$

Example 13: Find $f(x)$ if $f'(x) = \sin x - \cos x$ and $f(\pi) = 3$.

Integrate $f'(x)$:

$$f(x) = \int (\sin x - \cos x)\,dx = -\cos x - \sin x + C.$$

Now solve for the constant:

$$3 = -\cos(\pi) - \sin(\pi) + C$$

$$C = 2.$$

Therefore, the equation becomes:

$$f(x) = -\cos x - \sin x + 2.$$

Now we've covered the basics of integration. However, integration is a very sophisticated topic and there are many types of integrals that will cause you trouble. We will need several techniques to learn how to evaluate these integrals. The first and most important is called u-substitution, which we will cover in the second half of this chapter.

In the meantime, here are some solved problems. Do each problem, covering the answer first, then checking your answer.

PROBLEM 1. Evaluate $\int x^{\frac{3}{5}}dx$.

Answer: Here's the Power Rule again:

$$\int x^n dx = \frac{x^{n+1}}{n+1} + C.$$

Using the rule:

$$\int x^{\frac{3}{5}}dx = \frac{x^{\frac{8}{5}}}{\frac{8}{5}} + C.$$

You can rewrite it as $\dfrac{5x^{\frac{8}{5}}}{8} + C.$

PROBLEM 2. Evaluate $\int \left(5x^3 + x^2 - 6x + 4\right)dx$.

Answer: We can break this up into several integrals:

$$\int \left(5x^3 + x^2 - 6x + 4\right)dx = \int 5x^3 dx + \int x^2 dx - \int 6x\,dx + 4\int dx.$$

Each of these can be integrated according to the Power Rule:

$$\frac{5x^4}{4}+C+\frac{x^3}{3}+C-\frac{6x^2}{2}+C+4x+C.$$

This can be rewritten:

$$\frac{5x^4}{4}+\frac{x^3}{3}-3x^2+4x+C.$$

Notice that we combine the constant terms into one constant term C.

PROBLEM 3. Evaluate $\int\left(3-x^2\right)^2 dx$.

Answer: First, expand the integrand:

$$\int\left(9-6x^2+x^4\right)dx.$$

Break this up into several integrals:

$$\int 9dx-6\int x^2dx+\int x^4dx.$$

And integrate according to the Power Rule:

$$9x-2x^3+\frac{x^5}{5}+C.$$

PROBLEM 4. Evaluate $\int\left(4\sin x-3\cos x\right)dx$.

Answer: Break this problem into two integrals:

$$4\int\sin x\,dx-3\int\cos x\,dx.$$

Each of these trig integrals can be evaluated according to its rule:

$$-4\cos x-3\sin x+C.$$

PROBLEM 5. $\int\left(2\sec^2 x-5\csc^2 x\right)dx$.

Answer: Break the integral in two:

$$2\int\sec^2 x\,dx-5\int\csc^2 x\,dx.$$

Each of these trig integrals can be evaluated according to its rule:

$$2\tan x+5\cot x+C.$$

PRACTICE PROBLEMS

Now evaluate the following integrals. The answers are in chapter 21.

1. $\displaystyle\int \frac{1}{x^4}\,dx$

2. $\displaystyle\int \frac{5}{\sqrt{x}}\,dx$

3. $\displaystyle\int \frac{x^5+7}{x^2}\,dx$

4. $\displaystyle\int \left(5x^4-3x^2+2x+6\right)dx$

5. $\displaystyle\int \left(3x^{-3}-2x^{-2}+x^4+16x^7\right)dx$

6. $\displaystyle\int \left(1+x^2\right)(x-2)dx$

7. $\displaystyle\int x^{\frac{1}{3}}(2+x)dx$

8. $\displaystyle\int \left(x^3+x\right)^2 dx$

9. $\displaystyle\int \frac{x^6-2x^4+1}{x^2}\,dx$

10. $\displaystyle\int x(x-1)^3\,dx$

11. $\displaystyle\int (\cos x-5\sin x)dx$

12. $\displaystyle\int \sec x(\sec x+\tan x)dx$

13. $\displaystyle\int \left(\sec^2 x+x\right)dx$

14. $\displaystyle\int \frac{\sin x}{\cos^2 x}\,dx$

15. $\displaystyle\int \frac{\cos^3 x+4}{\cos^2 x}\,dx$

16. $\int \dfrac{\sin 2x}{\cos x} dx$

17. $\int \left(1 + \cos^2 x \sec x\right) dx$

18. $\int \left(\tan^2 x\right) dx$

19. $\int \dfrac{1}{\csc x} dx$

20. $\int \left(x - \dfrac{2}{\cos^2 x}\right) dx$

U-SUBSTITUTION

When we discussed differentiation, one of the most important techniques we mastered was the Chain Rule. Now, you'll learn the integration corollary of the Chain Rule (called *u*-substitution), which we use when the integrand is a composite function. All you do is replace the function with *u*, and then you can integrate the simpler function using the Power Rule (as shown below):

$$\int u^n du = \frac{u^{n+1}}{n+1} + C$$

Suppose you have to integrate $\int (x-4)^{10} dx$. You could expand out this function and integrate each term, but that'll take a while. Instead, you can follow these four steps:

Step 1: Let $u = x - 4$. Then $\dfrac{du}{dx} = 1$ (rearrange this to get $du = dx$).

Step 2: Substitute $u = x - 4$ and $du = dx$ into the integrand:

$$\int u^{10} du.$$

Step 3: Integrate:

$$\int u^{10} du = \frac{u^{11}}{11} + C.$$

Step 4: Substitute back for *u*:

$$/ \frac{(x-4)^{11}}{11} + C.$$

That's *u*-substitution. The main difficulty you'll have will be picking the appropriate function to set equal to *u*. The best way to get better is to practice. The object is to pick a function and replace it with *u*, then take the derivative of *u* to find *du*. If we can't replace *all* of the terms in the integrand, we *can't* do the substitution.

Let's do some examples.

Example 1: $\int 10x\left(5x^2 - 3\right)^6 dx =$

Once again, you could expand this out and integrate each term, but that would be difficult. Use *u*-substitution.

Let $u = 5x^2 - 3$. Then $\dfrac{du}{dx} = 10x$ and $du = 10x\,dx$. Now substitute:

$$\int u^6 du.$$

And integrate:

$$\int u^6 du = \frac{u^7}{7} + C.$$

Substituting back gives you:

$$\frac{\left(5x^2 - 3\right)^7}{7} + C.$$

Confirm that this is the integral by differentiating $\dfrac{\left(5x^2 - 3\right)^7}{7} + C$:

$$\frac{d}{dx}\left[\frac{\left(5x^2 - 3\right)^7}{7} + C\right] = \frac{7\left(5x^2 - 3\right)^6}{7}(10x) = \left(5x^2 - 3\right)^6(10x).$$

Example 2: $\int 2x\sqrt{x^2 - 5}\,dx =$

If $u = x^2 - 5$, then $\dfrac{du}{dx} = 2x$ and $du = 2x\,dx$. Substitute *u* into the integrand:

$$\int u^{\frac{1}{2}}\,du.$$

Integrate:

$$\int u^{\frac{1}{2}}\,du = \frac{u^{3/2}}{3/2} + C = \frac{2u^{\frac{3}{2}}}{3} + C.$$

And substitute back:

$$\frac{2(x^2-5)^{\frac{3}{2}}}{3}+C.$$

Note: from now on, we're not going to rearrange $\frac{du}{dx}$; we'll go directly to "$du =$" format. You should be able to do that step on your own.

Example 3: $\int 3\sin(3x-1)dx$

Let $u = 3x - 1$. Then $du = 3dx$. Substitute the u in the integral:

$$\int \sin u\, dx.$$

Figure out the integral:

$$\int \sin u\, du = -\cos u + C.$$

And throw the x's back in:

$$-\cos(3x - 1) + C.$$

So far, this is only the simplest kind of u-substitution; naturally, the process can get worse when the substitution isn't as easy. Usually, you'll have to insert a constant term to put the integrand into a workable form. For example:

Example 4: $\int (5x+7)^{20}dx =$

Let $u = 5x + 7$. Then $du = 5dx$. Notice that we can't do the substitution immediately because we need to substitute for dx and we have $5dx$. No problem: because 5 is a constant, just solve for dx:

$$\frac{1}{5}du = dx.$$

Now you can substitute:

$$\int (5x+7)^{20}dx = \int u^{20}\left(\frac{1}{5}\right)du.$$

Rearrange the integral and solve:

$$\int u^{20}\left(\frac{1}{5}\right)du = \frac{1}{5}\int u^{20}du = \frac{1}{5}\frac{u^{21}}{21}+C = \frac{u^{21}}{105}+C$$

And now it's time to substitute back:

$$\frac{u^{21}}{105} + C = \frac{(5x+7)^{21}}{105} + C.$$

These examples are more typical of the ones you'll see on the exam.

Example 5: $\int x\cos(3x^2 + 1)dx =$

Let $u = 3x^2 + 1$. Then $du = 6x\,dx$. We need to substitute for $x\,dx$, so we can rearrange the du term:

$$\frac{1}{6}du = x\,dx.$$

Now substitute:

$$\int \frac{1}{6}\cos u\,du.$$

Evaluate the integral:

$$\int \frac{1}{6}\cos u\,du = \frac{1}{6}\sin u + C.$$

And substitute back:

$$\frac{1}{6}\sin u + C = \frac{1}{6}\sin(3x^2 + 1) + C.$$

Example 6: $\int x\sec^2(x^2)dx =$

Let $u = x^2$. Then $du = 2x\,dx$ and $\frac{1}{2}du = x\,dx$.

Substitute:

$$\int \frac{1}{2}\sec^2 u\,du.$$

Evaluate the integral:

$$\int \frac{1}{2}\sec^2 u\,du = \frac{1}{2}\tan u + C$$

Now the original function goes back in:

$$\frac{1}{2}\tan(x^2) + C.$$

That's all there is to it. This is a good technique to master, so practice on the following solved problems. Do each problem, covering the answer first, then checking your answer.

PROBLEM 1. Evaluate $\int \sec^2 3x \, dx$.

Answer: Let $u = 3x$ and $du = 3dx$. Then $\dfrac{1}{3} du = dx$.

Substitute and integrate:

$$\frac{1}{3} \int \sec^2 u \, du = \frac{1}{3} \tan u + C.$$

Then substitute back:

$$\frac{1}{3} \tan 3x + C.$$

PROBLEM 2. Evaluate $\int \sqrt{5x - 4} \, dx$.

Answer: Let $u = 5x - 4$ and $du = 5dx$. Then $\dfrac{1}{5} du = dx$.

Substitute and integrate:

$$\frac{1}{5} \int u^{\frac{1}{2}} \, du = \frac{2}{15} u^{\frac{3}{2}} + C.$$

Then substitute back:

$$\frac{2}{15} (5x - 4)^{\frac{3}{2}} + C.$$

PROBLEM 3. Evaluate $\int x \left(4x^2 - 7\right)^{10} dx$.

Answer: Let $u = 4x^2 - 7$ and $du = 8xdx$. Then $\dfrac{1}{8} du = x \, dx$.

Substitute and integrate:

$$\frac{1}{8} \int u^{10} \, du = \frac{1}{88} u^{11} + C.$$

Then substitute back:

$$\frac{1}{88} \left(4x^2 - 7\right)^{11} + C.$$

PROBLEM 4. Evaluate $\int \tan \dfrac{x}{3}\ \sec^2 \dfrac{x}{3} dx$.

Answer: Let $u = \tan \dfrac{x}{3}$ and $du = \dfrac{1}{3} \sec^2 \dfrac{x}{3} dx$. Then $3du = \sec^2 \dfrac{x}{3} dx$.

Substituting, we get:

$$3 \int u \ du = \frac{3}{2} u^2 + C.$$

Then substitute back:

$$\frac{3}{2} \tan^2 \frac{x}{3} + C.$$

PRACTICE PROBLEMS

Now evaluate the following integrals. The answers are in chapter 21.

1. $\int \sin 2x \cos 2x \ dx$

2. $\int \dfrac{3x \ dx}{\sqrt[3]{10 - x^2}}$

3. $\int x^3 \sqrt{5x^4 + 20} \ dx$

4. $\int \dfrac{dx}{(x-1)^2}$

5. $\int (x^2 + 1)(x^3 + 3x)^{-5} dx$

6. $\int \dfrac{1}{\sqrt{x}} \sin \sqrt{x} \ dx$

7. $\int x^2 \sec^2 x^3 \ dx$

8. $\int \dfrac{\cos\left(\dfrac{3}{x}\right)}{x^2} dx$

9. $\int \dfrac{\sin 2x}{(1 - \cos 2x)^3} dx$

10. $\int \sin(\sin x) \cos x \ dx$

12

DEFINITE INTEGRALS

It's time to learn one of the most important uses of the integral. We've already discussed how integration can be used to "antidifferentiate" a function; now you'll see that you can also use it to find the area under a curve. First, here's a little background about how to find the area without using integration.

Suppose you have to find the area under the curve $y = x^2 + 2$ from $x = 1$ to $x = 3$. The graph of the curve looks like this:

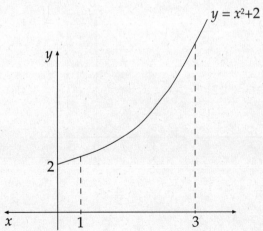

Don't panic yet. Nothing you've learned in geometry thus far has taught you how to find the area of something like this. You have learned how to find the area of a rectangle, though, and we're going to use rectangles to approximate the area between the curve and the x-axis.

Let's divide the region into two rectangles, one from $x = 1$ to $x = 2$ and one from $x = 2$ to $x = 3$, where the top of each rectangle comes just under the curve. It looks like this:

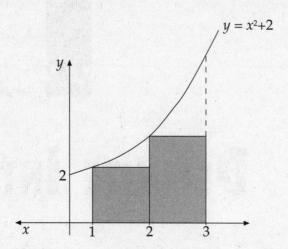

Notice that the width of each rectangle is 1. The height of the left rectangle is found by plugging 1 into the equation $y = x^2 + 2$ (yielding 3); the height of the right rectangle is found by plugging 2 into the same equation (yielding 6). The combined area of the two rectangles is $(1)(3) + (1)(6) = 9$. So we could say that the area under the curve is approximately 9 square units.

Naturally, this is a pretty rough approximation that significantly underestimates the area. Look at how much of the area we missed by using two rectangles! How do you suppose we make the approximation better? Divide the region into more, thinner rectangles.

This time, cut up the region into four rectangles, each with width of $\dfrac{1}{2}$. It looks like this:

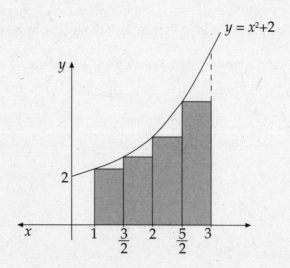

Now find the height of each rectangle the same way as before. Notice that the values we use are the left endpoints of each rectangle. The heights of the rectangles are, respectively:

$$(1)^2 + 2 = 3; \left(\frac{3}{2}\right)^2 + 2 = \frac{17}{4}; (2)^2 + 2 = 6 \text{ and } \left(\frac{5}{2}\right)^2 + 2 = \frac{33}{4}.$$

Now, multiply each height by the width of $\frac{1}{2}$ and add up the areas:

$$\left(\frac{1}{2}\right)(3) + \left(\frac{1}{2}\right)\left(\frac{17}{4}\right) + \left(\frac{1}{2}\right)(6) + \left(\frac{1}{2}\right)\left(\frac{33}{4}\right) = \frac{43}{4}.$$

This is a much better approximation of the area, but there's still a lot of space that isn't accounted for. We're still underestimating the area. The rectangles need to be thinner. But before we do that, let's do something else.

Notice how each of the rectangles is inscribed in the region. Suppose we used larger rectangles instead—that is, we could determine the height of each rectangle by the higher of the two y-values, not the lower. Then the region would look like this:

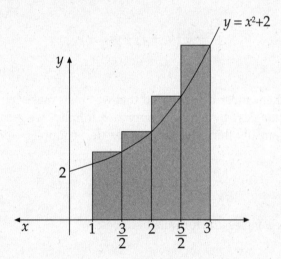

To find the area of the rectangles, we would still use the width of $\frac{1}{2}$, but the heights would change. The heights of each rectangle are now found by plugging in the right endpoint of each rectangle:

$$\left(\frac{3}{2}\right)^2 + 2 = \frac{17}{4}; (2)^2 + 2 = 6; \left(\frac{5}{2}\right)^2 + 2 = \frac{33}{4}; \text{ and } (3)^2 + 2 = 11.$$

Once again, multiply each height by the width of $\frac{1}{2}$ and add up the areas:

$$\left(\frac{1}{2}\right)\left(\frac{17}{4}\right) + \left(\frac{1}{2}\right)(6) + \left(\frac{1}{2}\right)\left(\frac{33}{4}\right) + \left(\frac{1}{2}\right)(11) = \frac{59}{4}.$$

The area under the curve using four inscribed rectangles is $\frac{43}{4}$, and the area using four circumscribed rectangles is $\frac{59}{4}$, so why not average the two? This gives us $\frac{51}{4}$, which is a better approximation of the area.

Now that we've found the area using the rectangles a few times, let's turn the method into a formula. Call the left endpoint of the interval a and the right endpoint of the interval b, and set the number of rectangles we use equal to n. Then the width of each rectangle is $\frac{b-a}{n}$. The height of the first inscribed rectangle is y_0, the height of the second rectangle is y_1, the height of the third rectangle is y_2, and so on, up to the last rectangle, which is y_{n-1}. If we use the left endpoint of each rectangle, the area under the curve is thus:

$$\left(\frac{b-a}{n}\right)\left[y_0 + y_1 + y_2 + y_3 \ldots + y_{n-1}\right]$$

If we use the right endpoint of each rectangle, then the formula is:

$$\left(\frac{b-a}{n}\right)\left[y_1 + y_2 + y_3 \ldots + y_n\right]$$

Intimidating, isn't it?

Now for the fun part. Remember how we said that we could make the approximation better by making more, thinner rectangles? By letting n approach infinity, we create an infinite number of rectangles that are infinitesimally thin. The formula for "left-endpoint" rectangles becomes:

$$\lim_{n\to\infty}\left(\frac{b-a}{n}\right)\left[y_0 + y_1 + y_2 + y_3 \ldots + y_{n-1}\right]$$

For "right-endpoint" rectangles, the formula becomes:

$$\lim_{n\to\infty}\left(\frac{b-a}{n}\right)\left[y_1 + y_2 + y_3 \ldots + y_n\right].$$

Note: Sometimes these are called "inscribed" and "circumscribed" rectangles, but that restricts the use of the formula. It's more exact to evaluate these rectangles using right and left endpoints.

When you txake the limit of this infinite sum, you get the integral. Surprise! (You knew the integral had to show up sometime, didn't you?) Actually, we write the integral like this:

$$\int_1^3 \left(x^2 + 2\right)dx$$

It is called a **definite integral**, and it means that we're finding the area under the curve $x^2 + 2$ from $x = 1$ to $x = 3$. (We'll discuss how to evaluate this in a moment.) On the AP exam, you'll only be asked to divide the region into a small number of rectangles, so it won't be very hard. Plus, you can use whichever type of rectangles that make you happy. Let's do an example.

Example 1: Approximate the area under the curve $y = x^3$ from $x = 2$ to $x = 3$ using four left-endpoint rectangles.

Draw four rectangles that look like this:

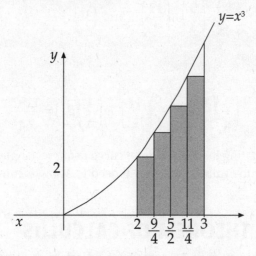

The width of each rectangle is $\dfrac{1}{4}$. The heights of the rectangles are:

$$2^3, \left(\frac{9}{4}\right)^3, \left(\frac{5}{2}\right)^3, \text{ and } \left(\frac{11}{4}\right)^3.$$

Therefore, the area is:

$$\left(\frac{1}{4}\right)(2^3) + \left(\frac{1}{4}\right)\left(\frac{9}{4}\right)^3 + \left(\frac{1}{4}\right)\left(\frac{5}{2}\right)^3 + \left(\frac{1}{4}\right)\left(\frac{11}{4}\right)^3 = \frac{893}{64}.$$

Example 2: Repeat example 1 using four right-endpoint rectangles.

Now draw four rectangles that look like this:

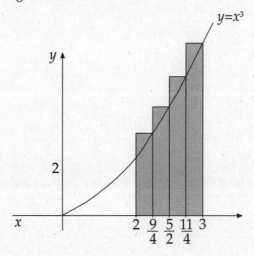

The width of each rectangle is still $\frac{1}{4}$, but the heights of the rectangles are now:

$$\left(\frac{9}{4}\right)^3, \left(\frac{5}{2}\right)^3, \left(\frac{11}{4}\right)^3, \text{ and } \left(3^3\right).$$

The area is now:

$$\left(\frac{1}{4}\right)\left(\frac{9}{4}\right)^3 + \left(\frac{1}{4}\right)\left(\frac{5}{2}\right)^3 + \left(\frac{1}{4}\right)\left(\frac{11}{4}\right)^3 + \left(\frac{1}{4}\right)\left(3^3\right) = \frac{1197}{64}.$$

That's all there is to approximating the area under a curve using rectangles. Now, let's learn how to find the area exactly. In order to evaluate this, you'll need to know(drum roll please)...

THE FUNDAMENTAL THEOREM OF CALCULUS

Above, we said that if you create an infinite number of infinitely thin rectangles, you'll get the integral of the curve. We wrote that the integral is:

$$\int_2^3 x^3 dx.$$

There is a rule to evaluating this integral. The rule is called the Fundamental Theorem of Calculus, and it says:

$$\int_a^b f(x)dx = F(b) - F(a); \text{ where } F(x) \text{ is the antiderivative of } f(x).$$

Using this rule, you can find $\int_2^3 x^3 dx$ by integrating it, and we get $\frac{x^4}{4}$. Now all you do is plug in 3 and 2 and take the difference. We use the following notation to symbolize this:

$$\left.\frac{x^4}{4}\right|_2^3$$

Thus we have:

$$\frac{3^4}{4} - \frac{2^4}{4} = \frac{81}{4} - \frac{16}{4} = \frac{65}{4}.$$

Since $\frac{65}{4} = \frac{1040}{64}$, you can see how close we were with our two earlier approximations.

Example 3: Find $\int_1^3 (x^2 + 2)dx$.

Using the Fundamental Theorem of Calculus:

$$\int_1^3 (x^2 + 2)dx = \left(\frac{x^3}{3} + 2x \right)\Big|_1^3.$$

If we evaluate this, we get:

$$\left(\frac{3^3}{3} + 2(3) \right) - \left(\frac{1^3}{3} + 2(1) \right) = \frac{38}{3}.$$

This is the first function for which we found the approximate area by using inscribed rectangles. Our final estimate, where we averaged the inscribed and circumscribed rectangles, was $\frac{51}{4}$, and as you can see, that was very close (off by $\frac{1}{12}$).

We're only going to do a few approximations using rectangles, because it's not a big part of the AP. On the other hand, definite integrals are a big part of the rest of this book.

Example 4: $\int_1^5 (x^2 - x)dx = \left(\frac{x^3}{3} - \frac{x^2}{2} \right)\Big|_1^5 = \left(\frac{125}{3} - \frac{25}{2} \right) - \left(\frac{1}{3} - \frac{1}{2} \right) = \frac{88}{3}.$

Example 5: $\int_0^{\frac{\pi}{2}} \sin x \, dx = (-\cos x)\Big|_0^{\frac{\pi}{2}} = \left(-\cos \frac{\pi}{2} \right) - (-\cos 0) = 1.$

Example 6: $\int_0^{\frac{\pi}{4}} \sec^2 x \, dx = \tan x\Big|_0^{\frac{\pi}{4}} = \tan \frac{\pi}{4} - \tan 0 = 1.$

THE TRAPEZOID RULE

This next section is only for BC students (AB students can go take a breather). There's another approximation method that's even better than the rectangle method. Essentially, all you do is divide the region into trapezoids instead of rectangles. Let's use the problem that we did at the beginning of the chapter. We get a picture that looks like this:

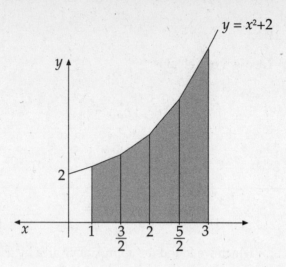

As you should recall from geometry, the formula for the area of a trapezoid is:

$$\frac{1}{2}(b_1 + b_2)h$$

where b_1 and b_2 are the two bases of the trapezoid. Notice that each of the shapes is a trapezoid on its side, so the height of each trapezoid is the length of the interval, $\frac{1}{2}$, and the bases are the y-values that correspond to each x-value. We found these earlier in the rectangle example; they are, in order: $3, \frac{17}{4}, 6, \frac{33}{4}$, and 11. We can find the area of each trapezoid and add them up:

$$\frac{1}{2}\left(3 + \frac{17}{4}\right)\left(\frac{1}{2}\right) + \frac{1}{2}\left(\frac{17}{4} + 6\right)\left(\frac{1}{2}\right) + \frac{1}{2}\left(6 + \frac{33}{4}\right)\left(\frac{1}{2}\right) + \frac{1}{2}\left(\frac{33}{4} + 11\right)\left(\frac{1}{2}\right) = \frac{51}{4}, \text{ or } 12.75.$$

Recall that the actual value of the area is $\frac{38}{3}$ or 12.67; the Trapezoid Rule gives a pretty good approximation.

Notice how each trapezoid shares a base with the trapezoid next to it, except for the end ones. This enables us to simplify the formula for using the Trapezoid Rule. Each trapezoid has a height equal to the length of the interval divided by the number of trapezoids we use. If the interval is from $x = a$ to $x = b$, and the number of trapezoids is n, then the height of each trapezoid is $\frac{b-a}{n}$. Then our formula is:

$$\left(\frac{1}{2}\right)\left(\frac{b-a}{2}\right)\left[y_0 + 2y_1 + 2y_2 + 2y_3 \ldots + 2y_{n-2} + 2y_{n-1} + y_n\right]$$

This is all you need to know about the Trapezoid Rule. Just follow the formula and you won't have any problems. Let's do one more example.

Example 7: Approximate the area under the curve $y = x^3$ from $x = 2$ to $x = 3$ using four inscribed trapezoids.

Following the rule, the height of each trapezoid is $\dfrac{3-2}{4} = \dfrac{1}{4}$. Thus, the approximate area is:

$$\left(\frac{1}{2}\right)\left(\frac{1}{4}\right)\left[2^3 + 2\left(\frac{9}{4}\right)^3 + 2\left(\frac{5}{2}\right)^3 + 2\left(\frac{11}{4}\right)^3 + 3^3\right] = \frac{1045}{64}.$$

Compare this answer to the actual value we found earlier—it's pretty close!

SIMPSON'S RULE

There's one last approximation method that doesn't really show up often on the AP exam, but since you're responsible for it, we'll show it to you briefly. Without going into proving anything, Simpson's Rule is similar to the Trapezoid Rule, except:

(a) we don't multiply by $\dfrac{1}{2}$; we divide each height by 3; and

(b) instead of multiplying each y-value by 2, we alternate multiplying by 4 and by 2.

The formula looks like this:

$$\left(\frac{b-a}{3n}\right)\left[y_0 + 4y_1 + 2y_2 + 4y_3 \ldots + 2y_{n-2} + 4y_{n-1} + y_n\right]$$

Note that the rule requires you to divide the region into an even number of intervals, and n must be at least 4.

Let's do the last trapezoid example using Simpson's Rule. The area is approximately:

$$\left(\frac{1}{12}\right)\left[2^3 + 4\left(\frac{9}{4}\right)^3 + 2\left(\frac{5}{2}\right)^3 + 4\left(\frac{11}{4}\right)^3 + 3^3\right] = \frac{65}{4}.$$

This is exactly what you would get if you took the definite integral. (This is because Simpson's Rule uses parabolas to approximate the area, so when the curve you're evaluating is a parabola, **Simpson's Rule** is exact.) Normally, the approximation is not quite so accurate, but Simpson's Rule is the best of the approximation methods that you'll learn. It's also the formula that many mathematicians use when they want a good approximation of the area under a curve.

Now that we've discussed how to approximate the area under a curve, and how to use definite integrals to find the exact area under a curve, we'll use these tools for most of the rest of this book. So, practice these problems carefully, and make sure that you know your stuff when it comes to taking definite integrals.

Practice, practice, practice.

PROBLEM 1. Find the area under the curve $y = 4 - x^2$ from $x = -1$ to $x = 1$ with $n = 4$ inscribed rectangles.

Answer: Draw four rectangles that look like this:

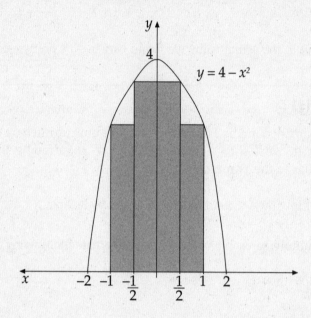

The width of each rectangle is $\dfrac{1}{2}$. The heights of the rectangles are found by evaluating $y = 4 - x^2$ at the appropriate endpoints:

$$\left(4-(-1)^2\right), \left(4-\left(-\frac{1}{2}\right)^2\right), \left(4-\left(\frac{1}{2}\right)^2\right), \text{ and } \left(4-(1)^2\right).$$

These can be simplified to $3, \dfrac{15}{4}, \dfrac{15}{4}$, and 3. Therefore the area is:

$$\left(\frac{1}{2}\right)(3)+\left(\frac{1}{2}\right)\left(\frac{15}{4}\right)+\left(\frac{1}{2}\right)\left(\frac{15}{4}\right)+\left(\frac{1}{2}\right)(3)=\frac{27}{4}$$

PROBLEM 2. Find the area under the curve $y = 4 - x^2$ from $x = -1$ to $x = 1$ with $n = 4$ circumscribed rectangles.

Answer: Draw four rectangles that look like this:

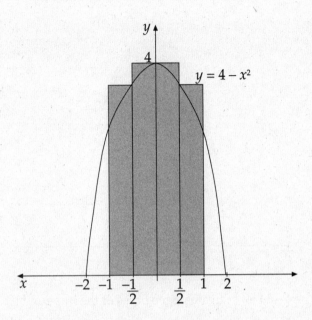

The width of each rectangle is $\frac{1}{2}$. The heights of the rectangles are found by evaluating $y = 4 - x^2$ at the appropriate endpoints:

$$\left(4 - \left(-\frac{1}{2}\right)^2\right), \ \left(4 - (0)^2\right), \ \left(4 - (0)^2\right), \ and \ \left(4 - \left(\frac{1}{2}\right)^2\right).$$

These can be simplified to $\frac{15}{4}$, 4, 4, and $\frac{15}{4}$. Therefore the area is:

$$\left(\frac{1}{2}\right)\left(\frac{15}{4}\right) + \left(\frac{1}{2}\right)(4) + \left(\frac{1}{2}\right)(4) + \left(\frac{1}{2}\right)\left(\frac{15}{4}\right) = \frac{31}{4}.$$

PROBLEM 3. Find the area under the curve $y = 4 - x^2$ from $x = -1$ to $x = 1$ using the Trapezoid Rule with $n = 4$.

Answer: Draw four trapezoids that look like this:

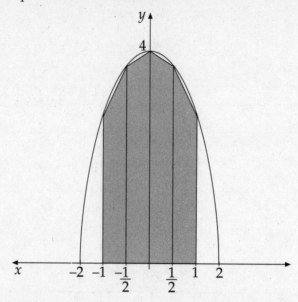

The width of each trapezoid is $\dfrac{1}{2}$. Evaluate the bases of the trapezoid by calculating $y = 4 - x^2$ at the appropriate endpoints. Following the rule, we get that the area is approximately:

$$\left(\frac{1}{2}\right)\left(\frac{1}{2}\right)\left[\left(4-(1)^2\right)+2\left(4-\left(-\frac{1}{2}\right)^2\right)+2\left(4-(0)^2\right)+2\left(4-\left(\frac{1}{2}\right)^2\right)+\left(4-(1)^2\right)\right] =$$
$$\left(\frac{1}{2}\right)\left(\frac{1}{2}\right)\left[3+\frac{15}{2}+8+\frac{15}{2}+3\right] = \frac{29}{4}.$$

PROBLEM 4. Find the area under the curve $y = 4 - x^2$ from $x = -1$ to $x = 1$ using Simpson's Rule with $n = 4$.

Answer: Using Simpson's Rule, we get that the area is approximately:

$$\left(\frac{1}{6}\right)\left[\left(4-(1)^2\right)+4\left(4-\left(-\frac{1}{2}\right)^2\right)+2\left(4-(0)^2\right)+4\left(4-\left(\frac{1}{2}\right)^2\right)+\left(4-(1)^2\right)\right] =$$
$$\left(\frac{1}{6}\right)\left[3+4\left(\frac{15}{4}\right)+2(4)+4\left(\frac{15}{4}\right)+(3)\right] = \frac{22}{3}$$

PROBLEM 5. Find the area under the curve $y = 4 - x^2$ from $x = -1$ to $x = 1$.

Answer: Now we can use the definite integral by evaluating

$$\int_{-1}^{1} \left(4 - x^2\right) dx.$$

This can be rewritten as:

$$\left(4x - \frac{x^3}{3}\right)\Bigg|_{-1}^{1}.$$

Follow the Fundamental Theorem of Calculus:

$$\left(4 - \frac{1}{3}\right) - \left(4(-1) - \frac{(-1)^3}{3}\right) = \frac{22}{3}.$$

A brief recap: Simpson's Rule gave us an exact value for the area, the Trapezoid Rule was very close, and the rectangles weren't as accurate.

PROBLEM 6. Find the area under the curve $y = 5 - 3x$ from $x = 0$ to $x = 1$.

Answer: Evaluate the integral $\int_{0}^{1} (5 - 3x) dx$:

$$\left(5x - \frac{3x^2}{2}\right)\Bigg|_{0}^{1} = \left(5 - \frac{3}{2}\right) - 0 = \frac{7}{2}.$$

PRACTICE PROBLEMS

Here's a great opportunity to practice finding the area beneath a curve and evaluating integrals. The answers are at the end of the unit.

1. Find the area under the curve $y = 2x - x^2$ from $x = 1$ to $x = 2$ with $n = 4$ inscribed rectangles.

2. Find the area under the curve $y = 2x - x^2$ from $x = 1$ to $x = 2$ with $n = 4$ circumscribed rectangles.

3. Find the area under the curve $y = 2x - x^2$ from $x = 1$ to $x = 2$ using the Trapezoid Rule with $n = 4$.

4. Find the area under the curve $y = 2x - x^2$ from $x = 1$ to $x = 2$ using Simpson's Rule with $n = 4$.

5. Find the area under the curve $y = 2x - x^2$ from $x = 1$ to $x = 2$.

6. Evaluate $\int_{-\frac{\pi}{2}}^{\frac{\pi}{2}} \cos x \, dx$.

7. Evaluate $\displaystyle\int_1^9 2x\sqrt{x}\,dx$.

8. Evaluate $\displaystyle\int_0^1 \left(x^4 - 5x^3 + 3x^2 - 4x - 6\right)dx$.

9. Evaluate $\displaystyle\int_{-4}^4 |x|dx$.

10. Evaluate $\displaystyle\int_{\frac{\pi}{2}}^{\frac{\pi}{2}} \sin x\,dx$.

THE MEAN VALUE THEOREM FOR INTEGRALS

As you recall, we did the Mean Value Theorem once before, in chapter 6, but this time we'll apply it to integrals, not derivatives. In fact, some books refer to it as the "Mean Value Theorem for Integrals" or MVTI. The most important aspect of the MVTI is that it enables you to find the average value of a function. In fact, the AP exam will often ask you to find the average value of a function, which is just its way of testing your knowledge of the MVTI.

Here's the theorem:

If $f(x)$ is continuous on a closed interval $[a, b]$, then at some point c in the interval $[a, b]$.

$$\int_a^b f(x)dx = f(c)(b-a)$$

This tells you that the area under the curve of $f(x)$ on the interval $[a, b]$, is equal to the value of the function at some value c (between a and b) times the length of the interval. If you look at this graphically, you can see that you're finding the area of a rectangle whose base is the interval and whose height is some value of $f(x)$ that creates a rectangle with the same area as the area under the curve.

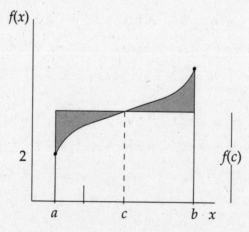

The number $f(c)$ gives us the average value of f on $[a, b]$. Thus, if we rearrange the theorem, we get the formula for finding the average value of $f(x)$ on $[a, b]$:

$$\frac{1}{b-a}\int_a^b f(x)dx.$$

There's all you need to know about finding average values. Try some examples.

Example 1: Find the average value of $f(x) = x^2$ from $x = 2$ to $x = 4$.

Evaluate the integral $\dfrac{1}{4-2}\displaystyle\int_2^4 x^2 dx$:

$$\frac{1}{4-2}\int_2^4 x^2 dx = \frac{1}{2}\left[\frac{x^3}{3}\bigg|_2^4\right] = \frac{1}{2}\left(\frac{64}{3} - \frac{8}{3}\right) = \frac{28}{3}.$$

Example 2: Find the average value of $f(x) = \sin x$ on $[0, \pi]$.

Evaluate $\dfrac{1}{\pi-0}\displaystyle\int_0^\pi \sin x\, dx$:

$$\frac{1}{\pi-0}\int_0^\pi \sin x\, dx = \frac{1}{\pi}(-\cos x)\bigg|_0^\pi = \frac{1}{\pi}(-\cos\pi + \cos 0) = \frac{2}{\pi}$$

THE FUNDAMENTAL THEOREM OF CALCULUS, PART TWO

As you saw in the last chapter, we've only half-learned the theorem. It has two parts, often referred to as the First and Second Fundamental Theorems of Calculus:

The First Fundamental Theorem of Calculus (which you've already seen):

If $f(x)$ is continuous at every point of $[a, b]$, and $F(x)$ is an antiderivative of $f(x)$ on

$[a, b]$, then $\displaystyle\int_a^b f(x)dx = F(b) - F(a)$.

The Second Fundamental Theorem of Calculus:

If $f(x)$ is continuous on $[a, b]$, then the derivative of the function $F(x) = \int_a^x f(t)dt$ is:

$$\frac{dF}{dx} = \frac{d}{dx}\int_a^x f(t)dt = f(x).$$

(Some books give the theorems in reverse order.)

We've already made use of the first theorem in evaluating definite integrals. In fact, we use the first Fundamental Theorem every time we evaluate a definite integral, so we're not going to give you any examples of that here. There is one aspect of the first Fundamental Theorem, however, that involves the area between curves (we'll discuss that in chapter 14).

But for now, you should know this:

If we have a point c in the interval $[a, b]$, then

$$\int_a^c f(x)dx + \int_c^b f(x)dx = \int_a^b f(x)dx.$$

In other words, we can divide up the region into parts, add them up, and the result is the area of the region. We'll get back to this in the chapter on the area between two curves.

THE SECOND FUNDAMENTAL THEOREM OF CALCULUS

The second theorem tells us how to find the derivative of an integral.

Example 3: Find $\dfrac{d}{dx}\displaystyle\int_1^x \cos t\, dt$.

The second Fundamental Theorem says that the derivative of this integral is just $\cos x$.

Example 4: Find $\dfrac{d}{dx}\displaystyle\int_2^x \left(1 - t^3\right) dt$.

Here, the theorem says that the derivative of this integral is just $(1 - x^3)$.

Isn't this easy? Let's add a couple of nuances. First, the constant term in the limits of integration is a "dummy term." Any constant will give the same answer. For example:

$$\frac{d}{dx}\int_2^x \left(1 - t^3\right) dt = \frac{d}{dx}\int_{-2}^x \left(1 - t^3\right)dt = \frac{d}{dx}\int_p^x \left(1 - t^3\right)dt = 1 - x^3.$$

In other words, all we're concerned with is the variable term.

Second, if the upper limit is a function of x, instead of just plain x, we multiply the answer by the derivative of that term. For example:

$$\frac{d}{dx}\int_2^{x^2}\left(1-t^3\right)dt=\left[1-\left(x^2\right)^3\right](2x)=\left(1-x^6\right)(2x).$$

Example 5: Find $\dfrac{d}{dx}\displaystyle\int_0^{3x^4}\left(t+4t^2\right)dt=\left[3x^4+4(3x^4)^2\right](12x^3).$

Try these solved problems on your own. You know the drill.

PROBLEM 1. Find the average value of $f(x)=\dfrac{1}{x^2}$ on the interval $[1, 3]$.

Answer: According to the Mean Value Theorem, the average value is found by evaluating

$$\frac{1}{3-1}\int_1^3\frac{dx}{x^2}.$$

Your result should be:

$$\frac{1}{2}\left(-\frac{1}{x}\right)\Bigg|_1^3=\frac{1}{2}\left(-\frac{1}{3}+1\right)=\frac{1}{3}.$$

PROBLEM 2. Find the average value of $f(x)=\sin x$ on the interval $[-\pi,\pi]$.

Answer: According to the Mean Value Theorem, the average value is found by evaluating

$$\frac{1}{2\pi}\left(\int_{-\pi}^{\pi}\sin x\,dx\right).$$

Integrating, we get:

$$\frac{1}{2\pi}(-\cos x)\Big|_{-\pi}^{\pi}=\frac{1}{2\pi}(-\cos\pi+\cos(-\pi))=0.$$

PROBLEM 3. Find $\dfrac{d}{dx}\displaystyle\int_1^x\frac{dt}{1-\sqrt[3]{t}}.$

Answer: According to the Second Fundamental Theorem of Calculus:

$$\frac{d}{dx}\int_1^x\frac{dt}{1-\sqrt[3]{t}}=\frac{1}{1-\sqrt[3]{x}}.$$

PROBLEM 4. Find $\dfrac{d}{dx}\displaystyle\int_1^{x^2}\dfrac{t\,dt}{\sin t}$.

Answer: According to the Second Fundamental Theorem of Calculus:

$$\frac{d}{dx}\int_1^{x^2}\frac{t\,dt}{\sin t}=\frac{x^2}{\sin\left(x^2\right)}(2x)=\frac{2x^3}{\sin x^2}.$$

PRACTICE PROBLEMS

Now try these problems. The answers are in chapter 21.

1. Find the average value of $f(x) = 4x\cos x^2$ on the interval $\left[0,\sqrt{\dfrac{\pi}{2}}\right]$.

2. Find the average value of $f(x) = \sqrt{x}$ on the interval $[0,16]$.

3. Find the average value of $f(x) = \sqrt{1-x}$ on the interval $[-1,1]$.

4. Find the average value of $f(x) = 2|x|$ on the interval $[-1,1]$.

5. Find $\dfrac{d}{dx}\displaystyle\int_1^x\sin^2 t\,dt$.

6. Find $\dfrac{d}{dx}\displaystyle\int_5^{3x}\left(t^2-t\right)dt$.

7. Find $\dfrac{d}{dx}\displaystyle\int_0^{x^2}|t|\,dt$.

8. Find $\dfrac{d}{dx}\displaystyle\int_1^x -2\cos t\,dt$.

13

Exponential and Logarithmic Functions, Part Two

You've learned how to integrate polynomials and some of the trig functions (there's more of them to come), and you have the first technique of integration: *u*-substitution. Now it's time to learn how to integrate some other functions—namely, exponential and logarithmic functions. Yes, the long-awaited second part of chapter 9. The first integral is the natural logarithm:

$$\int \frac{du}{u} = \ln |u| + C$$

Notice the absolute value in the logarithm. This ensures that you aren't taking the logarithm of a negative number. If you know that the term you're taking the log of is positive (for example, $x^2 + 1$), we can dispense with the absolute value marks. Let's do some examples.

Example 1: Find $\displaystyle\int \frac{5dx}{x+3}$.

Whenever an integrand contains a fraction, check to see if the integral is a logarithm. Usually, the process involves u-substitution. Let $u = x + 3$ and $du = dx$. Then:

$$\int \frac{5dx}{x+3} = 5\int \frac{du}{u} = 5\ln|u| + C.$$

Substituting back, the final result is:

$$5\ln|x+3| + C.$$

Example 2: Find $\displaystyle\int \frac{2x\ dx}{x^2+1}$.

Let $u = x^2 + 1$ $du = 2xdx$ and substitute into the integrand:

$$\int \frac{2x\ dx}{x^2+1} = \int \frac{du}{u} = \ln|u| + C.$$

Then substitute back:

$$\ln\left(x^2+1\right) + C.$$

MORE INTEGRALS OF TRIG FUNCTIONS

Remember when we started antiderivatives and we didn't do the integral of tangent, cotangent, secant, and cosecant? Well, their time has come.

Example 3: Find $\displaystyle\int \tan x\ dx$.

First, rewrite this integral as

$$\int \frac{\sin x}{\cos x}\ dx.$$

Now we let $u = \cos x$ $du = -\sin x\ dx$ and substitute:

$$\int -\frac{du}{u}.$$

Now, integrate and re-substitute:

$$\int -\frac{du}{u} = -\ln|u| = -\ln|\cos x| + C.$$

Thus, $\int \tan x \, dx = -\ln|\cos x| + C$.

Example 4: Find $\int \cot x \, dx$.

Just as before, rewrite this integral in terms of sine and cosine:

$$\int \frac{\cos x}{\sin x} \, dx.$$

Now we let $u = \sin x \quad du = \cos x \, dx$ and substitute:

$$\int \frac{du}{u}.$$

Now, integrate:

$$\ln|u| + C = \ln|\sin x| + C.$$

Therefore, $\int \cot x \, dx = \ln|\sin x| + C$. This looks a lot like the previous example, doesn't it?

Example 5: Find $\int \sec x \, dx$.

You could rewrite this integral as

$$\int \frac{1}{\cos x} \, dx,$$

but if you try u-substitution at this point, it won't work. So what should you do? You'll never guess, so we'll show you: multiply the $\sec x$ by $\dfrac{\sec x + \tan x}{\sec x + \tan x}$. This gives you:

$$\int \sec x \, dx = \int \sec x \frac{\sec x + \tan x}{\sec x + \tan x} dx = \int \frac{\sec^2 x + \sec x \tan x}{\sec x + \tan x} dx.$$

Now you can do u-substitution. Let $u = \sec x + \tan x \quad du = (\sec^2 x + \sec x \tan x) \, dx$. Then rewrite the integral as:

$$\int \frac{du}{u}.$$

Pretty slick, huh?

The rest goes according to plan as you integrate:

$$\int \frac{du}{u} = \ln |u| + C = \ln |\sec x + \tan x| + C.$$

Therefore, $\int \sec x \, dx = \ln |\sec x + \tan x| + C.$

Example 6: Find $\int \csc x \, dx.$

You guessed it! Multiply csc x by $\dfrac{\csc x + \cot x}{\csc x + \cot x} dx$. This gives you:

$$\int \csc x \frac{\csc x + \cot x}{\csc x + \cot x} dx = \int \frac{\csc^2 x + \csc x \cot x}{\csc x + \cot x} dx.$$

Let $u = \csc x + \cot x$ and $du = (-\csc^2 x - \csc x \cot x) \, dx$. And, just as in example 5, you can rewrite the integral as:

$$\int -\frac{du}{u}.$$

And integrate:

$$\int -\frac{du}{u} = -\ln |u| + C = -\ln |\csc x + \cot x| + C.$$

Therefore, $\int \csc x \, dx = -\ln |\csc x + \cot x| + C.$

As we do more integrals, the natural log will turn up over and over. It's important that you get good at recognizing when integrating requires the use of the natural log.

INTEGRATING e^x

Now let's learn how to find the integral of e^x. Remember that $\dfrac{d}{dx} e^x = e^x$? Well, you should be able to predict the formula below:

$$\int e^u \, du = e^u + C.$$

As with the natural logarithm, most of these integrals use u-substitution.

Example 7: Find $\int e^{7x}dx$.

Let $u = 7x$ $du = 7dx$ and $\frac{1}{7}du = dx$. Then you have:

$$\int e^{7x}dx = \frac{1}{7}\int e^u du = \frac{1}{7}e^u + C.$$

Substituting back, you get:

$$\frac{1}{7}e^{7x} + C.$$

In fact, whenever you see $\int e^{kx}dx$, where k is a constant, the integral is:

$$\int e^{kx}dx = \frac{1}{k}e^{kx} + C.$$

Example 8: Find $\int xe^{3x^2+1}\,dx$.

Let $u = 3x^2 + 1$, $du = 6x\,dx$ and $\frac{1}{6}\,du = x\,dx$. The result is:

$$\int xe^{3x+1}\,dx = \frac{1}{6}\int e^u du = \frac{1}{6}e^u + C.$$

Now it's time to put the x's back in:

$$\frac{1}{6}e^{3x^2+1} + C.$$

Example 9: Find $\int e^{\sin x}\cos x\,dx$.

Let $u = \sin x$ $du = \cos x\,dx$. The substitution here couldn't be simpler:

$$\int e^u du = e^u + C = e^{\sin x} + C.$$

As you can see, these integrals are pretty straightforward. The key is to use u-substitution to transform nasty-looking integrals into simple ones.

EXPONENTIAL FUNCTIONS

There's another type of exponential function whose integral you'll have to find occasionally:

$$\int a^u du.$$

As you should recall from your rules of logarithms and exponents, the term a^u can be written as $e^{u \ln a}$. Since $\ln a$ is a constant, we can transform $\int a^u du$ into $\int e^{u \ln a}$. If you integrate this, you'll get:

$$\int e^{u \ln a} du = \frac{1}{\ln a} e^{u \ln a} + C.$$

Now substituting back a^u for $e^{u \ln a}$:

$$\int a^u du = \frac{1}{\ln a} a^u + C.$$

Example 7: Find $\int 5^x dx$.

Follow the rule we just derived:

$$\int 5^x dx = \frac{1}{\ln 5} 5^x + C.$$

Because these integrals don't show up too often on the AP, this is the last you'll see of them in this book. You should, however, be able to integrate them using the rule, or by converting them into a form of $\int e^u du$.

Try these on your own. Do each problem with the answers covered, then check your answer.

PROBLEM 1. Evaluate $\int \frac{dx}{3x}$.

Answer: Move the constant term outside of the integral, like this:

$$\frac{1}{3} \int \frac{dx}{x}.$$

Now you can integrate:

$$\frac{1}{3} \int \frac{dx}{x} = \frac{1}{3} \ln|x| + C.$$

PROBLEM 2. Evaluate $\int \dfrac{3x^2 dx}{x^3 - 1}$.

Answer: Let $u = x^3 - 1$ and $du = 3x^2\, dx$, and substitute:

$$\int \frac{du}{u}.$$

Now integrate:

$$\ln|u| + C$$

And substitute back:

$$\ln|x^3 - 1| + C.$$

PROBLEM 3. Evaluate $\int e^{5x} dx$.

Answer: Let $u = 5x$ and $du = 5dx$. Then $\dfrac{1}{5}\, du = dx$. Substitute in:

$$\frac{1}{5}\int e^u du.$$

Integrate:

$$\frac{1}{5}e^u + C.$$

And substitute back:

$$\frac{1}{5}e^{5x} + C.$$

PROBLEM 4. Evaluate $\int 2^{3x} dx$.

Answer: Let $u = 3x$ and $du = 3dx$. Then $\dfrac{1}{3}\, du = dx$. Make the substitution:

$$\frac{1}{3}\int 2^u du.$$

Integrate according to the rule. Your result should be:

$$\frac{1}{3\ln 2}2^u + C.$$

Now get back to the expression as a function of x:

$$\frac{1}{3\ln 2}2^{3x}+C=\frac{1}{\ln 8}2^{3x}+C.$$

PRACTICE PROBLEMS

Evaluate the following integrals. The answers are in chapter 21.

1. $\displaystyle\int\frac{\sec^2 x}{\tan x}dx$

2. $\displaystyle\int\frac{\cos x}{1-\sin x}dx$

3. $\displaystyle\int\frac{1}{x\ln x}dx$

4. $\displaystyle\int\frac{1}{x}\cos(\ln x)dx$

5. $\displaystyle\int\frac{\sin x-\cos x}{\cos x}dx$

6. $\displaystyle\int\frac{dx}{\sqrt{x}\left(1+2\sqrt{x}\right)}$

7. $\displaystyle\int\frac{e^x dx}{1+e^x}$

8. $\displaystyle\int xe^{5x^2-1}dx$

9. $\displaystyle\int e^x\cos\left(2+e^x\right)dx$

10. $\displaystyle\int\frac{e^x+e^{-x}}{e^x-e^{-x}}dx$

11. $\displaystyle\int x4^{-x^2}dx$

12. $\displaystyle\int 7^{\sin x}\cos x\,dx$

14

THE AREA BETWEEN
TWO CURVES

These next two units discuss some of the most difficult topics you'll encounter in AP calculus. For some reason, students have terrible trouble setting up these problems, and the good old folks at ETS know it. Believe it or not, ETS actually only asks relatively simple versions of these problems on the exam.

Unfortunately, this unit and the next are always on the AP. We'll try to make them as simple as possible. You've already learned that if you want to find the area under a curve, you can integrate the function of the curve by using the endpoints as limits. So far, though, we've only talked about the area between a curve and the x-axis. What if you have to find the area between two curves?

VERTICAL SLICES

Suppose you wanted to find the area between the curve $y = x$ and the curve $y = x^2$ from $x = 2$ to $x = 4$. First, sketch the curves:

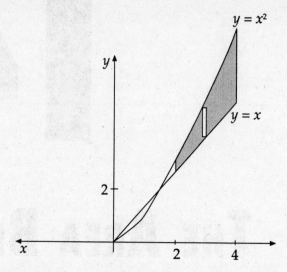

You can find the area by slicing up the region vertically, into a bunch of infinitely thin strips, and adding up the areas of all the strips. The height of each strip is $x^2 - x$, and the width of each strip is dx. Add up all the strips by using the integral:

$$\int_2^4 \left(x^2 - x\right)dx.$$

Then, evaluate it:

$$\left(\frac{x^3}{3} - \frac{x^2}{2}\right)\Bigg|_2^4 = \left(\frac{64}{3} - \frac{16}{2}\right) - \left(\frac{8}{3} - \frac{4}{2}\right) = \frac{38}{3}.$$

That wasn't so hard, was it? Don't worry. The process gets more complicated, but the idea remains the same. Now let's generalize this and come up with a rule:

> If a region is bounded by $f(x)$ above and $g(x)$ below at all points of the interval $[a, b]$, then the area of the region is given by:
>
> $$\int_a^b [f(x) - g(x)]dx.$$

Example 1: Find the area of the region between the parabola $y = 1 - x^2$ and the line $y = 1 - x$.

First, make a sketch of the region.

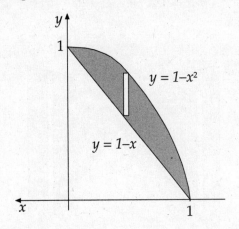

To find the points of intersection of the graphs, set the two equations equal to each other and solve for x:

$$1 - x^2 = 1 - x$$
$$x^2 - x = 0$$
$$x(x - 1) = 0$$
$$x = 0, 1$$

The left-hand edge of the region is $x = 0$ and the right-hand edge is $x = 1$, so the limits of integration are from 0 to 1.

Next, note that the top curve is always $y = 1 - x^2$ and the bottom curve is always $y = 1 - x$. (If the region has a place where the top and bottom curve switch, you need to make two integrals, one for each region. Fortunately, that's not the case here.) Thus, we need to evaluate:

$$\int_0^1 \left[(1 - x^2) - (1 - x) \right] dx.$$

$$\int_0^1 \left[(1 - x^2) - (1 - x) \right] dx = \int_0^1 \left(-x^2 + x \right) dx = \left(-\frac{x^3}{3} + \frac{x^2}{2} \right) \Bigg|_0^1 = \frac{1}{6}$$

Sometimes you're given the endpoints of the region; sometimes you have to find them on your own.

Example 2: Find the area of the region between the curve $y = \sin x$ and the curve $y = \cos x$ from 0 to $\dfrac{\pi}{2}$.

First, sketch the region.

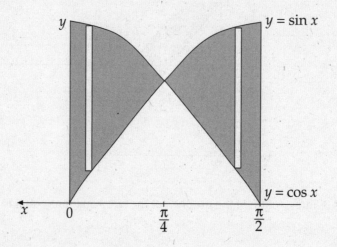

Notice that $\cos x$ is on top between 0 and $\dfrac{\pi}{4}$, then $\sin x$ is. The point where they cross is $\dfrac{\pi}{4}$, so you have to divide the area into two integrals: one from 0 to $\dfrac{\pi}{4}$, and the other from $\dfrac{\pi}{4}$ to $\dfrac{\pi}{2}$. In the first region, $\cos x$ is above $\sin x$, so the integral to evaluate is:

$$\int_0^{\frac{\pi}{4}} (\cos x - \sin x)\,dx$$

The integral of the second region is a little different, because $\sin x$ is above $\cos x$:

$$\int_{\frac{\pi}{4}}^{\frac{\pi}{2}} (\sin x - \cos x)\,dx.$$

If you add the two integrals, you'll get the area of the whole region:

$$\int_0^{\frac{\pi}{4}} (\cos x - \sin x)\,dx = (\sin x + \cos x)\Big|_0^{\frac{\pi}{4}} = \sqrt{2} - 1$$

$$\int_{\frac{\pi}{4}}^{\frac{\pi}{2}} (\sin x - \cos x)\,dx = (-\cos x - \sin x)\Big|_{\frac{\pi}{4}}^{\frac{\pi}{2}} = \sqrt{2} - 1$$

Adding these, we get that the area is $2\sqrt{2} - 2$.

HORIZONTAL SLICES

Now for the fun part. We can slice a region vertically when one function is at the top of our section and a different function is at the bottom. But what if the same function is both the top and the bottom of the slice (what we call a double-valued function)? You have to slice the region horizontally.

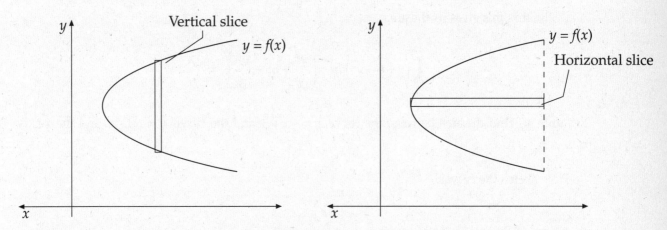

If we were to slice vertically, as in the left-hand picture, we'd have a problem. But if we were to slice horizontally, as in the right-hand picture, we don't have a problem. Instead of integrating an equation $f(x)$ with respect to x, we need to integrate an equation $f(y)$ with respect to y. As a result, our area formula changes a little:

> If a region is bounded by $f(y)$ on the right and $g(y)$ on the left at all points of the interval $[c, d]$, then the area of the region is given by:
>
> $$\int_c^d \left[f(y) - g(y)\right] dy.$$

Example 3: Find the area of the region between the curve $x = y^2$ and the curve $x = y + 6$ from $y = 0$ to $y = 3$.

First, sketch the region.

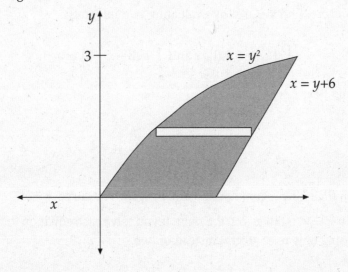

When you slice up the area horizontally, the right end of each section is the curve $x = y + 6$, and the left end of each section is always the curve $x = y^2$. Now set up our integral:

$$\int_0^3 \left(y + 6 - y^2\right)dy.$$

Evaluating this gives us the area:

$$\int_0^3 \left(y + 6 - y^2\right)dy = \left(\frac{y^2}{2} + 6y - \frac{y^3}{3}\right)\Bigg|_0^3 = \frac{27}{2}.$$

Example 4: Find the area between the curve $y = \sqrt{x+3}$ and the curve $y = \sqrt{3-x}$ and the x-axis from $x = -3$ to $x = 3$.

First, sketch the curves:

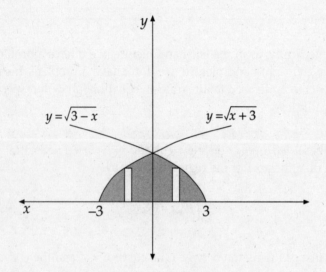

From $x = -3$ to $x = 0$, if you slice the region vertically, the curve $y = \sqrt{x+3}$ is on top, and the x-axis is on the bottom; from $x = 0$ to $x = 3$, the curve $y = \sqrt{3-x}$ is on top and the x-axis is on the bottom. Therefore, you can find the area by evaluating two integrals:

$$\int_{-3}^0 \left(\sqrt{x+3} - 0\right)dx \text{ and } \int_0^3 \left(\sqrt{3-x} - 0\right)dx.$$

Your results should be:

$$\frac{2}{3}(x+3)^{\frac{3}{2}}\Bigg|_{-3}^0 + \left(-\frac{2}{3}(3-x)^{\frac{3}{2}}\right)\Bigg|_0^3 = 4\sqrt{3}$$

Let's suppose you sliced the region horizontally instead. The curve $y = \sqrt{x+3}$ is always on the left, and the curve $y = \sqrt{3-x}$ is always on the right. If you solve each equation for x in terms of y, you save some time by using only one integral instead of two.

The two equations are $x = y^2 - 3$ and $x = 3 - y^2$. We also have to change the limits of integration from x-limits to y-limits. The two curves intersect at $y = \sqrt{3}$, so our limits of integration are from $y = 0$ to $y = \sqrt{3}$. The new integral is:

$$\int_0^{\sqrt{3}} \left[\left(3 - y^2\right) - \left(y^2 - 3\right)\right] dy = \int_0^{\sqrt{3}} \left(6 - 2y^2\right) dy = 6y - \left.\frac{2y^3}{3}\right|_0^{\sqrt{3}} = 4\sqrt{3}.$$

You get the same answer no matter which way you integrate (as long as you do it right!). The challenge of area problems is determining which way to integrate, and converting the equation to different terms. Unfortunately, there's no simple rule for how to do this. You have to look at the region and figure out its endpoints, as well as where the curves are with respect to each other.

Once you can do that, then the actual set-up of the integral(s) isn't that hard. Sometimes, evaluating the integrals isn't easy; however, if the integral of an AP question is difficult to evaluate, you'll only be required to set it up, not to evaluate it.

Here are some sample problems. On each, decide the best way to set up the integrals, and then evaluate them. Then check your answer.

PROBLEM 1. Find the area of the region between the curve $y = 3 - x^2$ and the line $y = 1 - x$ from $x = 0$ to $x = 2$.

Answer: You've been given a nice gift: the endpoints. First, make a sketch:

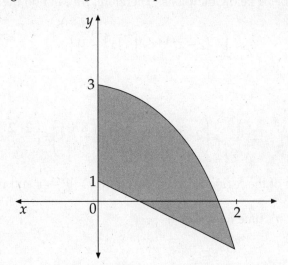

Since the curve $y = 3 - x^2$ is always above $y = 1 - x$ within the interval, you have to evaluate the following integral:

$$\int_0^2 \left[\left(3 - x^2\right) - (1 - x)\right] dx = \int_0^2 \left(2 + x - x^2\right) dx.$$

Therefore, the area of the region is:

$$\left.\left(2x + \frac{x^2}{2} - \frac{x^3}{3}\right)\right|_0^2 = \frac{10}{3}.$$

PROBLEM 2. Find the area between the x-axis and the curve $y = 2 - x^2$ from $x = 0$ to $x = 2$.

Answer: First, sketch this graph over the interval:

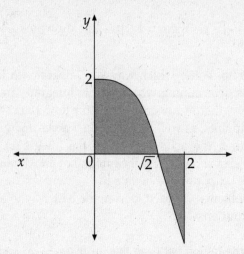

Since the curve crosses the x-axis at $\sqrt{2}$, you have to divide the region into two parts: from $x = 0$ to $x = \sqrt{2}$ and from $x = \sqrt{2}$ to $x = 2$. In the latter region, you'll need to integrate $y = -\left(2 - x^2\right) = x^2 - 2$ to adjust for the region's being below the x-axis. Therefore, we can find the area by evaluating:

$$\int_0^{\sqrt{2}} \left(2 - x^2\right)dx + \int_{\sqrt{2}}^2 \left(x^2 - 2\right)dx \,.$$

Integrating, we get:

$$\left(2x + -\frac{x^3}{3}\right)\Bigg|_0^{\sqrt{2}} + \left(\frac{x^3}{3} - 2x\right)\Bigg|_{\sqrt{2}}^2 = \left(2\sqrt{2} - \frac{2\sqrt{2}}{3}\right) - 0 + \left(\frac{8}{3} - 4\right) - \left(\frac{2\sqrt{2}}{3} - 2\sqrt{2}\right) = \frac{8\sqrt{2} - 4}{3}$$

PROBLEM 3. Find the area of the region between the curve $x = y^2 - 4y$ and the line $x = y$.

Answer: Once again, it's time for a sketch.

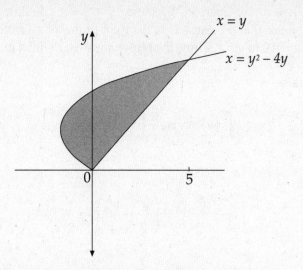

You don't have the endpoints this time, so you need to find where the two curves intersect. If you set them equal to each other, they intersect at $y = 0$ and at $y = 5$.

Since the curve $x = y^2$ is always to the left of $x = y$ over the interval we just found, we can evaluate the following integral:

$$\int_0^5 \left[y - \left(y^2 - 4y \right) \right] dy = \int_0^5 \left(5y - y^2 \right) dy .$$

The result of the integration should be:

$$\left(\frac{5y^2}{2} - \frac{y^3}{3} \right) \Bigg|_0^5 = \frac{125}{6} .$$

PROBLEM 4. Find the area between the curve $x = y^3 - y$ and the line $x = 0$ (the y-axis).

Answer: Sketch that sucker!

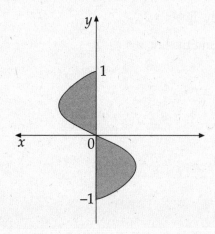

Next, find where the two curves intersect. By setting $y^3 - y = 0$, you'll find that they intersect at $y = -1$, $y = 0$, and $y = 1$. Notice that the curve is to the right of the y-axis from $y = -1$ to $y = 0$ and to the left of the y-axis from $y = 0$ to $y = 1$. Thus, the region must be divided into two parts: from $y = -1$ to $y = 0$ and from $y = 0$ to $y = 1$.

Set up the two integrals:

$$\int_{-1}^{0} \left(y^3 - y\right)dy + \int_{0}^{1} \left(y - y^3\right)dy .$$

And integrate them:

$$\left(\frac{y^4}{4} - \frac{y^2}{2}\right)\Bigg|_{-1}^{0} + \left(\frac{y^2}{2} - \frac{y^4}{4}\right)\Bigg|_{0}^{1} = \frac{1}{2} .$$

PRACTICE PROBLEMS

Find the area of the region between the two curves in each problem, and be sure to sketch each one. (We only gave you endpoints in one of them. Those are the breaks.) The answers are in chapter 21.

1. The curve $y = x^2 - 2$ and the line $y = 2$.

2. The curve $y = x^2$ and the curve $y = 4x - x^2$.

3. The curve $y = x^3$ and the curve $y = 3x^2 - 4$.

4. The curve $y = x^2 - 4x - 5$ and the curve $y = 2x - 5$.

5. The curve $y = x^3$ and the x-axis, from $x = -1$ to $x = 2$.

6. The curve $x = y^2$ and the line $x = y + 2$.

7. The curve $x = y^2$ and the curve $x = 3 - 2y^2$.

8. The curve $x = y^3 - y^2$ and the line $x = 2y$.

9. The curve $x = y^2 - 4y + 2$ and the line $x = y - 2$.

10. The curve $x = y^{\frac{2}{3}}$ and the curve $x = 2 - y^4$.

15

THE VOLUME OF A
SOLID OF REVOLUTION

Does the chapter title leave you in a cold sweat? Don't worry. You're not alone. This chapter covers the topic widely seen as the most difficult one on the AP exam. There is *always* a volume question on the test. The good news is that you're almost never asked to evaluate the integral—you usually only have to set it up. The difficulty with this chapter, as with chapter 14, is that there aren't any simple rules to follow. You have to draw the picture and figure it out.

In this chapter we're going to take the region between two curves, rotate it around a line (usually the x or y-axis), and find the volume of the region. There are two methods of doing this: the **washers method** and the **cylindrical shells method**. Sometimes you'll hear the washers method called the **disk method**, but a disk is only a washer without a hole in the middle.

THE DISK METHOD

Let's look at the region between the curve $y = \sqrt{x}$ and the x-axis (the curve $y = 0$), from $x = 0$ to $x = 1$, and revolve it about the x-axis. The picture looks like this:

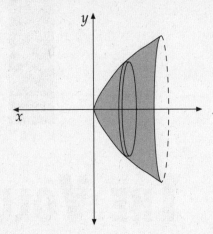

If you slice the resulting solid perpendicular to the x-axis, each cross-section of the solid is a circle, or disk (hence the phrase "disk method"). The radii of each disk vary from one value of x to the next, but you can find each radius by plugging it into the equation for y: each radius is $\sqrt{x}$. The area of each disk is therefore:

$$\pi\left(\sqrt{x}\right)^2 = \pi x.$$

Each disk is infinitesimally thin, so its thickness is dx; if you add up the volumes of all the disks, you'll get the entire volume. The way to add these up is by using the integral, with the endpoints of the interval as the limits of integration. Therefore, to find the volume, evaluate the integral:

$$\int_0^1 \pi x \, dx.$$

Now, let's generalize this. If you have a region whose area is bounded by the curve $y = f(x)$ and the x-axis on the interval $[a,b]$, each disk has a radius of $f(x)$, and the area of each disk will be:

$$\pi[f(x)]^2.$$

To find the volume, evaluate the integral:

$$\pi\int_a^b [f(x)]^2 \, dx.$$

This is the formula for finding the volume using disks.

Example 1: Find the volume of the solid that results when the region between the curve $y = x$ and the x-axis, from $x = 0$ to $x = 1$, is revolved about the x-axis.

As always, sketch the region to get a better look at the problem:

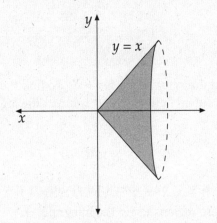

When you slice vertically, the top curve is $y = x$ and the limits of integration are from $x = 0$ to $x = 1$. Using our formula, we evaluate the integral:

$$\pi \int_0^1 x^2 dx .$$

The result is:

$$\pi \int_0^1 x^2 dx = \pi \frac{x^3}{3} \Big|_0^1 = \frac{\pi}{3}.$$

By the way, did you notice that the solid in the problem is a cone with a height and radius of 1? The formula for the volume of a cone is $\frac{1}{3} \pi r^2 h$, so you should expect to get $\frac{\pi}{3}$.

THE WASHER METHOD

Now, let's figure out how to find the volume of the solid that results when we revolve a region that does not touch the x-axis. Consider the region bounded above by the curve $y = x^3$ and below by the curve $y = x^2$, from $x = 2$ to $x = 4$, which is revolved about the x-axis. Sketch the region first:

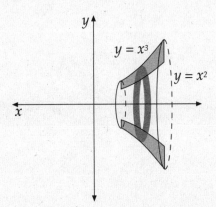

If you slice this region vertically, each cross-section looks like a washer (hence the phrase "washer method"):

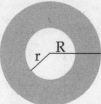

The outer radius is $R = x^3$ and the inner radius is $r = x^2$. To find the area of the region between the two circles, take the area of the outer circle, πR^2, and subtract the area of the inner circle, πr^2.

We can simplify this to:

$$\pi R^2 - \pi r^2 = \pi \left(R^2 - r^2 \right).$$

Because the outer radius is $R = x^3$ and the inner radius is $r = x^2$, the area of each region is $\pi \left(x^6 - x^4 \right)$. You can sum up these regions using the integral:

$$\pi \int_2^4 \left(x^6 - x^4 \right) dx.$$

Here's the general idea: In a region whose area is bounded above by the curve $y = f(x)$ and below by the curve $y = g(x)$, on the interval $[a, b]$, then each washer will have an area of

$$\pi \left[f(x)^2 - g(x)^2 \right].$$

To find the volume, evaluate the integral:

$$\pi \int_a^b \left[f(x)^2 - g(x)^2 \right] dx$$

This is the formula for finding the volume using washers when the region is rotated around the x-axis.

Example 2: Find the volume of the solid that results when the region bounded by $y = x$ and $y = x^2$, from $x = 0$ to $x = 1$, is revolved about the x-axis.

Sketch it first:

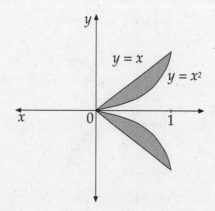

The top curve is $y = x$ and the bottom curve is $y = x^2$ throughout the region. Then our formula tells us that we evaluate the integral:

$$\pi \int_0^1 \left(x^2 - x^4\right) dx.$$

The result is:

$$\pi \int_0^1 \left(x^2 - x^4\right) dx = \pi \left(\frac{x^3}{3} - \frac{x^5}{5} \right) \Bigg|_0^1 = \frac{2\pi}{15}.$$

Suppose the region we're interested in is revolved around the y-axis instead of the x-axis. Now, to find the volume, you have to slice the region horizontally instead of vertically. We discussed how to do this in the previous unit on area.

Now, if you have a region whose area is bounded on the right by the curve $x = f(y)$ and on the left by the curve $x = g(y)$, on the interval $[c, d]$, then each washer has an area of

$$\pi \left[f(y)^2 - g(y)^2 \right].$$

To find the volume, evaluate the integral:

$$\pi \int_c^d \left[f(y)^2 - g(y)^2 \right] dy$$

This is the formula for finding the volume using washers when the region is rotated around the y-axis.

Example 3: Find the volume of the solid that results when the region bounded by the curve $x = y^2$ and the curve $x = y^3$, from $y = 0$ to $y = 1$ is revolved about the y-axis.

Sketch away:

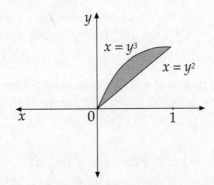

Since $x = y^2$ is always on the outside and $x = y^3$ is always on the inside, you have to evaluate the integral:

$$\pi \int_0^1 \left(x^4 - x^6\right) dx.$$

Here's what you should get:

$$\pi\int_0^1\left(x^4-x^6\right)dx = \pi\left[\frac{x^5}{5}-\frac{x^7}{7}\right]_0^1 = \frac{2\pi}{35}.$$

There's only one more nuance to cover. Sometimes you'll have to revolve the region about a line instead of one of the axes. If so, this will affect the radii of the washers; you'll have to adjust the integral to reflect the shift. Once you draw a picture, it usually isn't too hard to see the difference.

Example 4: Find the volume of the solid that results when the area bounded by the curve $y = x^2$ and the curve $y = 4x$ is revolved about the line $y = -2$. <u>Set up but do not evaluate the integral</u>. (This is how the AP exam will say it!)

You're not given the limits of integration here, so you need to find where the two curves intersect by setting the equations equal to each other:

$$x^2 = 4x$$

$$x^2 - 4x = 0$$

$$x = 0, 4.$$

These will be our limits of integration. Next, sketch the curve:

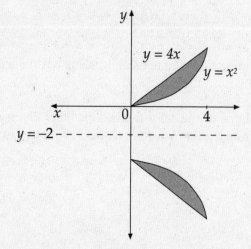

Notice that the distance from the axis of revolution is no longer found by just using each equation. Now, you need to add 2 to each equation to account for the shift in the axis. Thus, the radii are $x^2 + 2$ and $4x + 2$. This means that we need to evaluate the integral:

$$\pi\int_0^4\left[\left(x^2+2\right)^2-\left(4x+2\right)^2\right]dx.$$

Suppose instead that the region was revolved about the line $x = -2$. Sketch the region again:

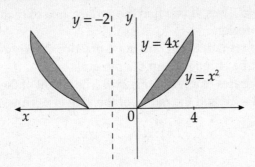

You'll have to slice the region horizontally this time; this means you're going to solve each equation for x in terms of y: $x = \sqrt{y}$ and $x = \dfrac{y}{4}$. We also need to find the y-coordinates of the intersection of the two curves: $y = 0, 16$.

Notice also that, again, each radius is going to be increased by 2 to reflect the shift in the axis of revolution. Thus we will have to evaluate the integral:

$$\pi \int_0^{16} \left[\left(\sqrt{y} + 2 \right)^2 - \left(\frac{y}{4} + 2 \right)^2 \right] dy \, .$$

Finding the volumes isn't that hard, once you've drawn a picture and figured out whether you need to slice vertically or horizontally, and whether the axis of revolution has been shifted. Sometimes, though, there will be times when you want to slice vertically yet revolve around the y-axis (or slice horizontally yet revolve around the x-axis). Here's the method for finding volumes in this way.

CYLINDRICAL SHELLS

Let's examine the region bounded above by the curve $y = 4 - x^2$ and below by the curve $y = x^2$, from $x = 0$ to $x = 1$. Suppose you had to revolve the region about the y-axis instead of the x-axis:

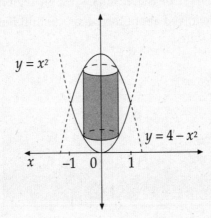

If you slice the region vertically and revolve the slice, you won't get a washer; you'll get a cylinder instead. Since each slice is an infinitesimally thin rectangle, the cylinder's "thickness" is also very, very thin, but real nonetheless. Thus, if you find the surface area of each cylinder and add them up, you'll get the volume of the region.

We know it's difficult to visualize, but you must practice drawing these pictures. If you can't draw the picture, you won't be able to set up the integral.

The formula for the surface area of a cylinder is $2\pi rh$. The height of the cylinder is the length of the vertical slice, $(4 - x^2) - x^2 = 4 - 2x^2$ and the radius of the slice is x. Thus, evaluate the integral:

$$2\pi \int_0^1 x\left(4 - 2x^2\right) dx.$$

The math goes like this:

$$2\pi \int_0^1 x\left(4 - 2x^2\right) dx = 2\pi \int_0^1 \left(4x - 2x^3\right) dx = 2\pi \left(2x^2 - \frac{x^4}{2} \right)\Bigg|_0^1 = 3\pi.$$

Suppose you tried to slice the region horizontally and use washers. You'd have to convert each equation and find the new limits of integration. Since the region is not bounded by the same pair of curves throughout, you would have to evaluate the region using several integrals. This is harder than anything the AP ever asks, so there's no need to go any further here. The cylindrical shells method was invented precisely so you can avoid severe headaches.

From a general standpoint: If we have a region whose area is bounded above by the curve $y = f(x)$ and below by the curve $y = g(x)$, on the interval $[a, b]$, then each cylinder will have a height of $f(x) - g(x)$, a radius of x, and an area of $2\pi x\left[f(x) - g(x)\right]$.

To find the volume, evaluate the integral:

$$2\pi \int_a^b x\left[f(x) - g(x)\right] dx.$$

This is the formula for finding the volume using cylindrical shells when the region is rotated around the y-axis.

Example 5: Find the volume of the region that results when the region bounded by the curve $y = \sqrt{x}$, the x-axis, and the line $x = 9$ is revolved about the y-axis. Set up but do not evaluate the integral.

Your sketch should look like this:

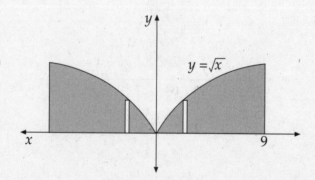

Notice that the limits of integration are from $x = 0$ to $x = 9$, and that each vertical slice is bounded from above by the curve $y = \sqrt{x}$ and from below by the x-axis $(y = 0)$. We need to evaluate the integral:

$$2\pi \int_0^9 x\left(\sqrt{x} - 0\right)dx = 2\pi \int_0^9 x\left(\sqrt{x}\right)dx.$$

Example 6: Find the volume that results when the region in example 5 is revolved about the line $x = -1$. <u>Set up but do not evaluate the integral</u>.

Sketch the figure:

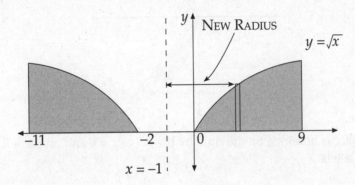

If you slice the region vertically, the height of the shell doesn't change because of the shift in axis of revolution, but you have to add 1 to each radius.

Our integral thus becomes:

$$2\pi \int_0^9 (x + 1)\left(\sqrt{x}\right)dx.$$

SHELLS AROUND THE X-AXIS

The last formula you need to learn involves slicing the region horizontally and revolving it about the x-axis. As you probably guessed, you'll get a cylindrical shell:

If you have a region whose area is bounded on the right by the curve $x = f(y)$ and on the left by the curve $x = g(y)$, on the interval $[c, d]$, then each cylinder will have a height of $f(y) - g(y)$, a radius of y, and an area of $2\pi y\left[f(y) - g(y)\right]$.

To find the volume, evaluate the integral:

$$2\pi \int_c^d y\left[f(y) - g(y)\right]dy.$$

This is the formula for finding the volume using cylindrical shells when the region is rotated around the x-axis.

Example 7: Find the volume of the region that results when the region bounded by the curve $x = y^3$ and the line $x = y$, from $y = 0$ to $y = 1$, is rotated about the x-axis. Set up but do not evaluate the integral.

Let your sketch be your guide:

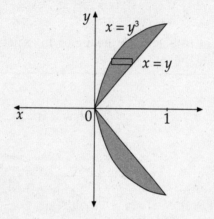

Each horizontal slice is bounded on the right by the curve $x = y$ and on the left by the line $x = y^3$. The integral to evaluate is:

$$2\pi \int_0^1 y\left(y - y^3\right)dy.$$

Suppose that you had to revolve this region about the line $y = -1$ instead. Now the region looks like this:

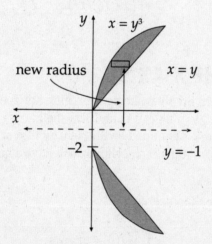

The radius of each cylinder is increased by 1 because of the shift in the axis of revolution, so the integral looks like this:

$$2\pi \int_0^1 (y + 1)\left(y - y^3\right)dy.$$

Wasn't this fun? Volumes of Solids of Revolution (their official name) require you to sketch the region carefully and to decide whether it'll be easier to slice the region vertically or horizontally. Once you figure out the slices' boundaries and the limits of integration (and you've adjusted for an axis of revolution, if necessary) it's just a matter of plugging into the integral. Usually, you won't be asked to evaluate the integral unless it's a simple one. Once you've conquered this topic, you're ready for anything.

Here are some solved problems. Do each problem, covering the answer first, then checking your answer.

PROBLEM 1. Find the volume of the solid that results when the region bounded by the curve $y = 16 - x^2$ and the curve $y = 16 - 4x$ is rotated about the x-axis. Use the washer method and <u>set up but do not evaluate the integral.</u>

Answer: First, sketch the region:

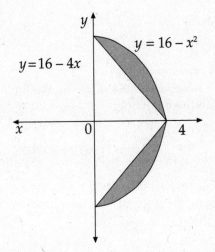

Next, find where the curves intersect by setting the two equations equal to each other:

$$16 - x^2 = 16 - 4x$$
$$x^2 = 4x$$
$$x^2 - 4x = 0$$
$$x = 0, 4$$

Slicing vertically, the top curve is always $y = 16 - x^2$ and the bottom is always $y = 16 - 4x$, so the integral looks like this:

$$\pi \int_0^4 \left[\left(16 - x^2\right)^2 - \left(16 - 4x\right)^2 \right] dx.$$

PROBLEM 2. Repeat problem 1, but revolve the region about the y-axis and use the cylindrical shells method. <u>Set up but do not evaluate the integral</u>.

Answer: Sketch the situation:

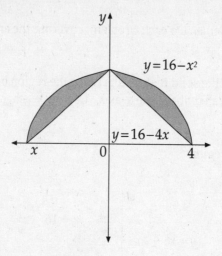

Slicing vertically, the top of each cylinder is $y = 16 - x^2$, the bottom is $y = 16 - 4x$, and the radius is x. Therefore, you should set up the following:

$$2\pi \int_0^4 x\left[\left(16 - x^2\right) - \left(16 - 4x\right)\right]dx$$

PROBLEM 3. Repeat problem 1 but revolve the region about the x-axis and use the cylindrical shells method. <u>Set up but do not evaluate the integral</u>.

Answer: Sketch it like this:

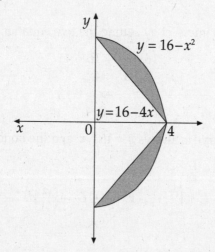

To slice horizontally, you have to solve each equation for x in terms of y and find the limits of integration with respect to y. First, solve for x in terms of y:

$$y = 16 - x^2 \text{ becomes } x = \sqrt{16 - y},$$

and

$$y = 16 - 4x \text{ becomes } x = \frac{16 - y}{4}.$$

Next, determine the limits of integration: they're from $y = 0$ to $y = 16$. Slicing horizontally, the curve $x = \sqrt{16 - y}$ is always on the right and the curve $x = \frac{16 - y}{4}$ is always on the left. The radius is y, so we evaluate:

$$2\pi \int_0^{16} y\left[\left(\sqrt{16 - y}\right) - \left(\frac{16 - y}{4}\right)\right] dy.$$

PROBLEM 4. Repeat problem 1 but revolve the region about the y-axis and use the washers method. <u>Set up but do not evaluate the integral</u>.

Answer: Get those pencils scribbling:

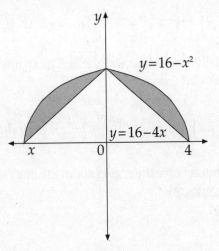

Slicing horizontally, the curve $x = \sqrt{16 - y}$ is always on the right and the curve $x = \frac{16 - y}{4}$ is always on the left. Therefore, your integral should look like this:

$$\pi \int_0^{16} \left[\left(\sqrt{16 - y}\right)^2 - \left(\frac{16 - y}{4}\right)^2\right] dy.$$

PROBLEM 5. Repeat problem 1, but revolve the region about the line $y = -3$. You may use either method. <u>Set up but do not evaluate the integral.</u>

Answer: Your sketch should resemble the one below (note that it's not drawn exactly to scale):

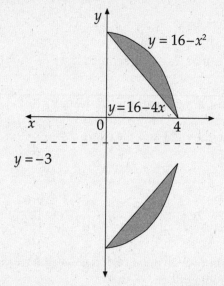

If you were to slice the region vertically, you would use washers. You'll need to add 3 to each radius to adjust for the axis of revolution. The integral to evaluate is:

$$\pi \int_0^4 \left[\left(16 - x^2 + 3\right)^2 - \left(16 - 4x + 3\right)^2 \right] dx.$$

To slice the region horizontally, use cylindrical shells. The radius of each shell would increase by 3, and you would evaluate:

$$2\pi \int_0^{16} (y+3) \left[\left(\sqrt{16-y}\right) - \left(\frac{16-y}{4}\right) \right] dy.$$

PROBLEM 6. Repeat problem 1, but revolve the region about the line $x = 8$. You may use either method. <u>Set up but do not evaluate the integral.</u>

Warning! This one is tricky!

Answer: First, sketch the region.

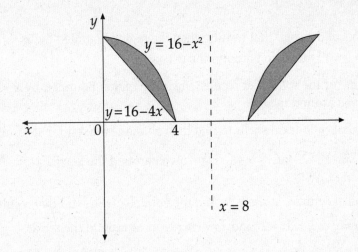

If you choose cylindrical shells, slice the region vertically; you'll need to adjust for the axis of revolution. Each radius can be found by subtracting x from 8. (Not 8 from x. That was the tricky part, in case you missed it.) The integral to evaluate is:

$$2\pi \int_0^4 (8-x)\left[(16-x^2)-(16-4x)\right]dx.$$

If you choose washers, slice the region horizontally. The radius of each washer is found by subtracting each equation from 8. Notice also that the curve $x = \dfrac{16-y}{4}$ is now the outer radius of the washer, and the curve $x = \sqrt{16-y}$ is the inner radius. The integral looks like:

$$\pi \int_0^{16}\left[\left(8-\left(\frac{16-y}{4}\right)\right)^2 - \left(8-\left(\sqrt{16-y}\right)\right)^2\right]dy.$$

PRACTICE PROBLEMS

Calculate the volumes below. The answers are in chapter 21.

1. Find the volume of the solid that results when the region bounded by $y = \sqrt{9-x^2}$ and the x-axis is revolved around the x-axis.

2. Find the volume of the solid that results when the region bounded by $y = \sec x$ and the x-axis from $x = -\dfrac{\pi}{4}$ to $x = \dfrac{\pi}{4}$ is revolved around the x-axis.

3. Find the volume of the solid that results when the region bounded by $x = 1 - y^2$ and the y-axis is revolved around the y-axis.

4. Find the volume of the solid that results when the region bounded by $x = \sqrt{5}y^2$ and the y-axis from $y = -1$ to $y = 1$ is revolved around the y-axis.

5. Find the volume of the solid that results when the region bounded by $y = x^3$, $x = 2$, and the x-axis is revolved around the line $x = 2$.

6. Use the method of cylindrical shells to find the volume of the solid that results when the region bounded by $y = x$, $x = 2$, and $y = -\dfrac{x}{2}$ is revolved around the y-axis.

7. Use the method of cylindrical shells to find the volume of the solid that results when the region bounded by $y = \sqrt{x}$, $y = 2x - 1$, and $x = 0$ is revolved around the y-axis.

8. Use the method of cylindrical shells to find the volume of the solid that results when the region bounded by $y = x^2$, $y = 4$, and $x = 0$ is revolved around the x-axis.

9. Use the method of cylindrical shells to find the volume of the solid that results when the region bounded by $y = 2\sqrt{x}$, $x = 4$, and $y = 0$ is revolved around the y-axis.

10. Use the method of cylindrical shells to find the volume of the solid that results when the region bounded by $y^2 = 8x$, $x = 2$, is revolved around the line $x = 4$.

16

INTEGRATION BY PARTS

This chapter is about one of the most important and powerful techniques of integral calculus. This always shows up on the AP exam. It's a small part of the AB exam and a large part of the BC exam, and you'll often find that an integral that otherwise seems to be undoable is simple once you use integration by parts.

THE FORMULA

Remember the Product Rule? It looks like this:

$$\frac{d}{dx}(uv) = u\frac{dv}{dx} + v\frac{du}{dx}.$$

If we write this in differential form (i.e., eliminate the dx), we get $d(uv) = udv + vdu$. Rearranging, we get the following: $udv = d(uv) - vdu$. Integrating both sides, we obtain the integration by parts formula:

$$\int u\,dv = uv - \int v\,du$$

This formula allows you to write one integral in terms of another integral. This will often help you turn a difficult integral into an easy one.

Example 1: Find $\int x \sin x \, dx$.

Upon first glance, we've got a problem. If you try u-substitution, neither term is even close to the derivative of the other. Thus, you can't substitute. What do you do? (Hint: look at the title of the chapter.)

Let $u = x$ and $dv = \sin x \, dx$. Then $du = dx$ and $v = -\cos x$. If you plug these parts into the formula, you get:

$$\int x \sin x \, dx = -x \cos x + \int \cos x \ dx.$$

Now we've turned the difficult left-side integral into an easy right-side integral:

$$\int x \sin x \, dx = -x \cos x + \sin x + C.$$

Notice that we let $u = x$ and $dv = \sin x \, dx$. If we had let $u = \sin x$ and $dv = x \, dx$, we would have calculated that:

$$du = \cos x \, dx \text{ and } v = \frac{x^2}{2},$$

and the formula would have been:

$$\int x \sin x \, dx = \frac{x^2 \sin x}{2} - \int \frac{x^2 \cos x}{2}.$$

This not only doesn't fix our problem, it makes it worse! This leads us to a general rule:

If one of the terms is a power of x, let that term be u and the other term be dv. (The main exception to this rule is when the other term is ln x.)

Let's assure ourselves that the formula works. If $\int x \sin x \, dx = -x \cos x + \sin x + C$, and you differentiate the right-hand side, the result is:

$$- x(-\sin x) + (-1) \cos x + \cos x = x \sin x.$$

As you can see, integration by parts is the Product Rule in reverse.

Example 2: Find $\int x^2 \sin x \, dx$.

As with Example 1, we let $u = x^2$ and $dv = \sin x \, dx$. Therefore:

$$du = 2x \, dx \text{ and } v = -\cos x.$$

Now plug these into the formula:

$$\int x^2 \sin x \, dx = -x^2 \cos x + 2 \int x \cos x \, dx.$$

Guess what you have to do now. You have to integrate by parts a second time!
Let $u = x$ and $dv = \cos x \, dx$, so $du = dx$ and $v = \sin x$. Now you have:

$$\int x^2 \sin x \, dx = -x^2 \cos x + 2x \sin x - 2 \int \sin x \, dx.$$

The final integral becomes:

$$\int x^2 \sin x \, dx = -x^2 \cos x + 2x \sin x + 2 \cos x + C$$

Often, you'll have to do integration by parts twice. Good news: the AP exam never asks you to do it three times.

Back when we did the integrals of exponential functions, we omitted the integral of the natural logarithm and we told you we would do it in this chapter. Are you ready?

Example 3: Find $\int \ln x \, dx$.

Let $u = \ln x$ and $dv = dx$. (Yes, we're allowed to do this!) Then $du = \dfrac{1}{x} dx$ and $v = x$.

Plugging into the formula, you get:

$$\int \ln x \, dx = x \ln x - \int dx.$$

Now the right side becomes:

$$\int \ln x \, dx = x \ln x - x + C.$$

You should add this integral to the others you've memorized or are comfortable with deriving:

$$\int \ln x \, dx = x \ln x - x + C$$

There's one last type of integration by parts technique you need to know for the AP exam. It requires you to perform two rounds of integration by parts (followed by some simple algebra), and it helps you solve for an unknown integral.

Example 4: Find $\int e^x \cos x \, dx$.

First, we let $u = e^x$ and $dv = \cos x \, dx$. Then, $du = e^x \, dx$ and $v = \sin x$. When you plug this information into the formula, you get:

$$\int e^x \cos x \, dx = e^x \sin x - \int e^x \sin x \, dx \ .$$

Now do integration by parts again. This time we let $u = e^x$ and $dv = \sin x \, dx$, so that $du = e^x \, dx$ and $v = -\cos x$. Now the formula looks like this:

$$\int e^x \cos x \, dx = e^x \sin x - \left(-e^x \cos x + \int e^x \sin x \, dx \right) = e^x \sin x + e^x \cos x - \int e^x \cos x \, dx.$$

It looks as if you're back to square one, but notice that the unknown integral is on both the left- and right-hand sides of the equation. If you now add $\int e^x \cos x \, dx$ to both sides (isn't algebra wonderful?), you get:

$$2\int e^x \cos x \, dx = e^x \sin x + e^x \cos x.$$

Now, all you do is divide both sides by 2 and throw in the constant:

$$\int e^x \cos x \, dx = \frac{e^x \sin x + e^x \cos x}{2} + C.$$

This is why integration by parts is so powerful. You can evaluate a difficult integral by repeatedly rewriting it in terms of integrals you do know.

The AP always has integration by parts problems, and they're generally just like these. In fact, if you can do all of the examples and problems in this chapter, you should be able to do any integration by parts problem that will appear on the AP.

PROBLEM 1. Evaluate $\int x \ln x \, dx$.

Answer: If you let $u = \ln x$ and $dv = x \, dx$, then $du = \frac{1}{x} dx$ and $v = \frac{x^2}{2}$. Notice that this is an exception to our rule about setting the x-term equal to u.

Now you have:

$$\int x \ln x \, dx = \left(\frac{x^2}{2} \right) \ln x - \int \frac{x^2}{2} \frac{1}{x} dx = \frac{x^2 \ln x}{2} - \frac{1}{2} \int x \, dx.$$

Now you can evaluate the right-hand integral:

$$\int x \ln x \, dx = \frac{x^2 \ln x}{2} - \frac{x^2}{4} + C.$$

PROBLEM 2. Evaluate $\int x^2 e^x \, dx$.

Answer: Let $u = x^2$ and $dv = e^x \, dx$. Then $du = 2x \, dx$ and $v = e^x$. The formula becomes:

$$\int x^2 e^x \, dx = x^2 e^x - 2 \int x e^x dx.$$

We need to use integration by parts a second time to evaluate the right-hand integral. Let $u = x$ and $dv = e^x dx$, so $du = dx$ and $v = e^x$. Now it looks like this:

$$\int x^2 e^x \, dx = x^2 e^x - 2\left[x e^x - \int e^x dx \right].$$

Now evaluate the right-hand integral:

$$\int x^2 e^x \, dx = x^2 e^x - 2x e^x + 2e^x + C.$$

PROBLEM 3. Evaluate $\int \sin x \; e^{-x} dx$.

Answer: Let $u = \sin x$ and $dv = e^x \, dx$. Then $du = \cos x \, dx$ and $v = -e^{-x}$. Now the integral looks like this:

$$\int \sin x \; e^{-x} dx = -\sin x e^{-x} + \int \cos x e^{-x} dx.$$

It's time for another round of integration by parts to evaluate the right-hand integral. Let $u = \cos x$ and $dv = e^x dx$. Then $du = -\sin x \, dx$ and $v = -e^{-x}$. This gives you:

$$\int \sin x e^{-x} dx = -\sin x e^{-x} - \cos x e^{-x} - \int \sin x e^{-x} dx.$$

Now add $\int \sin x e^{-x} dx$ to both sides:

$$2 \int \sin x e^{-x} dx = -\sin x e^{-x} - \cos x e^{-x} + C.$$

PRACTICE PROBLEMS

Evaluate each of the following integrals. The answers are in chapter 21.

1. $\int x \csc^2 x \, dx$

2. $\int x e^{2x} dx$

3. $\int \dfrac{\ln x}{x^2}\,dx$

4. $\int x^2 \cos x\,dx$

5. $\int x^2 \ln x\,dx$

6. $\int x \sin 2x\,dx$

7. $\int \ln^2 x\,dx$

8. $\int x \sec^2 x\,dx$

17

TRIG FUNCTIONS

INVERSE TRIG FUNCTIONS

In this unit, you'll learn a broader set of integration techniques. So far, you only know how to do a few types of integrals: polynomials, some trig functions, and logs. Now we'll turn our attention to a set of integrals that result in **inverse trigonometric functions**. But first, we need to go over the derivatives of the inverse trigonometric functions. (You might want to refer back to chapter 10 on derivatives of inverse functions.)

Suppose you have the equation $\sin y = x$. If you differentiate both sides with respect to x, you get:

$$\cos y \frac{dy}{dx} = 1.$$

Now divide both sides by $\cos y$:

$$\frac{dy}{dx} = \frac{1}{\cos y}.$$

This is an implicit derivative, as we saw in Unit One. In chapter 5, we used implicit derivatives; here, we'll find the explicit derivatives by a little substitution.

Because $\sin^2 y + \cos^2 y = 1$, we can replace $\cos y$ with $\sqrt{1 - \sin^2 y}$:

$$\frac{dy}{dx} = \frac{1}{\sqrt{1 - \sin^2 y}}.$$

Finally, because $x = \sin y$, replace $\sin y$ with x. The derivative equals:

$$\frac{1}{\sqrt{1 - x^2}}.$$

Now go back to the original equation $\sin y = x$ and solve for y in terms of x: $y = \sin^{-1} x$. If you differentiate both sides with respect to x, you get:

$$\frac{dy}{dx} = \frac{d}{dx} \sin^{-1} x.$$

Replace $\frac{dy}{dx}$, and you get the final result:

$$\frac{d}{dx} \sin^{-1} x = \frac{1}{\sqrt{1 - x^2}}.$$

This is the derivative of inverse sine. By similar means, you can find the derivatives of all six inverse trig functions. They're not difficult to derive, and they're also not difficult to memorize. The choice is up to you. We use u instead of x to account for the Chain Rule.

$$\frac{d}{dx}\left(\sin^{-1} u\right) = \frac{1}{\sqrt{1 - u^2}} \frac{du}{dx}; \quad -1 < u < 1 \qquad \frac{d}{dx}\left(\cos^{-1} u\right) = \frac{-1}{\sqrt{1 - u^2}} \frac{du}{dx}; \quad -1 < u < 1$$

$$\frac{d}{dx}\left(\tan^{-1} u\right) = \frac{1}{1 + u^2} \frac{du}{dx} \qquad\qquad \frac{d}{dx}\left(\cot^{-1} u\right) = \frac{-1}{1 + u^2} \frac{du}{dx}$$

$$\frac{d}{dx}\left(\sec^{-1} u\right) = \frac{1}{|u|\sqrt{u^2 - 1}} \frac{du}{dx}; \quad |u| > 1 \qquad \frac{d}{dx}\left(\csc^{-1} u\right) = \frac{-1}{|u|\sqrt{u^2 - 1}} \frac{du}{dx}; \quad |u| > 1$$

Notice the domain restrictions for inverse sine, cosine, secant, and cosecant.

Example 1: $\dfrac{d}{dx}\sin^{-1}x^2 = \dfrac{2x}{\sqrt{1-\left(x^2\right)^2}} = \dfrac{2x}{\sqrt{1-x^4}}$.

Example 2: $\dfrac{d}{dx}\tan^{-1}5x = \dfrac{5}{1+(5x)^2} = \dfrac{5}{1+25x^2}$.

Example 3: $\dfrac{d}{dx}\sec^{-1}\left(x^2-x\right) = \dfrac{2x-1}{\left|x^2-x\right|\sqrt{\left(x^2-x\right)^2-1}} = \dfrac{2x-1}{\left|x^2-x\right|\sqrt{x^4-2x^3+x^2-1}}$.

Finding the derivatives of inverse trig functions is just a matter of following the formulas. They're very rarely tested on the AP in this form and are not terribly important. In addition, the AP exam almost only tests inverse sine and tangent, and then usually only as integrals, not as derivatives.

Therefore, it's time to learn the integrals. The derivative formulas lead directly to the integral formulas, but because the only difference between the derivatives of inverse sine, tangent, and secant and those of inverse cosine, cotangent, and cosecant is the negative sign, only the former functions are generally used.

$$\int \frac{du}{\sqrt{1-u^2}} = \sin^{-1}u + C \qquad \left(\text{valid for } u^2 < 1\right)$$

$$\int \frac{du}{1+u^2} = \tan^{-1}u + C$$

$$\int \frac{du}{u\sqrt{u^2-1}} = \sec^{-1}u + C \qquad \left(\text{valid for } u^2 > 1\right)$$

The first two always show up on the AP exam, so you should learn to recognize the pattern. Most of the time, the integration requires some sticky algebra to get the integral into the proper form.

Example 4: $\displaystyle\int \frac{x\,dx}{\sqrt{1-x^4}} =$

Don't forget about u-substitution. Let $u = x^2$ and $du = 2x\,dx$. Therefore, $\dfrac{1}{2}du = x\,dx$. Substituting and integrating, we get:

$$\frac{1}{2}\int \frac{du}{\sqrt{1-u^2}} = \frac{1}{2}\sin^{-1}u + C.$$

Now substitute back:

$$\frac{1}{2}\sin^{-1}(x^2)+C.$$

Example 5: $\displaystyle\int \frac{dx}{\sqrt{4-x^2}} =$

Anytime you see an integral where you have the square root of a constant minus a function, try to turn it into an inverse sine. You can do that here with a little simple algebra:

$$\int \frac{dx}{\sqrt{4-x^2}} = \int \frac{dx}{\sqrt{4\left(1-\dfrac{x^2}{4}\right)}} = \frac{1}{2}\int \frac{dx}{\sqrt{1-\dfrac{x^2}{4}}}.$$

Now use u-substitution: let $u = \dfrac{x}{2}$ and $du = \dfrac{1}{2}dx$, so $2du = dx$, and substitute:

$$\frac{1}{2}\int \frac{dx}{\sqrt{1-\dfrac{x^2}{4}}} = \frac{1}{2}\int \frac{2du}{\sqrt{1-u^2}} = \sin^{-1}u+C.$$

When you substitute back, you get:

$$\sin^{-1}\frac{x}{2}+C.$$

Example 6: $\displaystyle\int \frac{e^x dx}{1+e^{2x}} =$

Again, use u-substitution. Let $u = e^x$ and $du = e^x\, dx$, and substitute in:

$$\int \frac{du}{1+u^2} = \tan^{-1}u+C.$$

Then substitute back:

$$\tan^{-1}e^x + C.$$

Let's do one last type that's slightly more difficult. For this one, you need to remember how to complete the square.

Example 7: $\displaystyle\int \frac{dx}{x^2+4x+5} =$

Complete the square in the denominator (your algebra teacher warned you about this).

$$x^2 + 4x + 5 = (x+2)^2 + 1.$$

Now rewrite the integral:

$$\int \frac{dx}{1+(x+2)^2}.$$

Using u-substitution, let $u = x + 2$ and $du = dx$. Then do the substitution:

$$\int \frac{du}{1+u^2} = \tan^{-1} u + C.$$

When you put the function of x back in to the integral, it reads:

$$\tan^{-1}(x + 2) + C.$$

Evaluating these integrals involves looking for a particular pattern. If it's there, all you have to do is use algebra and u-substitution to make the integrand conform to the pattern. Once that's accomplished, the rest is easy. These integrals will show up again when we do partial fractions. Otherwise, as far as the AP is concerned, this is all you need to know.

Try these solved problems and don't forget to look for those patterns. Get out your index card and cover the answers while you work.

PROBLEM 1. Find the derivative of $y = \sin^{-1}\left(\dfrac{x}{2}\right)$.

Answer: The rule is:

$$\frac{d}{dx}\left(\sin^{-1} u\right) = \frac{1}{\sqrt{1-u^2}} \frac{du}{dx}.$$

The algebra looks like this:

$$\frac{d}{dx}\left(\sin^{-1}\left(\frac{x}{2}\right)\right) = \frac{1}{\sqrt{1-\left(\frac{x}{2}\right)^2}}\frac{1}{2} = \frac{1}{2}\left(\frac{1}{\sqrt{1-\frac{x^2}{4}}}\right) = \frac{1}{2}\left(\frac{1}{\sqrt{\frac{4-x^2}{4}}}\right) = \frac{1}{2}\left(\frac{2}{\sqrt{4-x^2}}\right) = \frac{1}{\sqrt{4-x^2}}.$$

PROBLEM 2. Evaluate $\int \dfrac{dx}{4+x^2}$.

Answer: First, you need to do a little algebra. Factor 4 out of the denominator to obtain this:

$$\int \frac{dx}{4\left(1+\dfrac{x^2}{4}\right)}.$$

Next, rewrite the integrand as:

$$\frac{1}{4}\int \frac{dx}{\left(1+\left(\dfrac{x}{2}\right)^2\right)}.$$

Now you can use u-substitution. Let $u = \dfrac{x}{2}$ and $du = \dfrac{1}{2}dx$, and so $2du = dx$:

$$\frac{1}{2}\int \frac{du}{1+u^2}.$$

Now integrate it:

$$\frac{1}{2}\tan^{-1}u + C.$$

And re-substitute:

$$\frac{1}{2}\tan^{-1}\frac{x}{2} + C.$$

PROBLEM 3. Evaluate $\int \dfrac{dx}{\sqrt{-x^2+4x-3}}$.

Answer: This time, you need to use some algebra by completing the square of the polynomial under the square root sign;

$$-x^2 + 4x - 3 = -\left(x^2 - 4x + 3\right) = -\left[(x-2)^2 - 1\right] = \left[1 - (x-2)^2\right].$$

Now rewrite the integrand as:

$$\int \frac{dx}{\sqrt{1-(x-2)^2}}.$$

And use *u*-substitution. Let $u = x - 2$ and $du = dx$:

$$\int \frac{du}{\sqrt{1 - u^2}}.$$

Now, this looks familiar. Once you integrate, you get:

$$\sin^{-1} u + C$$

which becomes:

$$\sin^{-1}(x - 2) + C$$

after you substitute back.

PRACTICE PROBLEMS

Here is some more practice work on derivatives and integrals of inverse trig functions. The answers are in chapter 21.

1. Find the derivative of $\frac{1}{4}\tan^{-1}\left(\frac{x}{4}\right)$.

2. Find the derivative of $\sin^{-1}\left(\frac{1}{x}\right)$.

3. Find the derivative of $\tan^{-1}\left(e^x\right)$.

4. Evaluate $\int \frac{dx}{x\sqrt{x^2 - \pi}}$.

5. Evaluate $\int \frac{dx}{7 + x^2}$.

6. Evaluate $\int \frac{dx}{x\left(1 + \ln^2 x\right)}$.

7. Evaluate $\int \frac{\sec^2 x \, dx}{\sqrt{1 - \tan^2 x}}$.

8. Evaluate $\int \frac{dx}{\sqrt{9 - 4x^2}}$.

9. Evaluate $\int \frac{e^{3x} dx}{1 + e^{6x}}$.

INTEGRALS OF ADVANCED TRIGONOMETRIC FUNCTIONS

In chapter 11, you learned how to find the integrals of some of the trigonometric functions. Now it's time to figure out how to find the integrals of some of the more complicated trig expressions. First, recall some of the basic trigonometric integrals:

$$\int \sin x \, dx = -\cos x + C \qquad \int \cos x \, dx = \sin x + C$$

$$\int \sec^2 x \, dx = \tan x + C \qquad \int \csc^2 dx = -\cot x + C$$

$$\int \sec x \tan x \, dx = \sec x + C \qquad \int \csc x \cot x \, dx = -\csc x + C$$

$$\int \tan x \, dx = -\ln |\cos x| + C \qquad \int \cot x \, dx = \ln |\sin x| + C$$

$$\int \sec x \, dx = \ln |\sec x + \tan x| + C \qquad \int \csc x \, dx = -\ln |\csc x + \cot x| + C$$

Some of these are derived by reversing differentiation, others by *u*-substitution.

Example 1: Evaluate $\int \sin^2 x \, dx$.

You can start by replacing sin²x with $\dfrac{1 - \cos 2x}{2}$. This comes from taking the double angle formula cos 2x = 1 – 2 sin²x and solving for sin x. After you make that replacement, the integral looks like this:

$$\int \frac{1 - \cos 2x}{2} dx = \frac{1}{2} \int dx - \frac{1}{2} \int \cos 2x \, dx = \frac{x}{2} - \frac{\sin 2x}{4} + C.$$

Remember this substitution:

$$\sin^2 x = \frac{1 - \cos 2x}{2}.$$

You could use the substitution:

$$\cos^2 x = \frac{1 + \cos 2x}{2}$$

(which comes from taking the double angle formula cos 2x = 2 cos²x – 1 and solving for x) to find $\int \cos^2 x \, dx$ as well.

Now let's do some variations.

Example 2: Evaluate $\int \sin^2 x \cos x\, dx$.

Here, use some simple u-substitution. Let $u = \sin x$ and $du = \cos x\, dx$ and substitute:

$$\int \sin^2 x \cos x\, dx = \int u^2 du = \frac{u^3}{3} + C.$$

When you substitute back, you get:

$$\frac{\sin^3 x}{3} + C.$$

In fact, you can do any integral of the form $\int \sin^n x \cos x\, dx$ using u-substitution, and the result will be:

$$\frac{\sin^{n+1} x}{n+1} + C.$$

Similarly,

$$\int \cos^n x \sin x\, dx = -\frac{\cos^{n+1} x}{n+1} + C$$

How about higher powers of sine and cosine?

Example 3: Evaluate $\int \sin^3 x\, dx$.

First, break the integrand into this:

$$\int (\sin x)(\sin^2 x)\, dx.$$

Next, using trig substitution, you get:

$$\int (\sin x)(\sin^2 x)\, dx = \int (\sin x)(1 - \cos^2 x)\, dx.$$

You can turn this into two integrals:

$$\int \sin x\, dx - \int (\sin x \cos^2 x)\, dx.$$

You can do both of these integrals:

$$-\cos x + \frac{\cos^3 x}{3} + C.$$

Example 4: Evaluate $\int \sin^3 x \cos^2 x\, dx$.

Here, break the integrand into:

$$\int (\sin x)(\sin^2 x)(\cos^2 x)\, dx.$$

Next, using trig substitution, you get:

$$\int (\sin x)(\sin^2 x)(\cos^2 x)\, dx = \int (\sin x)(1 - \cos^2 x)(\cos^2 x)\, dx.$$

Now you can use u-substitution. Let $u = \cos x$, $du = -\sin x\, dx$, and substitute:

$$-\int (1 - u^2) u^2\, du = \int (u^4 - u^2)\, du = \frac{u^5}{5} - \frac{u^3}{3} + C.$$

Now substitute back, and you're done:

$$\frac{\cos^5 x}{5} - \frac{\cos^3 x}{3} + C$$

As you can see, these integrals all require you to know trig substitutions and u-substitution. The AP doesn't ask too many variations on these integrals, but let's do just a few other types, just in case.

Example 5: Evaluate $\int \tan^2 x\, dx$.

First, use the trigonometric substitution $\tan^2 x = \sec^2 x - 1$:

$$\int \tan^2 x\, dx = \int (\sec^2 x - 1)\, dx.$$

Now break this into two integrals that are easy to evaluate:

$$\int \sec^2 x\, dx - \int dx.$$

And integrate:

$$\tan x - x + C.$$

Example 6: Evaluate $\int \tan^3 x\, dx$.

Recognize what to do here? Right! Break up the integrand!

$$\int (\tan x)(\tan^2 x)\, dx.$$

Next, use the trig substitution $\tan^2 x = \sec^2 x - 1$ in the expression:

$$\int (\tan x)(\sec^2 x - 1) dx.$$

Break this into two integrals:

$$\int \tan x \sec^2 x \, dx - \int \tan x \, dx.$$

Tackle the first integral using u-substitution. Let $u = \tan x$, $du = \sec^2 x \, dx$:

$$\int u \, du = \frac{u^2}{2} + C.$$

Substituting back gives you:

$$\frac{\tan^2 x}{2} + C.$$

We've done the second integral before:

$$\int \tan x \, dx = \int \frac{\sin x}{\cos x} \, dx = -\ln |\cos x| + C.$$

Thus the integral is:

$$\frac{\tan^2 x}{2} - \ln |\cos x| + C.$$

Example 7: Evaluate $\int \sec^3 x \, dx$.

Now it's time for integration by parts. Let $u = \sec x$, $dv = \sec^2 x \, dx$. Therefore, $du = \sec x \tan x \, dx$ and $v = \tan x$. Plug these into the formula, and you get:

$$\int \sec^3 x \, dx = \sec x \tan x - \int \tan^2 x \sec x \, dx.$$

Next, substitute $\tan^2 x = \sec^2 x - 1$:

$$\int \tan^2 x \sec x \, dx = \int (\sec^2 x - 1)(\sec x) \, dx = \int \sec^3 x \, dx - \int \sec x \, dx.$$

So, the integral has become

$$\int \sec^3 x \, dx = \sec x \tan x - \int \sec^3 x \, dx + \int \sec x \, dx.$$

Now, add $\int \sec^3 x \, dx$ to both sides:

$$2\int \sec^3 x \, dx = \sec x \tan x + \int \sec x \, dx.$$

Finally, recall that the integral on the right is:

$$\int \sec x \, dx = \ln |\sec x + \tan x| + C.$$

The combined integral is:

$$2\int \sec^3 x \, dx = \sec x \tan x + \ln |\sec x + \tan x| + C.$$

When you divide by 2, you get your final answer:

$$\int \sec^3 x \, dx = \frac{\sec x \tan x + \ln |\sec x + \tan x|}{2} + C.$$

Whew! This is about as difficult as the AP ever gets. Believe it or not, trigonometric integrals can get even more complicated, involving higher powers or more difficult combinations of trig functions. Thankfully, the AP tends to limit itself to the simpler ones.

If you can integrate any power of a trig function up to 4, as well as some of the basic combinations, you'll be able to handle any trigonometric integral that appears on the AP exam.

Walk yourself through the step-by-step process by trying the some solved problems below. Cover each answer first, then check your answer.

PROBLEM 1. $\int \cos^2 x \, dx$

Answer: Use the substitution $\frac{1 + \cos 2x}{2}$. Now the integral looks like this:

$$\int \frac{1 + \cos 2x}{2} \, dx.$$

Divide this into two integrals:

$$\int \frac{1}{2} \, dx + \int \frac{\cos 2x}{2} \, dx$$

When you evaluate each one separately, you get:

$$\frac{x}{2} + \frac{\sin 2x}{4} + C.$$

PROBLEM 2. $\int \cos^3 x \, dx$

Answer: Here, split $\cos^3 x$ into $\cos x \cos^2 x$ and replace the second term with $(1 - \sin^2 x)$:

$$\int \cos x (1 - \sin^2 x) \, dx.$$

This can be rewritten as:

$$\int \cos x \, dx - \int \sin^2 x \cos x \, dx.$$

The left-hand integral is sin x. Do the right-hand integral using u-substitution. Let $u = \sin x$ and $du = \cos x \, dx$. You get:

$$\int u^2 du = \frac{u^3}{3}.$$

When you substitute back, the second integral becomes:

$$\frac{\sin^3 x}{3} + C.$$

Thus, the complete answer is:

$$\sin x - \frac{\sin^3 x}{3} + C.$$

PROBLEM 3. $\int \tan^2 x \sec^2 x \, dx$

Answer: This won't take long. Use u-substitution, by letting $u = \tan x$ and $du = \sec^2 x \, dx$:

$$\int u^2 du = \frac{u^3}{3}.$$

The final result when you substitute back is:

$$\frac{\tan^3 x}{3} + C.$$

PROBLEM 4. $\int (\tan x + \sec x)^2 dx$

Answer: First, expand the integrand:

$$\int \left(\tan^2 x + 2 \sec x \tan x + \sec^2 x \right) dx.$$

Next, use the trig substitution $\tan^2 x = \sec^2 x - 1$ to get:

$$\int \left(\sec^2 x - 1 + 2 \sec x \tan x + \sec^2 x \right) dx = \int \left(2 \sec^2 x - 1 + 2 \sec x \tan x \right) dx.$$

Separate this into three (yes, three) integrals, all of which you know how to evaluate:

$$\int 2 \sec^2 x \, dx - \int dx + \int 2 \sec x \tan x \, dx = 2 \tan x - x + 2 \sec x + C.$$

PRACTICE PROBLEMS

Evaluate the following integrals. The answers are in chapter 21.

1. $\displaystyle\int \sin^4 x \, dx$

2. $\displaystyle\int \cos^4 x \, dx$

3. $\displaystyle\int \cos^4 x \sin x \, dx$

4. $\displaystyle\int \sin^3 x \cos^5 x \, dx$

5. $\displaystyle\int \sin^2 x \cos^2 x \, dx$

6. $\displaystyle\int \tan^3 x \sec^4 x \, dx$

7. $\displaystyle\int \tan^5 x \, dx$

8. $\displaystyle\int \cot^2 x \sec x \, dx$

18

OTHER APPLICATIONS OF THE INTEGRAL

Each of the three topics discussed in this chapter appears only on the BC exam. If you're in an AB course, move on to chapter 19. And have a nice day ☺.

LENGTH OF A CURVE

Another way to apply integrals is to find the length of a curve. It's usually a pretty simple problem. All you have to do is use a formula.

Suppose you want to find the length of a curve of a function $y = f(x)$ from $x = a$ to $x = b$. Call this length L. You could divide L into a set of line segments, Δl add up the lengths of each of them, and you would find L. The formula for the length of each line segment can be found with the Pythagorean Theorem:

$$\Delta l = \sqrt{(\Delta x)^2 + (\Delta y)^2}$$

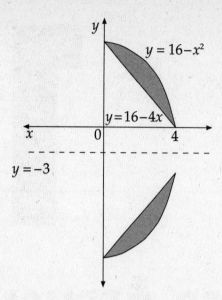

If we make Δl, Δx and Δy infinitesimally small, they become dx and dy. If we then add up all of these line segments, we get: $L = \int_a^b dl = \int_a^b \sqrt{dx^2 + dy^2} = \int_{x_1}^{x_2} \sqrt{1 + \left(\dfrac{dy}{dx}\right)^2}\, dx = \int_{y_1}^{y_2} \sqrt{1 + \left(\dfrac{dx}{dy}\right)^2}\, dy$

Here's the rule:

If the function $f(x)$ is continuous and differentiable on [a,b], then the length of the curve $y = f(x)$ from a to b is:

$$L = \int_a^b \sqrt{1 + \left(\frac{dy}{dx}\right)^2}\, dx$$

Example 1: Find the length of the curve $y = x^{\frac{3}{2}}$ from $x = 0$ to $x = 4$.

First, find the first derivative of the function: $\dfrac{dy}{dx} = \dfrac{3}{2} x^{\frac{1}{2}}$

Next, plug into the formula:

$$L = \int_0^4 \sqrt{1 + \left(\frac{3}{2} x^{\frac{1}{2}}\right)^2}\, dx.$$

Now evaluate the integral:

$$L = \int_0^4 \sqrt{1 + \left(\frac{3}{2} x^{\frac{1}{2}}\right)^2}\, dx = \int_0^4 \sqrt{1 + \frac{9}{4} x}\, dx.$$

Using u-substitution, let $u = 1 + \dfrac{9x}{4}$, $du = \dfrac{9}{4} dx$ and $\dfrac{4}{9} du = dx$. Then substitute:

$$\frac{4}{9} \int u^{\frac{1}{2}} du = \frac{8}{27} u^{\frac{3}{2}}.$$

Once you substitute back, the result is: $\dfrac{8}{27} \left(1 + \dfrac{9}{4} x\right)^{\frac{3}{2}} \Bigg|_0^4 = \dfrac{8}{27} \left(10^{\frac{3}{2}} - 1\right)$

As you can see, the formula isn't terribly difficult, but it can often lead to a really ugly integral. Fortunately, the AP will either give you a curve where the integral works out easily, as it did here, or you'll only be asked to set up the integral, not to evaluate it.

Sometimes you'll be given the curve as a function of y instead of x. Then the formula for the length of the curve $x = f(y)$ on the interval [c,d] is:

$$L = \int_a^b \sqrt{\left(\frac{dx}{dt}\right)^2 + \left(\frac{dy}{dt}\right)^2}\, dt$$

Example 2: Find the length of the curve $x = \sin y$, from $y = 0$ to $y = \dfrac{\pi}{2}$. Set up but do not evaluate the integral.

Since $\dfrac{dx}{dy} = \cos y$, we can plug it into the formula: $L = \int_0^{\frac{\pi}{2}} \sqrt{1 + \cos^2 y}\, dy$

PARAMETRIC FUNCTIONS

Sometimes the curve will be defined parametrically, usually in terms of t (for time). Then the formula for the length of the curve from $t = a$ to $t = b$ is:

$$L = \int_a^b \sqrt{\left(\frac{dx}{dt}\right)^2 + \left(\frac{dy}{dt}\right)^2}\, dt$$

Example 3: Find the length of the curve defined by $x = \sin t$ and $y = \cos t$, from $t = 0$ to $t = \pi$.

Take the derivatives of the two t-functions: $\dfrac{dx}{dt} = \cos t$ and $\dfrac{dy}{dt} = -\sin t$. Then use the formula:

$$L = \int_0^{\pi} \sqrt{\cos^2 t + \sin^2 t}\, dt$$

If you evaluate the integral, you get:

$$L = \int_0^\pi \sqrt{1}\, dt = \int_0^\pi dt = \pi$$

That's all there is to finding the length of a curve. As you've seen, many applications of the integral involve simple formulas. All you have to do is to plug in and evaluate the integral. If the integral is a difficult one, you probably just have to set it up.

Try these solved problems. Do each problem, covering the answer first, then checking your answer.

PROBLEM 1. Find the length of the curve $y = \frac{1}{3}(x^2 + 2)^{\frac{3}{2}}$ from $x = 0$ to $x = 3$.

Answer: First, we find the first derivative $\dfrac{dy}{dx} = \frac{1}{3} \cdot \frac{3}{2}(x^2 + 2)^{\frac{1}{2}}(2x) = x(x^2 + 2)^{\frac{1}{2}}$

Next, plug into the formula:

$$L = \int_a^b \sqrt{1 + \left(\frac{dy}{dx}\right)^2}\, dx = \int_0^3 \sqrt{1 + x^2(x^2 + 2)}\, dx.$$

Now you just have to evaluate the integral.

$$L = \int_0^3 \sqrt{1 + x^4 + 2x^2}\, dx = \int_0^3 \sqrt{(x^2 + 1)^2}\, dx = \left(\frac{x^3}{3} + x\right)\Bigg|_0^3 = 12.$$

PROBLEM 2. Find the length of the curve $x = \dfrac{y^4}{4} + \dfrac{1}{8y^2}$ from $y = 1$ to $y = 2$.

Answer: Here, you have x in terms of y so, first, we find $\dfrac{dx}{dy}$:

$$\frac{dx}{dy} = y^3 - \frac{1}{4y^3}.$$

Next, find $\left(\dfrac{dx}{dy}\right)^2$:

$$\left(\frac{dx}{dy}\right)^2 = \left(y^3 - \frac{1}{4y^3}\right)^2 = y^6 - \frac{1}{2} + \frac{1}{16y^6}.$$

Plug this into the formula:

$$L = \int_c^d \sqrt{1+\left(\frac{dx}{dy}\right)^2}\,dy = \int_1^2 \sqrt{1+\left(y^6 - \frac{1}{2} + \frac{1}{16y^6}\right)}\,dy = \int_1^2 \sqrt{y^6 + \frac{1}{2} + \frac{1}{16y^6}}\,dy.$$

And evaluate the integral:

$$L = \int_1^2 \sqrt{y^6 + \frac{1}{2} + \frac{1}{16y^6}}\,dy = \int_1^2 \sqrt{\left(y^3 + \frac{1}{4y^3}\right)^2}\,dy = \left(\frac{y^4}{4} - \frac{1}{8y^2}\right)\Big|_1^2 = \frac{123}{32}.$$

Problem 3. Find the length of the curve $x = \tan t$ and $y = \sec t$ from $t = 0$ to $t = \frac{\pi}{4}$. Set up but do not evaluate the integral.

Answer: To use the formula: $L = \int_a^b \sqrt{\left(\frac{dx}{dt}\right)^2 + \left(\frac{dy}{dt}\right)^2}\,dt$,

you need to determine that $\frac{dx}{dt} = \sec^2 t$ and $\frac{dy}{dt} = \sec t\tan t$. Plug these into the formula:

$$L = \int_0^{\frac{\pi}{4}} \sqrt{\sec^4 t + \sec^2 t\tan^2 t}\,dt.$$

PRACTICE PROBLEMS

Find the length of the following curves between the specified intervals. Evaluate the integrals only when you're asked to. The answers are in chapter 21.

1. $y = \frac{x^3}{12} + \frac{1}{x}$ from $x = 1$ to $x = 2$

2. $y = \tan x$ from $x = -\frac{\pi}{6}$ to $x = 0$ (<u>Set up but do not evaluate the integral.</u>)

3. $y = \sqrt{1-x^2}$ from $x = 0$ to $x = \frac{1}{4}$ (<u>Set up but do not evaluate the integral.</u>)

4. $x = \frac{y^3}{18} + \frac{3}{2y}$ from $y = 2$ to $y = 3$

5. $x = \sqrt{1-y^2}$ from $y = -\frac{1}{2}$ to $y = \frac{1}{2}$ (<u>Set up but do not evaluate the integral.</u>)

6. $x = \sin y - y \cos y$ from $y = 0$ to $y = \pi$ (<u>Set up but do not evaluate the integral.</u>)

7. $x = \cos t$ and $y = \sin t$ from $t = \dfrac{\pi}{6}$ to $t = \dfrac{\pi}{3}$

8. $x = \sqrt{t}$ and $y = \dfrac{1}{t^3}$ from $t = 1$ to $t = 4$ (<u>Set up but do not evaluate the integral.</u>)

9. $x = 3t^2$ and $y = 2t$ from $t = 1$ to $t = 2$ (<u>Set up but do not evaluate the integral.</u>)

THE METHOD OF PARTIAL FRACTIONS

This is the last technique you'll learn to evaluate integrals. There are many, many more types of integrals and techniques to learn; in fact, there are courses in college primarily concerned with integrals and their uses! Fortunately for you, they're not on the AP exam (and therefore, not in this book). The BC exam usually has a partial fractions integral or two, and the concept isn't terribly hard.

We use the method of partial fractions to evaluate certain types of integrals that contain rational expressions. First, let's discuss the type of algebra you'll be doing.

If you wanted to add the expressions $\dfrac{3}{x-1}$ and $\dfrac{5}{x+2}$, you would do the following:

$$\frac{3}{x-1} + \frac{5}{x+2} = \frac{3(x+2) + 5(x-1)}{(x-1)(x+2)} = \frac{8x+1}{(x-1)(x+2)}.$$

Now, suppose you had to do this in reverse; you were given the fraction on the right and you wanted to determine what two fractions were added to give you that fraction. Another way of asking this is: What constants A and B exist, such that $\dfrac{A}{x-1} + \dfrac{B}{x+2} = \dfrac{8x+1}{(x-1)(x+2)}$?

(We know you know already. Just play along.)

How do you go about solving for A and B? First, multiply through by $(x-1)(x+2)$ to clear the denominator:

$$A(x+2) + B(x-1) = 8x+1.$$

Next, simplify the left side:

$$Ax + 2A + Bx - B = 8x + 1.$$

Now, if you group the terms on the left, you get:

$$(A+B)x + (2A-B) = 8x+1.$$

Therefore, $A + B = 8$ and $2A - B = 1$

If you solve this pair of simultaneous equations, you get $A = 3$ and $B = 5$. Surprised? We hope not. Why would you need to know this method? Suppose you wanted to find:

$$\int \frac{8x+1}{(x-1)(x+2)}\,dx.$$

You now know you can rewrite this integral as:

$$\int \frac{3}{x-1}dx + \int \frac{5}{x-2}dx.$$

These integrals are easily evaluated:

$$\int \frac{3}{x-1}dx + \int \frac{5}{x-2}dx = 3\ln|x-1| + 5\ln|x-2| + C.$$

Example 1: Evaluate $\int \frac{x+18}{(3x+5)(x+4)}dx$.

You need to find A and B such that:

$$\frac{A}{3x+5} + \frac{B}{x+4} = \frac{x+18}{(3x+5)(x+4)}.$$

First, multiply through by $(3x + 5)(x + 4)$:

$$A(x + 4) + B(3x + 5) = x + 18.$$

Next, simplify and group the terms:

$$Ax + 4A + 3Bx + 5B = x + 18$$

$$(A + 3B)x + (4A + 5B) = x + 18$$

You now have two simultaneous equations: $A + 3B = 1$ and $4A + 5B = 18$. If we solve the equations, we get $A = 7$ and $B = -2$. Thus, you can rewrite the integral as:

$$\int \frac{7}{3x+5}dx - \int \frac{2}{x+4}dx.$$

These are both logarithmic integrals. The solution is:

$$\frac{7}{3}\ln\left|x+\frac{5}{3}\right| - 2\ln|x+4| + C.$$

There are three main types of partial fractions that appear on the AP. You've just seen the first type: one with two linear factors in the denominator. The second type has a repeated linear term in the denominator.

Example 2: Evaluate $\int \frac{2x+4}{(x-1)^2}dx$.

Now you need to find two constants A and B, such that:

$$\frac{A}{x-1} + \frac{B}{(x-1)^2} = \frac{2x+4}{(x-1)^2}.$$

Multiplying through by $(x - 1)^2$, we get: $A(x - 1) + B = 2x + 4$. Now simplify:

$$Ax - A + B = 2x + 4$$

$$Ax + (B - A) = 2x + 4$$

Thus, $A = 2$ and $B - A = 4$, so $B = 6$. Now we can rewrite the integral as:

$$\int \frac{2}{x-1}\,dx + \int \frac{6}{(x-1)^2}\,dx.$$

The solution is:

$$2\ln|x-1| - \frac{6}{x-1} + C.$$

The third type has an irreducible quadratic factor in the denominator.

Example 3: Evaluate $\displaystyle\int \frac{3x+5}{(x^2+1)(x+2)}\,dx$.

For a quadratic factor, you need to find A, B, and C such that:

$$\frac{Ax+B}{x^2+1} + \frac{C}{x+2} = \frac{3x+5}{(x^2+1)(x+2)}.$$

Now you have a term $Ax + B$ over the quadratic term. Whenever you have a quadratic factor, you need to use a linear numerator, not a constant numerator.

Multiply through by $(x^2 + 1)(x + 2)$ and you get:

$$(Ax + B)(x + 2) + C(x^2 + 1) = 3x + 5.$$

Simplify and group the terms:

$$Ax^2 + 2Ax + Bx + 2B + Cx^2 + C = 3x + 5$$

$$(A + C)x^2 + (2A + B)x + (2B + C) = 3x + 5.$$

This gives you three equations:

$$A + C = 0$$
$$2A + B = 3$$
$$2B + C = 5.$$

If you solve these three equations, the values are:

$$A = \frac{1}{5}, \ B = \frac{13}{5}, \text{ and } C = -\frac{1}{5}.$$

Now you can rewrite the integral:

$$\int \frac{\frac{1}{5}x + \frac{13}{5}}{x^2 + 1}\, dx - \int \frac{\frac{1}{5}}{x+1}\, dx.$$

Break the first integral into two integrals:

$$\frac{1}{5}\int \frac{x}{x^2+1}\, dx + \frac{13}{5}\int \frac{dx}{x^2+1} - \frac{1}{5}\int \frac{1}{x+2}\, dx.$$

You can evaluate the first integral using u-substitution; the second integral is inverse tangent, and the third integral is a natural logarithm:

$$\frac{1}{10}\ln\left|x^2+1\right| + \frac{13}{5}\tan^{-1}(x) - \frac{1}{5}\ln\left|x+2\right| + C.$$

Notice how all of these integrals contained a natural logarithm term? That's typical of a partial fractions integral. Now try these solved problems. Do each problem, covering the answer first, then checking your answer.

PROBLEM 1. Evaluate $\int \dfrac{4}{x^2 - 6x + 5}\, dx$.

Answer: First, factor the denominator:

$$\int \frac{4}{(x-1)(x-5)}\, dx.$$

Find A and B such that:

$$\frac{A}{x-5} + \frac{B}{x-1} = \frac{4}{(x-5)(x-1)}.$$

First, multiply through by $(x - 5)(x - 1)$:

$$A(x - 1) + B(x - 5) = 4.$$

Next, simplify and group the terms:

$$Ax - A + Bx - 5B = 4$$

$$(A + B)x + (-A - 5B) = 4$$

You now have two simultaneous equations: $A + B = 0$ and $-A - 5B = 4$. If you solve the equations, we get $A = 1$ and $B = -1$. Thus, we can rewrite the integral as:

$$\int \frac{1}{x-5}\, dx - \int \frac{1}{x-1}\, dx.$$

These are both logarithmic integrals. The solution is: $\ln|x-5|-\ln|x-1|+C$.

PROBLEM 2. Evaluate $\int \dfrac{(x+1)dx}{x(x^2+1)}$.

Answer: You need to find A, B, and C, such that:

$$\frac{A}{x}+\frac{Bx+C}{x^2+1}=\frac{x+1}{x(x^2+1)}.$$

Multiply through by $x(x^2+1)$:

$$A(x^2+1)+(Bx+C)(x)=x+1.$$

Simplify and group the terms:

$$Ax^2+A+Bx^2+Cx=x+1$$

$$(A+B)x^2+Cx+A=x+1$$

Now you have three equations:

$$A+B=0$$

$$C=1$$

$$A=1$$

You only have to solve for B: Since $A=1$, $B=-1$. Now rewrite the integral as:

$$\int\frac{1}{x}dx+\int\frac{-x+1}{x^2+1}dx.$$

Break this into:

$$\int\frac{1}{x}dx-\int\frac{x\,dx}{x^2+1}+\int\frac{1}{x^2+1}dx$$

The first integral is a natural logarithm, the second integral requires u-substitution, and the third integral is an inverse tangent. The answer is:

$$\ln|x|-\frac{1}{2}\ln(x^2+1)+\tan^{-1}(x)+C.$$

PROBLEM 3. Evaluate $\int\dfrac{6x^2-17x+25}{(x-1)(x-2)(x-3)}dx$.

Answer: You need to find A, B, and C such that:

$$\frac{A}{x-1}+\frac{B}{x-2}+\frac{C}{x-3}=\frac{6x^2-17x+25}{(x-1)(x-2)(x-3)}.$$

First, multiply through by $(x-1)(x-2)(x-3)$:

$$A(x - 2)(x - 3) + B(x - 1)(x - 3) + C(x - 1)(x - 2) = 6x^2 - 17x + 25.$$

Next, simplify and group the terms:

$$A(x^2 - 5x + 6) + B(x^2 - 4x + 3)x + C(x^2 - 3x + 2) = 6x^2 - 17x + 25$$

$$(A + B + C)x^2 + (-5A - 4B - 3C)x + (6A + 3B + 2C) = 6x^2 - 17x + 25$$

Now you have three simultaneous equations:

$$A + B + C = 6$$

$$-5A - 4B - 3C = -17$$

$$6A + 3B + 2C = 25$$

The three constants become: $A = 7$, $B = -15$, and $C = 14$. Thus, you can rewrite the integral as:

$$\int \frac{7}{x-1} dx - \int \frac{15}{x-2} dx + \int \frac{14}{x-3} dx.$$

These are all logarithmic integrals. The solution is:

$$7\ln |x-1| - 15 \ln|x-2| + 14\ln |x-3| + C.$$

PRACTICE PROBLEMS

Evaluate the following integrals. The answers are in chapter 21.

1. $\int \dfrac{x+4}{(x-1)(x+6)} dx$

2. $\int \dfrac{x}{(x-3)(x+1)} dx$

3. $\int \dfrac{1}{x^3 + x^2 - 2x} dx$

4. $\int \dfrac{2x+1}{x^2 - 7x + 12} dx$

5. $\int \dfrac{2x-1}{(x-1)^2} dx$

6. $\int \dfrac{1}{(x+1)(x^2+1)} dx$

7. $\int \dfrac{2x+1}{(x+1)(x^2+2)}\,dx$

8. $\int \dfrac{x^2+3x-1}{x^3-1}\,dx$

IMPROPER INTEGRALS

There's one last type of integral that shows up on the BC exam, but not on the AB. These are improper integrals, which are integrals evaluated over an open interval, rather than a closed one. More formally, an integral is improper if: (a) its integrand becomes infinite at one or more points in the interval of integration; or (b) one or both of the limits of integration is infinite.

We integrate an improper integral by evaluating the integral for a constant (*a*) and then taking the limit of the resulting integral as *a* approaches the limit in question. If the limit exists, the integral **converges** and the value of the integral is the limit. If the limit doesn't exist, the integral **diverges** and the integral cannot be evaluated there.

Confusing, huh. Let's clear the air with an example.

Example 1: Evaluate $\displaystyle\int_0^4 \sqrt{\dfrac{4+x}{4-x}}\,dx$.

At $x = 4$, the denominator is zero and the integrand is undefined. Therefore, replace the 4 in the limit of the integral with *a* and evaluate:

$$\lim_{a \to 4}\int_0^a \sqrt{\dfrac{4+x}{4-x}}\,dx.$$

Multiply the numerator and the denominator of the integrand by $\sqrt{4+x}$:

$$\int \sqrt{\dfrac{4+x}{4-x}}\,\dfrac{\sqrt{4+x}}{\sqrt{4+x}}\,dx = \int \dfrac{4+x}{\sqrt{16-x^2}}\,dx = \int \dfrac{4}{\sqrt{16-x^2}}\,dx + \int \dfrac{x}{\sqrt{16-x^2}}\,dx.$$

Evaluate the first integral by factoring 16 out of the denominator:

$$\int \dfrac{4\,dx}{4\sqrt{1-\left(\dfrac{x}{4}\right)^2}} = \int \dfrac{dx}{\sqrt{1-\left(\dfrac{x}{4}\right)^2}}.$$

Next, use *u*-substitution. Let $u = \dfrac{x}{4}$ and $du = \dfrac{1}{4}\,dx$:

$$\int \dfrac{4\,du}{\sqrt{1-u^2}} = 4\sin^{-1}u = 4\sin^{-1}\left(\dfrac{x}{4}\right)$$

Evaluate the second integral with u-substitution. Let $u = 16 - x^2$ and $du = -2xdx$:

$$-\frac{1}{2}\int u^{-\frac{1}{2}} = -u^{\frac{1}{2}} = -\sqrt{16-x^2}\,.$$

The final result is:

$$\int \frac{4}{\sqrt{16-x^2}}\,dx + \int \frac{x\,dx}{\sqrt{16-x^2}} = 4\sin^{-1}\frac{x}{4} - \sqrt{16-x^2}\,.$$

Now you have to evaluate the integral at the limits of integration:

$$4\sin^{-1}\frac{x}{4} - \sqrt{16-x^2}\,\Big|_0^a = \left(4\sin^{-1}\frac{a}{4} - \sqrt{16-a^2}\right) + (4).$$

Finally, take the limit as a tends to 4.

$$\lim_{a\to 4}\left(\left(4\sin^{-1}\frac{a}{4} - \sqrt{16-a^2}\right) + (4)\right) = \left(\left(4\sin^{-1}(1)\right) + (4)\right) = 2\pi + 4.$$

This may look complicated, but it was actually a very straightforward process. Because the integrand didn't exist at $x = 4$, replace 4 with a and then integrate. When you evaluated the limits of integration using a instead of 4, you then took the limit as a tends to 4.

Example 2: Evaluate $\displaystyle\int_0^\infty \frac{dx}{1+x^2}$.

Replace the upper limit of integration with a and take the limit as a tends to infinity:

$$\lim_{a\to\infty}\int_0^a \frac{dx}{1+x^2}\,.$$

Integrate:

$$\int_0^a \frac{dx}{1+x^2} = \tan^{-1}x\,\Big|_0^a = \tan^{-1}a\,.$$

Now, take the limit:

$$\lim_{a\to\infty}\tan^{-1}a = \frac{\pi}{2}\,.$$

Let's do an integral that becomes undefined in the middle of the limits of integration.

Example 3: Evaluate $\displaystyle\int_0^2 \frac{dx}{\sqrt[3]{x-1}}$.

Note that this integral becomes undefined at $x = 1$. Because of this, you have to divide the integral into two parts: one for $x < 1$ and one for $x > 1$:

$$\lim_{a\to1^-}\int_0^a \frac{dx}{\sqrt[3]{x-1}} \text{ and } \lim_{b\to1^+}\int_b^2 \frac{dx}{\sqrt[3]{x-1}}.$$

If either of these limits fails to exist, then the integral from 0 to 2 diverges. Evaluate them both:

$$\lim_{a\to1^-}\left[\frac{3}{2}(x-1)^{\frac{2}{3}}\Big|_0^a\right] + \lim_{b\to1^+}\left[\frac{3}{2}(x-1)^{\frac{2}{3}}\Big|_b^2\right].$$

$$= \lim_{a\to1^-}\left[\frac{3}{2}(a-1)^{\frac{2}{3}} - \frac{3}{2}\right] + \lim_{b\to1^+}\left[\frac{3}{2} - (b-1)^{\frac{2}{3}}\right] = -\frac{3}{2} + \frac{3}{2} = 0.$$

Because both limits exist and are finite, the integral converges and the value is 0.

There are several other facets of evaluating improper integrals that we won't explore here because they don't appear on the AP exam. You'll only have to do the most simple of these, usually one with infinity as a limit of integration.

Evaluate these integrals and check your work.

PROBLEM 1. Evaluate $\displaystyle\int_1^\infty \frac{dx}{x}$.

Answer: First, evaluate the integral:

$$\int_1^a \frac{dx}{x} = \ln x\Big|_1^a = \ln a.$$

Next, we evaluate the limit: $\displaystyle\lim_{a\to\infty} \ln a = \infty$.

Therefore $\displaystyle\int_1^\infty \frac{dx}{x} = \infty$, and the integral diverges.

PROBLEM 2. Evaluate $\displaystyle\int_1^\infty \frac{dx}{x^2}$.

Answer: Evaluate the integral:

$$\int_1^a \frac{dx}{x^2} = -\frac{1}{x}\Big|_1^a = -\frac{1}{a} + 1.$$

Next, evaluate the limit:

$$\lim_{a \to \infty}\left(-\frac{1}{a}+1\right)=1.$$

Therefore $\int_1^\infty \frac{dx}{x^2}=1$, and the integral converges.

PROBLEM 3. Evaluate $\int_0^1 \frac{dx}{\sqrt{1-x^2}}$.

Answer: Evaluate the integral:

$$\int_0^a \frac{dx}{\sqrt{1-x^2}}=\sin^{-1}x\Big|_0^a=\sin^{-1}(a)-\sin^{-1}(0)=\sin^{-1}(a).$$

Next evaluate:

$$\lim_{a \to 1^-}\left(\sin^{-1}a\right)=\frac{\pi}{2}.$$

Therefore $\int_0^1 \frac{dx}{\sqrt{1-x^2}}=\frac{\pi}{2}$, and the integral converges.

PROBLEM 4. Evaluate $\int_{-2}^2 \frac{dx}{x^2}$.

Answer: The integrand becomes undefined at $x = 0$, so you have to divide the integral into:

$$\lim_{a \to 0^-}\int_{-2}^a \frac{dx}{x^2} \text{ and } \lim_{b \to 0^+}\int_b^2 \frac{dx}{x^2}.$$

Now, evaluating both integrals:

$$\lim_{a \to 0^-}\left[-\frac{1}{x}\Big|_{-2}^a\right]+\lim_{b \to 0^+}\left[-\frac{1}{x}\Big|_b^2\right]=\lim_{a \to 0^-}\left[-\frac{1}{a}-\frac{1}{2}\right]+\lim_{b \to 0^+}\left[-\frac{1}{2}+\frac{1}{b}\right]=\infty.$$

The integral diverges.

PRACTICE PROBLEMS

Evaluate the following integrals. The answers are in chapter 21.

1. $\displaystyle\int_0^1 \frac{dx}{\sqrt{x}}$.

2. $\displaystyle\int_{-1}^1 \frac{dx}{x^{2/3}}$.

3. $\displaystyle\int_{-\infty}^0 e^x\,dx$.

4. $\displaystyle\int_{-\infty}^{\frac{\pi}{2}} \tan\theta\,d\theta$.

5. $\displaystyle\int_1^4 \frac{dx}{1-x}$.

6. $\displaystyle\int_0^1 \frac{x+1}{\sqrt{x^2+2x}}\,dx$.

7. $\displaystyle\int_{-\infty}^0 \frac{dx}{(2x-1)^3}$.

8. $\displaystyle\int_{-\infty}^{\infty} x^3\,dx$.

9. $\displaystyle\int_0^3 \frac{dx}{x-2}$.

10. $\displaystyle\int_{-1}^8 \frac{dx}{\sqrt[3]{x}}$.

19

DIFFERENTIAL
EQUATIONS

There are many types of differential equations, but only a very small number of them appear on the AP exam. There are courses devoted to learning how to solve a wide variety of differential equations, but AP calculus provides only a very basic introduction to the topic.

If you're given an equation in which the derivative of a function is equal to some other function, you can determine the original function by integrating both sides of the equation and then solving for the constant term.

Example 1: If $\dfrac{dy}{dx} = \dfrac{4x}{y}$ and $y(0) = 5$, find an equation for y in terms of x.

The first step in solving these is to put all of the terms that contain y on the left side of the equals sign and all of the terms that contain x on the right side. We then have: $y \, dy = 4x \, dx$. The second step is to integrate both sides:

$$\int y \, dy = \int 4x \, dx.$$

And then you integrate:

$$\frac{y^2}{2} = 2x^2 + C.$$

You're not done yet. The final step is to solve for the constant by plugging in $x = 0$ and $y = 5$:

$$\frac{5^2}{2} = 2(0^2) + C, \text{ so } C = \frac{25}{2}. \text{ The solution is } \frac{y^2}{2} = 2x^2 + \frac{25}{2}.$$

That's all there is to it. Separate the variables, integrate both sides, and solve for the constant.

EXPONENTIAL GROWTH

Often, the equation will involve a logarithm. Let's do an example.

Example 2: If $\dfrac{dy}{dx} = 3x^2 y$ and $y(0) = 2$, find an equation for y in terms of x.

First, put the y terms on the left and the x terms on the right:

$$\frac{dy}{y} = 3x^2 dx.$$

Next, integrate both sides:

$$\int \frac{dy}{y} = \int 3x^2 dx.$$

The result is: $\ln y = x^3 + C$.

It's customary to solve this equation for y. You can do this by putting both sides into exponent form:

$$y = e^{x^3 + C}.$$

This can be rewritten as $y = e^{x^3} e^c$ and, because e^c is a constant, the equation becomes:

$$y = Ce^{x^3}.$$

This is the preferred form of the equation. Now, solve for the constant. Plug in $x = 0$ and $y = 2$, and you get $2 = Ce^0$.

Because $e^0 = 1$, $C = 2$. The solution is $y = 2e^{x^3}$.

This is the typical differential equation that you'll see on the AP exam. The other common problem type involves position, velocity, and acceleration. We did several problems of this type in chapter 8, before you knew how to use integrals. In a sample problem, you're given the velocity and acceleration and told to find distance (the reverse of what we did before).

Example 3: If the acceleration of a particle is given by $a(t) = -32$ ft/sec^2, and the velocity of the particle is 64 ft/sec and the height of the particle is 32 ft at time $t = 0$, find: (a) the equation of the particle's velocity at time t; (b) the equation for the particle's height, h, at time t; and (c) the maximum height of the particle.

Part A: Because acceleration is the rate of change of velocity with respect to time, you can write that $\dfrac{dv}{dt} = -32$. Now separate the variables and integrate both sides:

$$\int dv = \int -32 \, dt.$$

Integrating this expression, we get $v = -32t + C$. Now we can solve for the constant by plugging in $t = 0$ and $v = 64$. We get: $64 = -32(0) + C$ and $C = 64$. Thus, velocity is $v = -32t + 64$.

Part B: Because velocity is the rate of change of displacement with respect to time, you can write that

$$\frac{dh}{dt} = -32t + 64.$$

Separate the variables and integrate both sides:

$$\int dh = \int (-32t + 64) \, dt.$$

Integrate the expression: $h = -16t^2 + 64t + C$. Now solve for the constant by plugging in $t = 0$ and $h = 32$:

$$32 = -16(0^2) + 64(0) + C \text{ and } C = 32.$$

Thus, the equation for height is $h = -16t^2 + 64t + 32$.

Part C: In order to find the maximum height, you need to take the derivative of the height with respect to time and set it equal to zero. Notice that the derivative of height with respect to time is the velocity; just set the velocity equal to zero and solve for t:

$$-32t + 64 = 0, \text{ so } t = 2.$$

Thus, at time $t = 2$, the height of the particle is a maximum. Now, plug $t = 2$ into the equation for height:

$$h = -16(2)^2 + 64(2) + 32 = 96.$$

Therefore, the maximum height of the particle is 96 feet.

You have now seen all that the AP tests on differential equations (also known as "Diff-EQ's"). Fortunately, the AP barely scratches the surface of the subject, so if you can handle the questions in this section, you'll be ready for any differential equation question the AP asks.

Here are some solved problems. Do each problem, covering the answer first, then checking your answer.

PROBLEM 1. If $\dfrac{dy}{dx} = \dfrac{3x}{2y}$ and $y(0) = 10$, find an equation for y in terms of x.

Answer: First, separate the variables: $2y \, dy = 3x \, dx$. Then, we take the integral of both sides: $\displaystyle\int 2y \, dy = \int 3x \, dx$

Next, integrate both sides:

$$y^2 = \frac{3x^2}{2} + C.$$

Finally, solve for the constant:

$$10^2 = \frac{3(0)^2}{2} + C \text{ so } C = 100.$$

The solution is $y^2 = \dfrac{3x^2}{2} + 100$.

PROBLEM 2. If $\dfrac{dy}{dx} = 4xy^2$ and $y(0) = 1$, find an equation for y in terms of x.

Answer: First, separate the variables: $\dfrac{dy}{y^2} = 4x \, dx$ and take the integral of both sides:

$$\int \frac{dy}{y^2} = \int 4x \, dx.$$

Next, integrate both sides: $-\dfrac{1}{y} = 2x^2 + C$. You can rewrite this as: $y = -\dfrac{1}{2x^2 + C}$.

Finally, solve for the constant:

$$1 = -\frac{1}{2(0)^2 + C} = \frac{1}{C}, \text{ so } C = -1.$$

The solution is $y = -\dfrac{1}{2x^2 - 1}$.

PROBLEM 3. If $\dfrac{dy}{dx} = \dfrac{y^2}{x}$ and $y(1) = \dfrac{1}{3}$, find an equation for y in terms of x.

Answer: This time, separating the variables gives us this: $\dfrac{dy}{y^2} = \dfrac{dx}{x}$.

Then, take the integral of both sides: $\displaystyle\int \dfrac{dy}{y^2} = \int \dfrac{dx}{x}$

Next, integrate both sides: $-\dfrac{1}{y} = \ln x + C$ and rearrange:

$$y = \dfrac{-1}{\ln x + C}.$$

Finally, solve for the constant: $\dfrac{1}{3} = \dfrac{-1}{C}$, so $C = -3$. The solution is $y = \dfrac{-1}{\ln x - 3}$.

PROBLEM 4. A city had a population of 10,000 in 1980 and 13,000 in 1990. Assuming an exponential growth rate, estimate the city's population in 2000.

Answer: The phrase "exponential growth rate" means that $\dfrac{dy}{dt} = ky$, where k is a constant. Take

the integral of both sides: $\displaystyle\int \dfrac{dy}{y} = \int k \, dt$.

Then, integrate both sides: ($\ln y = kt + C$) and put both sides in exponential form:
$$y = e^{kt+c} = Ce^{kt}.$$

Next, use the information about the population to solve for the constants. If you treat 1980 as $t = 0$ and 1990 as $t = 10$, then:

$$10{,}000 = Ce^{k(0)} \text{ and } 13{,}000 = Ce^{k(10)}.$$

So, $C = 10{,}000$ and $k = \dfrac{1}{10}\ln 1.3 \approx .0262$.

The equation for population growth is approximately $y = 10{,}000e^{.0262t}$. We can estimate that the population in 2000 will be:

$$y = 10{,}000e^{.0262(20)} = 16{,}900.$$

EULER'S METHOD

This is a new topic on the BC exam. As far as we can tell, there will probably be one question that will require you to know Euler's method, so if you find this confusing, skip it. It won't make any real difference in your score.

THE METHOD

You are going to use your calculator to find an approximate answer to a differential equation. The method is quite simple. First, you need a starting point and an initial slope.

Next, we use increments of h to come up with approximations. Each new approximation will use the following rules

$$x_n = x_{n-1} + h$$

$$y_n = y_{n-1} + h \cdot y'_{n-1}$$

Repeat for $n = 1, 2, 3, \ldots$

This is much easier to understand if we do an example.

Example 1: Use Euler's Method, with $h = 0.2$, to estimate $y(1)$ if $y' = y - 2$ and $y(0) = 4$.

We are given that the curve goes through the point $(0, 4)$. We will call the coordinates of this point $x_0 = 0$ and $y_0 = 4$. The slope is found by plugging $y_0 = 4$ into $y' = y - 2$, so we have an initial slope of $y'_0 = 4 - 2 = 2$.

Now we need to find the next set of points.

Step 1: Increase x_0 by h to get x_1.

$$x_1 = 0.2$$

Step 2: Multiply h by y'_0 and add to y_0 to get y_1.

$$y_1 = 4 + 0.2(2) = 4.4$$

Step 3: Find y'_1 by plugging y_1 into the equation for y'

$$y'_1 = 4.4 - 2 = 2.4$$

Repeat until you get to the desired point (in this case $x = 1$).

Step 1: Increase x_1 by h to get x_2.

$$x_2 = 0.4$$

Step 2: Multiply h by y'_1 and add to y_1 to get y_2.

$$y_2 = 4.4 + 0.2(2.4) = 4.88$$

Step 3: Find y'_2 by plugging y_2 into the equation for 9

$$y'_2 = 4.88 - 2 = 2.88$$

Step 1: $x_3 = x_2 + h$.

$$x_3 = 0.6$$

Step 2: $y_3 = y_2 + h(y_2')$

$$y_3 = 4.88 + 0.2(2.88) = 5.456$$

Step 3: $y_3' = y_3 - 2$

$$y_3' = 5.456 - 2 = 3.456$$

Step 1: $x_4 = x_3 + h$

$$x_4 = 0.8$$

Step 2: $y_4 = y_3 + h(y_3')$

$$y_4 = 5.456 + 0.2(3.456) = 6.1472$$

Step 3: $y_4' = y_4 - 2$

$$y_4' = 6.1472 - 2 = 4.1472$$

Step 1: $x_5 = x_4 + h$

$$x_5 = 1.0$$

Step 2: $y_5 = y_4 + h(y_4')$

$$y_5 = 6.1472 + 0.2(4.1472) = 6.97644$$

We don't need to go any farther because we are asked for the value of y when $x = 1$.

The answer is $y = 6.97644$.

Let's do another example.

Example 2: Use Euler's method, with $h = 0.1$, to estimate $y(0.5)$ if $y' = y - 1$ and $y(0) = 3$.

We start with $x_0 = 0$ and $y_0 = 3$. The slope is found by plugging $y_0 = 3$ into $y' = y - 1$, so we have an initial slope of $y_0' = 3 - 1 = 2$.

Step 1: Increase x_0 by h to get x_1.

$$x_1 = 0.1$$

Step 2: Multiply h by y_0' and add to y_0 to get y_1.

$$y_1 = 3 + 0.1(2) = 3.2$$

Step 3: Find y_1' by plugging y_1 into the equation for y'

$$y_1' = 3.2 - 1 = 2.2$$

Step 1: $x_2 = x_1 + h$.

$$x_2 = 0.2$$

Step 2: $y_2 = y_1 + h(y_1')$

$$y_2 = 3.2 + 0.1(2.2) = 3.42$$

Step 3: $y_2' = y_2 - 1$

$$y_2' = 3.42 - 1 = 2.42$$

Step 1: $x_3 = x_2 + h$.

$$x_3 = 0.3$$

Step 2: $y_3 = y_2 + h(y_2')$

$$y_3 = 3.42 + 0.1(2.42) = 3.662$$

Step 3: $y_3' = y_3 - 1$

$$y_3' = 3.662 - 1 = 2.662$$

Step 1: $x_4 = x_3 + h$.

$$x_4 = 0.4$$

Step 2: $y_4 = y_3 + h(y_3')$

$$y_4 = 3.662 + 0.1(2.662) = 3.9282$$

Step 3: $y_4' = y_4 - 1$

$$y_4' = 3.9282 - 1 = 2.9282$$

Step 1: $x_5 = x_4 + h$.

$$x_5 = 0.5$$

Step 2: $y_5 = y_4 + h(y_4')$

$$y_5 = 3.9282 + 0.1(2.9282) = 4.22102$$

The answer is $y = 4.22102$.

Now try these on your own. Do each problem first with the answer covered, then check your answer.

Problem 1: Use Euler's method, with $h = 0.2$, to estimate $y(2)$ if $y' = 2y + 1$ and $y(1) = 5$.

We start with $x_0 = 1$ and $y_0 = 5$. The slope is found by plugging $y_0 = 5$ into $y' = 2y + 1$, so we have an initial slope of $y_0' = 2(5) + 1 = 11$.

Step 1: Increase x_0 by h to get x_1.

$$x_1 = 1.2$$

Step 2: Multiply h by y_0' and add to y_0 to get y_1.

$$y_1 = 5 + 0.2(11) = 7.2$$

Step 3: Find y_1' by plugging y_1 into the equation for y'

$$y_1' = 2(7.2) + 1 = 15.4$$

Step 1: $x_2 = x_1 + h$.

$$x_2 = 1.4$$

Step 2: $y_2 = y_1 + h(y_1')$

$$y_2 = 7.2 + 0.2(15.4) = 10.28$$

Step 3: $y_2' = 2y_2 + 1$

$$y_2' = 2(10.28) + 1 = 21.56$$

Step 1: $x_3 = x_2 + h$.

$$x_3 = 1.6$$

Step 2: $y_3 = y_2 + h(y_2')$

$$y_3 = 10.28 + 0.2(21.56) = 14.592$$

Step 3: $y_3' = 2y_3 + 1$

$$y_3' = 2(14.592) + 1 = 30.184$$

Step 1: $x_4 = x_3 + h$.

$$x_4 = 1.8$$

Step 2: $y_4 = y_3 + h(y_3')$

$$y_4 = 14.592 + 0.2(30.184) = 20.6288$$

Step 3: $y_4' = 2y_4 + 1$

$$y_4' = 2(20.6288) + 1 = 42.2576$$

Step 1: $x_5 = x_4 + h$.

$$x_5 = 2$$

Step 2: $y_5 = y_4 + h(y_4')$

$$y_5 = 20.6288 + 0.2(42.2576) = 29.08032$$

The answer is $y = 29.08032$.

Problem 2: Use Euler's Method, with $h = 0.1$, to estimate $y(0.5)$ if $y' = y^2 + 1$ and $y(0) = 0$.

We start with $x_0 = 0$ and $y_0 = 0$. The slope is found by plugging $y_0 = 0$ into $y' = y^2 + 1$, so we have an initial slope of $y_0' = 1$.

Step 1: Increase x_0 by h to get x_1.

$$x_1 = 0.1$$

Step 2: Multiply h by y_0' and add to y_0 to get y_1.

$$y_1 = 0 + 0.1(1) = 0.1$$

Step 3: Find y_1' by plugging y_1 into the equation for y'

$$y_1' = (0.1)^2 + 1 = 1.01$$

Step 1: $x_2 = x_1 + h$.

$$x_2 = 0.2$$

Step 2: $y_2 = y_1 + h(y_1')$

$$y_2 = 0.1 + 0.1(1.01) = 0.201$$

Step 3: $y_2' = (y_2)^2 + 1$

$$y_2' = (0.201)^2 + 1 = 1.040$$

Step 1: $x_3 = x_2 + h$.

$$x_3 = 0.3$$

Step 2: $y_3 = y_2 + h(y_2')$

$$y_3 = 0.201 + 0.1(1.040) = 0.305$$

Step 3: $y_3' = (y_3)^2 + 1$

$$y_3' = (0.305)^2 + 1 = 1.093$$

Step 1: $x_4 = x_3 + h$.

$$x_4 = 0.4$$

Step 2: $y_4 = y_3 + h(y_3')$

$$y_4 = 0.305 + 0.1(1.093) = 0.414$$

Step 3: $y_4' = (y_4)^2 + 1$

$$y_4' = (0.414)^2 + 1 = 1.171$$

Step 1: $x_5 = x_4 + h$.

$$x_5 = 0.5$$

Step 2: $y_5 = y_4 + h(y_4')$

$$y_5 = 0.414 + 0.1(1.171) = 0.531$$

The answer is $y = 0.531$.

PRACTICE PROBLEMS

1. Use Euler's Method, with $h = 0.25$, to estimate $y(1)$ if $y' = y - x$ and $y(0) = 2$.

2. Use Euler's Method, with $h = 0.2$, to estimate $y(1)$ if $y' = -y$ and $y(0) = 1$.

3. Use Euler's Method, with $h = 0.1$, to estimate $y(0.5)$ if $y' = 4x^3$ and $y(0) = 0$.

SOLUTIONS

1. 4.441

2. 0.328

3. 0.04

PRACTICE PROBLEMS

Now try these problems. The answers are in chapter 21.

1. If $\dfrac{dy}{dx} = \dfrac{7x^2}{y^3}$ and $y(3) = 2$, find an equation for y in terms of x.

2. If $\dfrac{dy}{dx} = 5x^2\,y$ and $y(0) = 6$, find an equation for y in terms of x.

3. If $\dfrac{dy}{dx} = \dfrac{1}{y + x^2 y}$ and $y(0) = 2$, find an equation for y in terms of x .

4. If $\dfrac{dy}{dx} = \dfrac{e^x}{y^2}$ and $y(0) = 1$, find an equation for y in terms of x.

5. If $\dfrac{dy}{dx} = \dfrac{y^2}{x^3}$ and $y(1) = 2$, find an equation for y in terms of x.

6. If $\dfrac{dy}{dx} = \dfrac{\sin x}{\cos y}$ and $y(0) = \dfrac{3\pi}{2}$, find an equation for y in terms of x.

7. A colony of bacteria grows exponentially and the colony's population is 4,000 at time $t = 0$ and 6,500 at time $t = 3$. How big is the population at time $t = 10$?

8. A rock is thrown upward with an initial velocity, $v(t)$, of 18 m/s from a height, $h(t)$, of 45m. If the acceleration of the rock is a constant -9 m/s², find the height of the rock at time $t = 4$.

9. The rate of growth of the volume of a sphere is proportional to its volume. If the volume of the sphere is initially 36π ft³, and expands to 90π ft³ after 1 second, find the volume of the sphere after 3 seconds.

10. A radioactive element decays exponentially proportionally to its mass. One-half of its original amount remains after 5,750 years. If 10,000 grams of the element are present initially, how much will be left after 1,000 years?

20 INFINITE SERIES

The study of infinite series is another important and complex calculus topic. Fortunately for you, it's another minor part of BC calculus. However, you will be asked to do a few different things with series.

THE SEQUENCE

First, let's learn some terminology. A **sequence** of numbers is an infinite succession of numbers that

follow a pattern: For example: $1, 2, 3, 4, \ldots$; or $\dfrac{1}{2}, \dfrac{1}{3}, \dfrac{1}{4} \ldots$; or $1, -1, 1, -1 \ldots$

Terms in a sequence are usually denoted with a subscript; the first term of a sequence is a_1, the second term of a sequence is a_2, the nth term of a sequence is a_n, and so on.

Usually, the terms in a sequence are generated by a formula. For example, the sequence $a_n = \dfrac{n-1}{n}$ beginning with $n=1$ is: $0, \dfrac{1}{2}, \dfrac{2}{3}, \dfrac{3}{4}, \ldots$. This is found by plugging in 1 for n, then 2, then 3, and so on. Notice that n is always an integer.

A sequence converges to a number if, as n gets bigger, the terms get closer to a certain number; that number is the limit of the sequence. Here's the official math jargon:

A sequence has a limit L if, for any $\varepsilon > 0$ there is an associated positive integer N such that $|a_n - L| < \varepsilon$ for all $n \geq N$. If so, the sequence **converges** to L and we write: $\lim_{n \to \infty} a_n = L$.

If the sequence has no finite limit, it **diverges**.

For example: the sequence $a_n = \dfrac{n-1}{n}$ converges, because $\lim_{n \to \infty} \dfrac{n-1}{n} = 1$. As n gets bigger, the terms of the sequence get closer and closer to 1. The sequence $a_2 = 2^n$ diverges, because $\lim_{n \to \infty} 2^n = \infty$. As n gets bigger, the terms of the sequence get bigger.

THE SERIES

Now that we've defined a sequence, we can define a series. A **series** is an expression of the form $a_1 + a_2 + a_3 + a_4 + \ldots a_n + \ldots$ If the series stops at some final term a_n, the series is finite. If the series continues indefinitely, it's an infinite series. (Makes sense, right?) These are the ones that primarily show up on the AP. Since a series is the sum of all the terms of a sequence, we often use sigma notation like this:

$\sum_{n=1}^{\infty} a_n$. The Harmonic Series, for example, looks like this:

$$\sum_{n=1}^{\infty} \frac{1}{n} = 1 + \frac{1}{2} + \frac{1}{3} + \frac{1}{4} + \ldots$$

A **partial sum** of a series is the sum of the series up to a particular value of n. If the sequence of partial sums of a series converges to a limit L, then the series is said to converge to that limit L, and we write:

$$\sum_{n=1}^{\infty} a_n = L$$

If there is no such limit, (like the Harmonic Series, for example) then the series diverges.

Example 1: Does the series $\sum_{n=1}^{\infty} \dfrac{1}{3^n}$ converge and, if so, to what value?

The nth partial sum of this series is $\dfrac{1}{3} + \dfrac{1}{3^2} + \dfrac{1}{3^3} + \ldots \dfrac{1}{3^n}$. You can figure out the limit using the following method. First, write it out:

$$S_n = \frac{1}{3} + \frac{1}{3^2} + \frac{1}{3^3} + \dots \frac{1}{3^n}.$$

Next, multiply through by $\frac{1}{3}$:

$$\frac{1}{3}S_n = \frac{1}{3^2} + \frac{1}{3^3} + \dots \frac{1}{3^{n+1}}.$$

If you now subtract the second expression from the first, you get:

$$\frac{2}{3}S_n = \frac{1}{3} - \frac{1}{3^{n+1}}.$$

Now, multiply through by $\frac{3}{2}$:

$$S_n = \frac{3}{2}\left(\frac{1}{3} - \frac{1}{3^{n+1}}\right).$$

Finally, take the limit:

$$\lim_{n \to \infty} \frac{3}{2}\left(\frac{1}{3} - \frac{1}{3^{n+1}}\right), \text{ we get } \frac{3}{2}\left(\frac{1}{3} - 0\right) = \frac{1}{2}.$$

The series converges to $\frac{1}{2}$.

THE GEOMETRIC SERIES

The problem above contains an example of a series called a geometric series. Generally, this series takes the following form:

$$a + ar + ar^2 + ar^3 + \dots + ar^{n-1} + \dots = \sum_{n=1}^{\infty} ar^{n-1}.$$

These series often show up on the AP exam. You're usually asked to determine whether the series converges or diverges, and if it converges, to what limit. The test for convergence of a geometric series is very simple:

If $|r| < 1$ the series converges.

If $|r| \geq 1$, the series diverges.

So, for example, the series $\dfrac{1}{2} + \dfrac{1}{2^2} + \dfrac{1}{2^3} + \dots$ converges, whereas the series $2 + 2^2 + 2^3 + \dots$ diverges.

If a geometric series converges, you can always figure out its sum by doing the following:

$$\text{Let } S = a + ar + ar^2 + ar^3 + \dots ar^{n-1}$$

Multiply the expression by r:

$$rS = ar + ar^2 + ar^3 + \dots ar^n.$$

Now subtract the second expression from the first:

$$S - rS = a - ar^n.$$

Factor S out of the left side:

$$S(1 - r) = a - ar^n.$$

Divide through by $1 - r$:

$$S = \frac{a - ar^n}{(1-r)} = \frac{a(1-r^n)}{(1-r)}.$$

Thus, if we want to find the sum of the first n terms of a geometric series, we use the following formula:

$$S_n = \frac{a(1-r^n)}{(1-r)}$$

For example, the sum of the first four terms of the series in example 1 is:

$$S_4 = \frac{\dfrac{1}{3}\left(1-\left(\dfrac{1}{3}\right)^4\right)}{\left(1-\dfrac{1}{3}\right)} = \frac{40}{81}.$$

For an infinite geometric series, if it converges, $\lim\limits_{n\to\infty} r^n = 0$, and its sum is found this way:

$$S = \frac{a}{1-r}$$

In Example 1 above, the first term is $a = \dfrac{1}{3}$, and $r = \dfrac{1}{3}$. The sum is: $\dfrac{\dfrac{1}{3}}{1-\dfrac{1}{3}} = \dfrac{1}{2}$.

Many students find it is easier to work through the derivation of the sum than to actually memorize the formula. Either way, you should always be able to find the sum of an infinite geometric series.

Example 2: Find the limit to which the series $2 + \dfrac{2}{5} + \dfrac{2}{25} + \dfrac{2}{125} + \ldots$ converges.

The first term is $a = 2$ and $r = \dfrac{1}{5}$, so the limit is $\dfrac{2}{1 - \dfrac{1}{5}} = \dfrac{5}{2}$.

The AP's treatment of series doesn't get very complicated, and the exam usually confines itself to finding the limit of an infinite geometric series. Just remember that if $|r| \geq 1$, the geometric series diverges and it has no limit. No exceptions.

THE RATIO TEST

Sometimes you'll simply be asked to determine whether a series converges. With a geometric series, you already know what to do. If the series isn't a geometric one, you can usually use the ratio test:

Let $\sum a_n$ be a series where all of the terms are positive, and suppose that

$$\lim_{n \to \infty} \frac{a_{n+1}}{a_n} = \rho.$$

Then:

(a) If $\rho < 1$, the series converges.

(b) If $\rho > 1$, the series diverges.

(c) If $\rho = 1$, the test provides insufficient information and the series might converge or diverge.

Example 3: Determine whether the series $\left| \sum\limits_{n=1}^{\infty} \dfrac{n}{3^n} \right|$ converges.

Use the ratio test:

$$\rho = \lim_{n \to \infty} \frac{\dfrac{n+1}{3^{n+1}}}{\dfrac{n}{3^n}} = \lim_{n \to \infty} \frac{n+1}{3^{n+1}} \frac{3^n}{n} = \lim_{n \to \infty} \frac{n+1}{n} \frac{3^n}{3^{n+1}} = \frac{1}{3}$$

Because $\rho < 1$, the series converges. Notice that it does <u>not</u> converge to the limit $\dfrac{1}{3}$. That value simply tells us that the series is a convergent one.

Example 4: Determine whether the series $\sum_{n=1}^{\infty} \dfrac{1}{3n}$ converges.

By the ratio test, you discover that:

$$\rho = \lim_{n \to \infty} \dfrac{\dfrac{1}{3(n+1)}}{\dfrac{1}{3n}} = \lim_{n \to \infty} \dfrac{1}{3(n+1)} \dfrac{3n}{1} = \lim_{n \to \infty} \dfrac{3n}{3(n+1)} = 1.$$

Because $\rho = 1$, the test is insufficient and we don't know whether the series converges or diverges.

ALTERNATING SERIES

The next type of series to concern yourself with is the alternating series. This is a series whose terms are alternately positive and negative. For example,

$$1 - \dfrac{1}{2} + \dfrac{1}{3} - \dfrac{1}{4} + \ldots \dfrac{(-1)^{n+1}}{n} + \ldots.$$

Usually, you'll be asked to determine whether such a series converges. First, test the following:

The series $\sum_{n=1}^{\infty} (-1)^{n+1} b_n$ converges if all three of the following conditions are satisfied:

(a) $b_n > 0$ (which means that the terms must alternate in sign).

(b) $b_n > b_{n+1}$ for all n

(c) $b_n \to 0$ as $n \to \infty$

Example 5: Determine whether the series $1 - \dfrac{1}{2} + \dfrac{1}{3} - \dfrac{1}{4} + \ldots \dfrac{(-1)^{n+1}}{n} + \ldots$ converges.

Let's check the criteria:

(a) Each of the b_n's is positive.

(b) Each b_n is greater than its succeeding b_{n+1}.

(c) The b_n's are tending to zero as n approaches infinity.

Therefore, the series converges. This series is called the alternating harmonic series (a nifty combination of the harmonic and alternating series). Notice that this series converges, whereas the harmonic series diverges.

ABSOLUTE CONVERGENCE

Usually, you'll be asked to determine whether a series converges "absolutely".

A series $\sum a_n$ *converges absolutely* if the corresponding series $\sum |a_n|$ converges.

POWER SERIES

The most important type of series, and the one that the AP concentrates on, is the power series, which takes the form:

$$\sum_{n=0}^{\infty} c_n x^n = c_0 + c_1 x + c_2 x^2 + c_3 x^3 + \ldots + c_n x^n + \ldots.$$

Notice that we're now going from $n = 0$ to $n = \infty$, rather than from $n = 1$ to $n = \infty$. This is because the first term is a constant.

You'll be asked to determine whether a power series converges. The test is as follows:

For any power series in x, exactly one of the following is true:

(a) The series converges only for $x = 0$;

(b) The series converges absolutely for all x; or

(c) The series converges absolutely for all x in some open interval $(-R, R)$, and diverges if $x < -R$ or $x > R$. The series may converge or diverge at the endpoints of the interval.

For case (c), the number R is called the **radius of convergence**. In case (a), the radius of convergence is zero; in case (b) the radius is all x. The set of all values of x in the interval $(-R, R)$ is called the **interval of convergence**.

Example 6: Find the radius and interval of convergence for the power series $\sum_{n=0}^{\infty} x^n$.

This is a geometric series with $r = x$; thus, it converges if $r < 1$.
Therefore, the radius of convergence is R = 1, and the interval of convergence is $(-1, 1)$.

Example 7: Find the radius and interval of convergence for the power series $\displaystyle\sum_{n=0}^{\infty}\frac{x^n}{n!}$.

In order to determine convergence, apply the ratio test for absolute convergence:

$$\rho=\lim_{n\to\infty}\left|\frac{x^{n+1}}{(n+1)!}\frac{n!}{x^n}\right|=\lim_{n\to\infty}\left|\frac{x}{n+1}\right|=0.$$

Because $\rho=0$ for all x, the series converges absolutely for all x. Therefore, the radius of convergence is $R=\infty$ and the interval of convergence is $(-\infty,\infty)$.

TAYLOR SERIES

The final, and most important, topic in infinite series is the Taylor series. The power series and the Taylor series are the most frequent question topics on infinite series on the AP exam. A Taylor series expansion about a point $x=a$ is a power series expansion that's useful to approximate the function in the neighborhood of the point $x=a$.

If a function f has derivatives of all orders at a, then the Taylor series for f about $x=a$ is:

$$\sum_{k=0}^{\infty}\frac{f^{(k)}(a)}{k!}(x-a)^k=f(a)+f'(a)(x-a)+\frac{f''(a)}{2!}(x-a)^2+\ldots+\frac{f^{(n)}(a)}{n!}(x-a)^n+\ldots$$

The special case of the Taylor series for $a=0$ is called the **Maclaurin series:**

$$\sum_{k=0}^{\infty}\frac{f^{(k)}(0)}{k!}(x)^k=f(0)+f'(0)x+\frac{f''(0)}{2!}x^2+\ldots+\frac{f^{(n)}(0)}{n!}x^n+\ldots$$

The AP usually asks for a Taylor series at $x=0$. Note that this is the same thing as the Maclaurin series. Let's do an example to see what one of these looks like.

Example 8: Find the Taylor series about $a=1$ generated by $f(x)=\dfrac{1}{x}$.

First, find the derivatives of $\dfrac{1}{x}$ and compute their values at $x=1$:

$$
\begin{aligned}
f(x)&=x^{-1} & f(1)&=1\\
f'(x)&=-x^{-2} & f'(1)&=-1\\
f''(x)&=2x^{-3} & f''(1)&=2=2!\\
f'''(x)&=-6x^{-4} & f'''(1)&=-6=-3!\\
f^{(4)}(x)&=24x^{-5} & f^{(4)}(1)&=24=4!\\
f^{(n)}(x)&=(-1)^n\frac{n!}{x^{n+1}} & f^{(n)}(1)&=(-1)^n n!
\end{aligned}
$$

Next, plug into the formula to generate the Taylor series:

$$1-(x-1)+(x-1)^2+...=\sum_{k=0}^{\infty}(-1)^k(x-1)^k.$$

Example 9: Find the Taylor series about $a = 0$ generated by $f(x) = e^x$.

First, find the derivatives of e^x and compute their values at $x = 0$.

$$f(x) = e^x \qquad\qquad f(0) = 1$$
$$f'(x) = e^x \qquad\qquad f'(0) = 1$$
$$f''(x) = e^x \qquad\qquad f''(0) = 1$$
$$f'''(x) = e^x \qquad\qquad f'''(0) = 1$$
$$f^{(4)}(x) = e^x \qquad\qquad f^{(4)}(0) = 1$$
$$f^{(n)}(x) = e^x \qquad\qquad f^{(n)}(0) = 1$$

Next, plug into the formula to generate the Taylor series:

$$\sum_{k=0}^{\infty}\frac{x^k}{k!}=1+x+\frac{x^2}{2!}+\frac{x^3}{3!}+\frac{x^4}{4!}+...$$

This is a very important expansion to know. Memorize it and be able to use it on the AP. You're permitted to write the Taylor series expansion for e^x without explaining how to generate it.

That's all you need to know about Taylor series. There are four expansions you should prepare for on the AP. We already discussed the expansion for e^x. You'll also need to know the expansions for $\sin x$, $\cos x$, and $\ln(1 + x)$. You'll see them in the forthcoming practice problems.

PROBLEM 1. Find the sum of the series $4+\dfrac{4}{3}+\dfrac{4}{9}+\dfrac{4}{27}+...+\dfrac{4}{3^6}$.

Answer: This is a geometric series where $a = 4$, $r = \dfrac{1}{3}$, and $n = 7$. According to the formula, the sum is:

$$S_n = \frac{a(1-r^n)}{(1-r)}$$

Plugging in, you should get:

$$S_7 = \frac{4\left(1-\left(\dfrac{1}{3}\right)^7\right)}{\left(1-\dfrac{1}{3}\right)} = 5.997.$$

PROBLEM 2. Find the limit to which the series $4 + \dfrac{4}{3} + \dfrac{4}{9} + \dfrac{4}{27} + \ldots$ converges.

Answer: This is the same series, but now you want the limit at infinity, so use the formula:

$$S = \frac{a}{1-r}.$$

The result is:

$$S = \frac{4}{1 - \dfrac{1}{3}} = 6$$

PROBLEM 3. Determine whether the series $\displaystyle\sum_{n=1}^{\infty} \dfrac{n-1}{5^n}$ converges.

Answer: By the ratio test:

$$\rho = \lim_{n \to \infty} \frac{\dfrac{n}{5^{n+1}}}{\dfrac{n-1}{5^n}} = \lim_{n \to \infty} \frac{n}{5^{n+1}} \frac{5^n}{n-1} = \lim_{n \to \infty} \frac{n}{n-1} \frac{5^n}{5^{n+1}} = \frac{1}{5}$$

Because $\rho < 1$, the series converges.

PROBLEM 4. Find the Taylor series about $a = 0$ generated by $f(x) = \sin x$.

Answer: First, find the derivatives of $\sin x$ and compute their values at $x = 0$.

$$
\begin{aligned}
f(x) &= \sin x & f(0) &= 0 \\
f'(x) &= \cos x & f'(0) &= 1 \\
f''(x) &= -\sin x & f''(0) &= 0 \\
f'''(x) &= -\cos x & f'''(0) &= -1 \\
f^{(4)}(x) &= \sin x & f^{(4)}(0) &= 0 \\
f^{(5)}(x) &= \cos x & f^{(5)}(0) &= 1
\end{aligned}
$$

As you've seen us do before, plug into the formula to generate the Taylor Series:

$$x - \frac{x^3}{3!} + \frac{x^5}{5!} + \ldots = \sum_{k=0}^{\infty} (-1)^k \frac{x^{2k+1}}{(2k+1)!}$$

PROBLEM 5. Find the radius and interval of convergence for the series $\sum\limits_{n=1}^{\infty} \dfrac{x^n}{1+n^2}$.

Answer: In order to determine convergence, apply the ratio test for absolute convergence:

$$\rho = \lim_{n \to \infty} \left| \frac{x^{n+1}}{1+(n+1)^2} \frac{1+n^2}{x^n} \right| = \lim_{n \to \infty} \left| x \frac{(1+n^2)}{(1+(n+1)^2)} \right| = |x|$$

Thus, if $|x| < 1$, then $\rho < 1$ and the series converges.

If $|x| > 1$, then $\rho > 1$
and the series diverges.

If $x = 1$, then $\sum\limits_{n=0}^{\infty} \dfrac{x^n}{1+n^2} = \sum\limits_{n=0}^{\infty} \dfrac{1}{1+n^2}$; (which converges).

If $x = -1$, then $\sum\limits_{n=0}^{\infty} \dfrac{x^n}{1+n^2} = \sum\limits_{n=0}^{\infty} \dfrac{(-1)^n}{1+n^2}$; (which converges).

Therefore, the radius of convergence is 1 and the interval of convergence is [–1,1].

PRACTICE PROBLEMS

Now try these problems involving the series we've discussed in this chapter. The answers are in chapter 21 (which is next!).

1. Find the sum of the series $2 + \dfrac{2}{5} + \dfrac{2}{25} + \ldots + \dfrac{2}{5^4}$.

2. Find the sum of the series $8 + \dfrac{8}{7} + \dfrac{8}{49} + \dfrac{8}{343} + \dfrac{8}{7^4} + \ldots$.

3. Does the series $\sum\limits_{n=1}^{\infty} \dfrac{5^n}{(n-1)!}$ converge or diverge?

4. Does the series $\sum\limits_{n=1}^{\infty} \dfrac{5^n}{n^2}$ converge or diverge?

5. Find the Taylor series about $a = 0$ generated by $f(x) = \cos x$.

6. Find the Taylor series about $a = 0$ generated by $f(x) = \ln(1 + x)$.

7. Find the Taylor series about $a = 0$ generated by $f(x) = e^{-x}$.

8. Find the first three nonzero terms of the Taylor series about $a = \dfrac{\pi}{3}$ generated by $f(x) = \sin x$.

9. Find the radius and interval of convergence for the series $\displaystyle\sum_{n=0}^{\infty} 3^n x^n$.

10. Find the radius and interval of convergence for the series $\displaystyle\sum_{n=0}^{\infty} (-1)^n \frac{x^{2n}}{(2n)!}$.

21

ANSWERS TO PRACTICE PROBLEMS

LIMITS

1. 13

2. $\dfrac{4}{5}$

3. π^2

4. 4

5. 0

6. ∞

7. $\dfrac{1}{10}$

8. $\sqrt{5}$

9. ∞

10. $-\infty$

11. Does Not Exist.

12. 1

13. -1

14. ∞

15. Does Not Exist.

16. ∞

17. (a) 4; (b) 5; (c) Does Not Exist.

18. (a) 4; (b) 4; (c) 4

19. $\dfrac{3}{\sqrt{2}}$

20. 0

21. 3

22. $\dfrac{3}{8}$

23. $\dfrac{7}{5}$

24. Does Not Exist.

25. 0

26. 0

27. $\dfrac{49}{121}$

28. 6

29. $\cos x$

30. $-\dfrac{1}{x^2}$

CONTINUITY

1. Yes. It satisfies all three conditions.

2. No. It fails condition 3.

3. No. It fails condition 1.

4. No. It is discontinuous at any integral multiple of $\dfrac{\pi}{2}$.

5. No. It is discontinuous at the endpoints of the interval.

6. Yes.

7. The function is continuous for $k = \dfrac{9}{16}$.

8. The function is continuous for $k = 6$ and $k = -1$.

9. The removable discontinuity is at $\left(3, \dfrac{11}{5}\right)$.

10. (a) $\lim\limits_{x \to -\infty} f(x) = 0$

    (b) $\lim\limits_{x \to \infty} f(x) = 0$

    (c) $\lim\limits_{x \to 3^-} f(x) = 1$

    (d) $\lim\limits_{x \to 3^+} f(x) = 1$

    (e) $f(3)$ does not exist.

    (f) $f(x)$ has a jump discontinuity at $x = -3$ and an essential discontinuity at $x = 5$.

The Definition of the Derivative

1. 5

2. 4

3. 20

4. −10

5. $16x$

6. $−20x$

7. $40a$

8. 54

9. $−9x^2$

10. $4x^3$

11. $5x^4$

12. $\dfrac{1}{3}$

13. $\dfrac{5}{4}$

14. $\dfrac{1}{2}$

15. $−\sin x$

16. $2x + 1$

17. $3x^2 + 3$

18. $\dfrac{−1}{x^2}$

19. $2ax + b$

20. $−\dfrac{2}{x^3}$

Basic Differentiation

1. $64x^3 + 16x$

2. $10x^9 + 36x^5 + 18x$

3. $77x^6$

4. $80x^9$

5. $54x^2 + 12$

6. $6x^{11}$

7. $−3x^8 − 2x^2$

8. 0

9. $\dfrac{2}{ab}x − \dfrac{2}{a^2} + \dfrac{d}{ax^2}$

10. $64x^{−9} + \dfrac{6}{\sqrt{x}}$

11. $−42x^{−8} − \dfrac{2}{\sqrt{x}}$

12. $−\dfrac{5}{x^6} − \dfrac{8}{x^9}$

13. $\dfrac{1}{2\sqrt{x}} − \dfrac{3}{x^4}$

14. $216x^2 − 48x + 36$

15. $−14x^6 − 5x^4 + 12x^3 − 36x^2 − 6$

16. 0

17. $−\dfrac{16}{x^5} + \dfrac{10}{x^6} + \dfrac{36}{x^7}$

18. $\dfrac{1}{2\sqrt{x}}$

19. $−\dfrac{16}{x^2} + \dfrac{14}{x^3} − \dfrac{24}{x^4} + \dfrac{16}{x^5}$

20. 0

21. $3x^2 + 6x + 3$

22. $\dfrac{1}{2\sqrt{x}} + \dfrac{1}{3\sqrt[3]{x^2}} + \dfrac{2}{3\sqrt[3]{x}}$

23. $6x^2 + 6x − 14$

24. $\dfrac{5}{6}x^{-\frac{1}{6}} + \dfrac{7}{10}x^{-\frac{3}{10}}$

25. $5ax^4 + 4bx^3 + 3cx^2 + 2dx + e$

Product Rule, Quotient Rule, and Chain Rule

1. $\dfrac{-80x^9 + 75x^8 + 12x^2 - 6x}{\left(5x^7 + 1\right)^2}$

2. $3x^2 - 6x - 1$

3. $10(x+1)^9$

4. $\dfrac{16x^2 - 32}{\sqrt{x^2 - 4}}$

5. $\dfrac{3x^2 - 3x^4}{\left(x^2 + 1\right)^4}$

6. $\dfrac{1}{4}\left(\dfrac{2x-5}{5x+2}\right)^{-\frac{3}{4}}\left(\dfrac{29}{(5x+2)^2}\right)$

7. $\dfrac{32x^{11} + 7x^{\frac{7}{2}}}{16x^8}$

8. $3x^2 + 1 + \dfrac{1}{x^2} + \dfrac{3}{x^4}$

9. $\dfrac{4x^3}{(x+1)^5}$

10. $100\left(x^2 + x\right)^{99}(2x + 1)$

11. $\dfrac{-2x}{\left(x^2 + 1\right)^{\frac{1}{2}}\left(x^2 - 1\right)^{\frac{3}{2}}}$

12. $\dfrac{9}{64}$

13. 106

14. 0

15. $\dfrac{x^2 - 6x + 3}{(x-3)^2}$

16. 6

17. $-\dfrac{7}{4}$

18. $\dfrac{2x^3 + x}{\sqrt{x^4 + x^2}}$

19. $-\dfrac{1}{4}$

20. $\dfrac{-2}{(x-1)^3}$

21. -24

22. $\left(t^3 - 6t^{\frac{5}{2}}\right)\left(5 + \dfrac{1}{2\sqrt{t}}\right) + \left(3t^2 - 15t^{\frac{3}{2}}\right)\left(5t + \sqrt{t}\right)$

23. $-\dfrac{14\sqrt{2}}{32\sqrt{3}}$

24. 2

25. $\dfrac{48v^5}{\left(v^2 + 8\right)^4}$

Derivatives of Trig Functions

1. $2\sin x \cos x$, or $\sin 2x$

2. $-2x \sin x^2$

3. $\sec^3 x + \sec x \tan^2 x$

4. $-4 \csc^2 4x$

5. $\dfrac{3\cos 3x}{2\sqrt{\sin 3x}}$

6. $\dfrac{2\cos x}{(1 - \sin x)^2}$

7. $-4x \csc^2(x^2)\cot(x^2)$

8. $6\cos 3x \cos 4x - 8\sin 3x \sin 4x$

9. $16\sin 2x$

10. $\left[\cos(1+\cos^2 x)+\sin(1+\cos^2 x)\right](-2\cos x \sin x)$

11. $\dfrac{2\tan x \sec^2 x}{(1-\tan x)^3}$

12. $\sec\theta(2\sec^2 2\theta)+\sec\theta\tan\theta\tan 2\theta$

13. $-(\cos\theta)\sin(1+\sin\theta)$

14. $\dfrac{\sec\theta\tan\theta+\sec\theta\tan^2\theta-\sec^3\theta}{(1+\tan\theta)^2}$

15. $-\dfrac{4}{x^2}\csc^2\left(\dfrac{2}{x}\right)\left[1+\cot\left(\dfrac{2}{x}\right)\right]^{-3}$

16. $-\dfrac{\sin\sqrt{x}}{2\sqrt{x}}\cos\left(\cos\left(\sqrt{x}\right)\right)$

Implicit Differentiation

1. $\dfrac{dy}{dx}=\dfrac{3x^2}{3y^2+1}$

2. $\dfrac{dy}{dx}=\dfrac{8y-x}{y-8x}$

3. $\dfrac{dy}{dx}=\dfrac{1}{2}$

4. $\dfrac{dy}{dx}=-\dfrac{\sin x+\cos x}{\sin y+\cos y}$

5. $\dfrac{dy}{dx}=\dfrac{8}{7}$

6. $\dfrac{dy}{dx}=\dfrac{1}{7}$

7. $\dfrac{dy}{dx}=-1$

8. $\dfrac{d^2y}{dx^2}=\dfrac{-4y^2-x^2}{16y^3}$

9. $\dfrac{d^2y}{dx^2}=\dfrac{\sin x\sin^2 y-\cos y\cos^2 x}{\sin^3 y}$

10. $\dfrac{d^2y}{dx^2}=1$

Equations of Tangent Lines

1. $y=5x-3$

2. $y=24x-54$

3. $y=-x+6$

4. $3x+64y=25$

5. $-x+6y=38$

6. $y=-3x+4$

7. $y=0$

8. $y=-39x+166$

9. $-24x+7y=-47$

10. $y=0$

11. $x=\pm\sqrt{\dfrac{3}{2}}$

12. $-x+2y=11$

13. $x=9$

14. $x=-\dfrac{3}{2}$; $y=\dfrac{41}{4}$

15. $a=1$; $b=0$; $c=1$

Mean Value Theorem and Rolle's Theorem

1. $c = 0$

2. $c = \dfrac{4}{\sqrt{3}}$

3. $c = \dfrac{-12 + 8\sqrt{3}}{3}$ or approximately 0.62

4. $c = \sqrt{2}$

5. No solution.

6 $c = 4$

7. $c = \pm \dfrac{1}{\sqrt{3}}$

8. $c = \dfrac{1}{2}$

9. No solution.

10. $c = \dfrac{1}{8}$

Maxima and Minima

1. Area is 32.

2. $x \approx 1.69$

3. 16 by 24

4. Radius is $\sqrt[3]{\dfrac{256}{\pi}}$

5. $x = \approx 1350$ m

6. (1,0)

7. $r = \dfrac{288}{4 + \pi} \approx 40 \; in$

8. $\theta = \dfrac{\pi}{4}$ radians or 45 degrees

9. \$4,456 million

Curve Sketching

1. min. at $\left(\sqrt{3}, -6\sqrt{3} - 6 \right)$; max. at $\left(-\sqrt{3}, 6\sqrt{3} - 6 \right)$; inflection point at $(0, -6)$

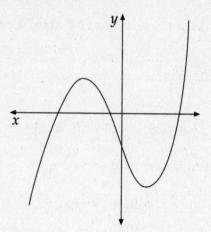

2. min. at $(-3, -4)$; max. at $(-1, 0)$; inflection point at $(-2, -2)$

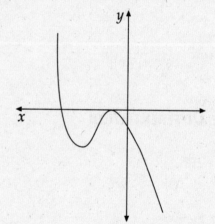

3. min. at $(0, -36)$; max. at $\left(\sqrt{\dfrac{13}{2}}, \dfrac{25}{4} \right)$ and $\left(-\sqrt{\dfrac{13}{2}}, \dfrac{25}{4} \right)$; inflection points at

$$\left(\sqrt{\frac{13}{6}}, -12.53\right) \text{ and } \left(-\sqrt{\frac{13}{6}}, -12.53\right)$$

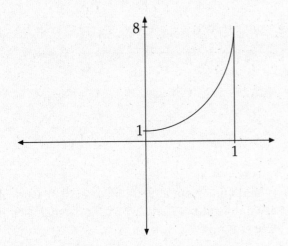

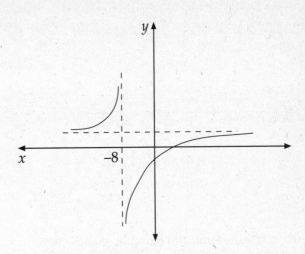

4. max. at (0,0); min. at (2, –4) and (–2, –4); inflection points at $\left(\dfrac{2}{\sqrt{3}}, -\dfrac{20}{9}\right)$ and $\left(-\dfrac{2}{\sqrt{3}}, -\dfrac{20}{9}\right)$

6. y-intercept at $\left(0, \dfrac{4}{3}\right)$; x-intercepts at (2,0) and (–2,0); vertical asymptote at $x = 3$; no horizontal asymptote; max. at (.76, 1.53); min. at (5.24, 10.47); no inflection points

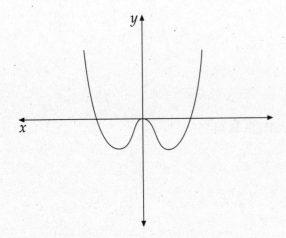

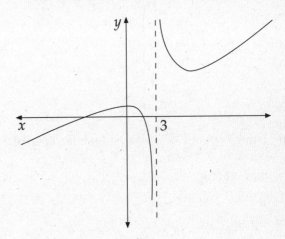

5. y-intercept at $\left(0, -\dfrac{3}{8}\right)$; x-intercept at (3, 0); vertical asymptote at $x = -8$; horizontal asymptote at $y = 1$; no max. or min; no inflection points

7. x- and y-intercept at (0,0); vertical asymptotes at $x = 5$ and $x = -5$; horizontal asymptote at $y = 0$; no max. or min.; inflection point at (0,0)

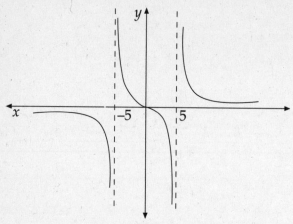

8. no max. or min.; no inflection points; cusp at (0,3)

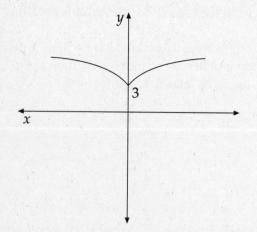

9. x-intercept at $\left(\frac{27}{8},0\right)$; x-and y-intercept (0,0); max. at (1,1).

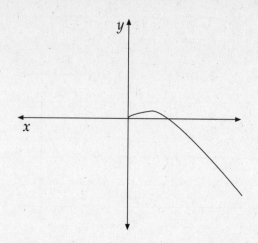

10. x– and y-intercept at (0,0); max. at (0,0); vertical asymptote at $x = 2$ and $x = -2$; horizontal asymptote at $y = 3$

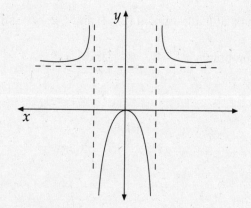

RELATED RATES

1. 2000 ft² / s

2. $\left(\frac{3}{4}\right)$ in / s

3. 100 km / hr

4. 3 m / s

5. $243\sqrt{3} \ in^2/s$

6. $\frac{5}{7}$ in² / s

7. $\frac{25}{6}$ ft / s

8. $\dfrac{4}{3}$ ft / s

9. 380 volts / s

10. $\dfrac{3\pi}{5}$ in²/min

POSITION, VELOCITY, AND ACCELERATION

1. $v(t) = 3t^2 - 18t + 24$; $a(t) = 6t - 18$

2. $v(t) = 2\cos(2t) - \sin t$; $a(t) = -4\sin(2t) - \cos t$

3. $t = 3$

4. $t = \pi, 3\pi$

5. Distance is 69.

6. Distance is 48.

7. Velocity is 0, acceleration is 0. This should come as no surprise because $2\sin^2 t + 2\cos^2 t = 2$, so the position is a constant.

8. $t = \dfrac{-16 + \sqrt{280}}{6} \approx 0.1222$

9. The velocity is never 0, so the particle does not change direction.

10. Distance is $2 + \sin^2 4 \approx 2.57$

EXPONENT AND LOGARITHMIC FUNCTIONS

1. $f'(x) = \dfrac{4x^3}{x^4 + 8}$.

2. $f'(x) = \dfrac{1}{x} + \dfrac{1}{6 + 2x}$

3. $f'(x) = \csc x$

4. $f'(x) = \ln(\cos 3x) - 3x \tan 3x - 3x^2$

5. $f'(x) = \dfrac{2}{x} - \dfrac{x}{5 + x^2}$.

6. $f'(x) = e^{x\cos x}(\cos x - x \sin x)$

7. $f'(x) = -3e^{-3x} \sin 5x + 5e^{-3x} \cos 5x$

8. $f'(x) = -\dfrac{1}{x^2}$

9. $f'(x) = e^{\tan 4x} \dfrac{4x \sec^2 4x - 1}{4x^2}$

10. $f'(x) = \pi e^{\pi x} - \pi$

11. $f'(x) = \dfrac{3}{x \ln 12}$

12. $f'(x) = \dfrac{1}{\ln 6}\left(\dfrac{1}{x} + \dfrac{\sec^2 x}{\tan x}\right)$

13. $f'(x) = \dfrac{1}{x e^{4x} \ln 4} - \dfrac{4 \log_4 x}{e^{4x}}$

14. $f'(x) = \dfrac{3}{2}$

15. $f'(x) = \dfrac{2 \ln x}{x \ln 10}$

16. $f'(x) = \left(3e^{3x}\right) - \left[3^{ex}(e \ln 3)\right]$

17. $f'(x) = 10^{\sin x}(\ln 10 \cos x)$

18. $f'(x) = (\sin x)^{\tan x}\left[(\sec^2 x)(\ln \sin x) + 1\right]$

19. $f'(x) = \ln 10$

20. $f'(x) = x^4\, 5^x\, (5 + x \ln 5)$

DERIVATIVES OF INVERSE FUNCTIONS

1. $\dfrac{16}{15}$

2. $-\dfrac{1}{12}$

3. $\dfrac{1}{e}$

4. $\dfrac{1}{3}$

5. $\dfrac{1}{4}$

6. 1

7. e

8. $\dfrac{15}{8}$

Newton's Method

1. $x_3 = 1.135$

2. $x_3 = -2.115$

3. $x = 3.148$

4. $x = 0.662 , 1.511$

5. $x = -1.425$

6. $x = 5.286$

7. $x = 1.192$

8. $x = 2.475$

Derivatives of Parametric Functions

1. $x = y^2$

2. $y = \sqrt{1-x^2}$

3. $y = x^2 - 6x$

4. $y - 48 = \dfrac{2}{3}(x - 40)$

5. $y - 1 = \sqrt{2}\left(x - \sqrt{2}\right)$

6. $(-2, 7)$

7. $(\ln 2, -4)$

8. $t = \dfrac{2\pi}{3}, \dfrac{4\pi}{3}$

L'Hopital's Rule

1. $\dfrac{3}{4}$

2. -1

3. $\dfrac{1}{6}$

4. -2

5. -2

6. 0

7. $\dfrac{1}{7}$

8. 1

9. $\dfrac{1}{2}$

10. 1

Differentials

1. 5.002

2. 3.999375

3. 1.802

4. 997

5. ± 2.16 in^3

6. 3 mm^3

7. -1.732 cm^2

8. (a) 1.963 m^3; (b) 15.71 m^3

Logarithmic Differentiation

1. $y\left[\dfrac{1}{x} - \dfrac{3x^2}{4(1-x^3)}\right]$

2. $\dfrac{y}{2}\left[\dfrac{1}{1+x} + \dfrac{1}{1-x}\right]$

3. $y\left[\dfrac{9x^2}{2(x^3+5)} - \dfrac{2x}{3(4-x^2)} - \dfrac{4x^3-2x}{(x^4-x^2+6)}\right]$

4. $y\left[\cot x - \tan x - \dfrac{3x^2}{2(x^3-4)}\right]$

5. $y\left[\dfrac{3(2x-2)}{(x^2-2x)} + \dfrac{4(-12x^3+7)}{(5-3x^4+7x)} - \dfrac{3(2x+1)}{(x^2+x)}\right]$

6. $y\left[\dfrac{1}{x-1} - \dfrac{1}{x} - \sec x \csc x\right]$

7. $y\left[\dfrac{2(1-2x)}{(x-x^2)} + \dfrac{3(3x^2+4x^3)}{(x^3+x^4)} + \dfrac{4(6x^5-5x^4)}{(x^6-x^5)}\right]$

8. $\dfrac{y}{4}\left[\dfrac{1}{x} - \dfrac{1}{1-x} + \dfrac{1}{1+x} - \dfrac{2x}{x^2-1} + \dfrac{1}{5-x}\right]$

THE ANTIDERIVATIVE

1. $-\dfrac{1}{3x^3} + C$

2. $10x^{\frac{1}{2}} + C$

3. $\dfrac{x^4}{4} - \dfrac{7}{x} + C$

4. $x^5 - x^3 + x^2 + 6x + C$

5. $-\dfrac{3}{2x^2} + \dfrac{2}{x} + \dfrac{x^5}{5} + 2x^8 + C$

6. $\dfrac{x^4}{4} - \dfrac{2x^3}{3} + \dfrac{x^2}{2} - 2x + C$

7. $\dfrac{3x^{\frac{4}{3}}}{2} + \dfrac{3x^{\frac{7}{3}}}{7} + C$

8. $\dfrac{x^7}{7} + \dfrac{2x^5}{5} + \dfrac{x^3}{3} + C$

9. $\dfrac{x^5}{5} - \dfrac{2x^3}{3} - \dfrac{1}{x} + C$

10. $\dfrac{x^5}{5} - \dfrac{3x^4}{4} + x^3 - \dfrac{x^2}{2} + C$

11. $\sin x + 5\cos x + C$

12. $\tan x + \sec x + C$

13. $\tan x + \dfrac{x^2}{2} + C$

14. $\sec x + C$

15. $\sin x + 4\tan x + C$

16. $-2\cos x + C$

17. $x + \sin x + C$

18. $\tan x - x + C$

19. $-\cos x + C$

20. $\dfrac{x^2}{2} - 2\tan x + C$

U-SUBSTITUTION

1. $\dfrac{\sin^2 2x}{4} + C$

2. $-\dfrac{9}{4}\left(10-x^2\right)^{\frac{2}{3}} + C$

3. $\dfrac{1}{30}\left(5x^4+20\right)^{\frac{3}{2}} + C$

4. $-\dfrac{1}{x-1} + C$

5. $-\dfrac{1}{12}\left(x^3+3x\right)^{-4} + C$

6. $-2\cos\sqrt{x}+C$

7. $\dfrac{1}{3}\tan(x^3)+C$

8. $-\dfrac{1}{3}\sin\left(\dfrac{3}{x}\right)+C$

9. $-\dfrac{1}{4}(1-\cos 2x)^{-2}+C$

10. $-\cos(\sin x)+C$

DEFINITE INTEGRALS

1. $\dfrac{17}{32}$

2. $\dfrac{25}{32}$

3. $\dfrac{21}{32}$

4. $\dfrac{2}{3}$

5. $\dfrac{2}{3}$

6. 2

7. $\dfrac{968}{5}$

8. $-\dfrac{161}{20}$,

9. 16

10. 0

MVTI AND THE FUNDAMENTAL THEOREMS OF CALCULUS

1. $\dfrac{4}{\sqrt{2\pi}}$

2. $\dfrac{8}{3}$

3. $\dfrac{2\sqrt{2}}{3}$

4. $1.$

5. $\sin^2 x$

6. $27x^2-9x$

7. $2x^3$

8. $-2\cos x$

EXPONENTIAL AND LOGARITHMIC FUNCTIONS

1. $\ln|\tan x|+C$

2. $-\ln|1-\sin x|+C$

3. $\ln|\ln x|+C$

4. $\sin(\ln x)+C$

5. $-\ln|\cos x|-x+C$

6. $\ln\left|1+2\sqrt{x}\,\right|+C$

7. $\ln|1+e^x|+C$

8. $\dfrac{1}{10}e^{5x^2-1}+C$

9. $\sin(2+e^x)+C$

10. $\ln|e^x-e^{-x}|+C$

11. $\dfrac{4^{-x^2}}{\ln 16} + C$

12. $\dfrac{7^{\sin x}}{\ln 7} + C$

AREAS BETWEEN TWO CURVES

1. $\dfrac{32}{3}$

2. $\dfrac{8}{3}$

3. $\dfrac{27}{4}$

4. 36

5. $\dfrac{17}{4}$

6. $\dfrac{9}{2}$

7. 4

8. $\dfrac{37}{12}$

9. $\dfrac{9}{2}$

10. $\dfrac{12}{5}$

THE VOLUME OF A SOLID OF REVOLUTION

1. 36π

2. 2π

3. $\dfrac{16\pi}{15}$

4. 2π

5. $\dfrac{16\pi}{5}$

6. 8π

7. $\dfrac{7\pi}{15}$

8. $\dfrac{128\pi}{5}$

9. $\dfrac{x}{8} - \dfrac{\sin 4x}{32} + C$

10. $\dfrac{896\pi}{15}$

INTEGRATION BY PARTS

1. $-x\cot x + \ln|\sin x| + C$

2. $\dfrac{x}{2}e^{2x} - \dfrac{1}{4}e^{2x} + C$

3. $-\dfrac{1}{x}\ln x - \dfrac{1}{x} + C$

4. $x^2\sin x + 2x\cos x - 2\sin x + C$

5. $\dfrac{x^3 \ln x}{3} - \dfrac{x^3}{9} + C$

6. $-\dfrac{x\cos 2x}{2} - \dfrac{\sin 2x}{4} + C$

7. $x\ln^2 x - 2x \ln x + 2x + C$

8. $x\tan x + \ln|\cos x| + C$

Inverse Trig Functions

1. $\dfrac{1}{16+x^2}$

2. $\dfrac{-1}{\left| x\sqrt{x^2-1} \right|}$

3. $\dfrac{e^x}{1+e^{2x}}$

4. $;\ x^2 > \pi$

5. $\dfrac{1}{\sqrt{\pi}}\sec^{-1}\left(\dfrac{x}{\sqrt{\pi}}\right)+C;\, x^2 > \pi$

6. $\tan^{-1}(\ln x) + C$

7. $\sin^{-1}(\tan x) + C$

8. $\dfrac{1}{2}\sin^{-1}\left(\dfrac{2x}{3}\right)+C$

9. $-\dfrac{1}{3}\sin\left(\dfrac{3}{x}\right)+C$

Integrals of Advanced Trig Functions

1. $\dfrac{3x}{8}-\dfrac{\sin 2x}{4}+\dfrac{\sin 4x}{32}+C$

2. $\dfrac{3x}{8}+\dfrac{\sin 2x}{4}+\dfrac{\sin 4x}{32}+C$

3. $-\dfrac{\cos^5 x}{5}+C$

4. $-\dfrac{\cos^6 x}{6}+\dfrac{\cos^8 x}{8}+C$

5. $\dfrac{x}{8}-\dfrac{\sin 4x}{32}-+C$

Length of a Curve

6. $\dfrac{\tan^4 x}{4}+\dfrac{\tan^6}{}$

7. $\dfrac{\tan^4 x}{4}-\dfrac{\tan^2 x}{2}-\ln|\cos x|+C$

8. $-\csc x + C$

Length of a Curve

1. $\dfrac{13}{12}$

2. $\int_{-\frac{\pi}{6}}^{0}\sqrt{1+\sec^4 x}\,dx$

3. $\int_0^{\frac{1}{4}}\sqrt{\dfrac{1}{1-x^2}}\,dx$

4. $\dfrac{47}{36}$

5. $\int_{\frac{1}{2}}^{\frac{1}{2}}\sqrt{\dfrac{1}{1-y^2}}\,dy$

6. $\int_0^{\pi}\sqrt{1+y^2\sin^2 y}\,dy$

7. $\dfrac{\pi}{6}$

8. $\dfrac{1}{2}\int_1^4 \dfrac{\sqrt{t^7+36}}{t^4}\,dt$

9. $2\int_1^2\sqrt{1+9t^2}\,dt$

Partial Fractions

1. $\dfrac{1}{7}\left[5\ln|x-1|+2\ln|x+6|\right]+C$

2. $\dfrac{1}{4}\left[3\ln|x-3|+\ln|x+1|\right]+C$

3. $-\dfrac{1}{2}\ln|x| + \dfrac{1}{3}\ln|x+2| + \dfrac{1}{6}\ln|x-1| + C$

4. $-7\ln|x-3| + 9\ln|x-4| + C$

5. $2\ln|x-1| - \dfrac{1}{x-1} + C$

6. $\dfrac{1}{2}\ln|x+1| - \dfrac{1}{4}\ln(x^2+1) + \dfrac{1}{2}\tan^{-1}x + C$

7. $-\dfrac{1}{3}\ln|x+1| + \dfrac{1}{6}\ln(x^2+2) + \dfrac{5\sqrt{2}}{6}\tan^{-1}\left(\dfrac{x}{\sqrt{2}}\right) + C$

8. $\ln|x-1| + \dfrac{4}{\sqrt{3}}\tan^{-1}\dfrac{2x+1}{\sqrt{3}} + C$

IMPROPER INTEGRALS

1. 2

2. 6

3. 1

4. Diverges

5. Diverges

6. $\sqrt{3}$

7. $-\dfrac{1}{4}$

8. Diverges

9. Diverges

10. $\dfrac{9}{2}$

DIFFERENTIAL EQUATIONS

1. $y = \sqrt[4]{\dfrac{28x^3}{3} - 236}$

2. $y = 6e^{\frac{5x^3}{3}}$

3. $y = \sqrt{2\tan^{-1}x + 4}$

4. $y = \sqrt[3]{3e^x - 2}$

5. $y = 2x^2$

6. $y = \sin^{-1}(-\cos x)$

7. 20,000 (approximately)

8. 45 m

9. $\dfrac{1125\pi}{2}\,\text{ft}^3$

10. 8,900 grams (approximately)

SERIES

1. Sum is 2.499.

2. Sum is $\dfrac{28}{3}$.

3. It converges.

4. It diverges.

5. $\cos x = 1 - \dfrac{x^2}{2!} + \dfrac{x^4}{4!} - \dfrac{x^6}{6!} + \ldots = \sum\limits_{k=0}^{\infty}(-1)^k\dfrac{x^{2k}}{(2k)!}$

6. $\ln(1+x) = x - \dfrac{x^2}{2} + \dfrac{x^3}{3} - \dfrac{x^4}{4} + \ldots \sum\limits_{k=0}^{\infty}(-1)^k\dfrac{x^{k+1}}{k+1}$

7. $e^{-x} = 1 - x + \dfrac{x^2}{2!} - \dfrac{x^3}{3!} - \dfrac{x^4}{4} + \ldots \sum\limits_{k=0}^{\infty}(-1)^k\dfrac{x^k}{k!}$

8. $\dfrac{\sqrt{3}}{2} + \dfrac{1}{2}\left(x - \dfrac{\pi}{3}\right) - \dfrac{\sqrt{3}}{4}\left(x - \dfrac{x}{3}\right)^2$

9. The radius of convergence is $\dfrac{1}{3}$; the interval of convergence is $\left(-\dfrac{1}{3}, \dfrac{1}{3}\right)$.

10. The radius of convergence is ∞ ; the interval of convergence is $(-\infty, \infty)$.

22

THE PRINCETON REVIEW AP CALCULUS AB DIAGNOSTIC EXAM

CALCULUS AB

SECTION I

Time—1 hour and 30 minutes.

Number of questions—40

Percent of total grade—50

Part A consists of 28 questions that will be answered on side 1 of the answer sheet. Following are the directions for Section I, Part A.

Directions: Solve each of the following problems, using the available space for scratchwork. After examining the form of the choices, decide which is the best of the choices given and fill in the corresponding oval on the answer sheet. No credit will be given for anything written in the test book. Do not spend too much time on any one problem.

In this test:

Unless otherwise specified, the domain of a function f is assumed to be the set of all real numbers x for which $f(x)$ is a real number.

1. If $f(x) = 5x^{\frac{4}{3}}$, then $f'(8) =$

(A) 10   (B) $\dfrac{40}{3}$ (C) 40   (D) 80   (E) $\dfrac{160}{3}$

2. $\lim\limits_{x \to \infty} \dfrac{5x^2 - 3x + 1}{4x^2 + 2x + 5}$ is

(A) 0   (B) $\dfrac{4}{5}$ (C) $\dfrac{3}{11}$ (D) $\dfrac{5}{4}$ (E)

GO ON TO THE NEXT PAGE

3. If $f(x) = \dfrac{3x^2 + x}{3x^2 - x}$ then $f'(x)$ is

(A) 1

(B) $\dfrac{6x^2 + 1}{6x^2 - 1}$

(C) $\dfrac{-6}{(3x - 1)^2}$

(D) $\dfrac{-2x^2}{(x^2 - x)^2}$

(E) $\dfrac{36x^3 - 2x}{(x^2 - x)^2}$

---

4. If the function f is continuous for all real numbers and if $f(x) = \dfrac{x^2 - 7x + 12}{x - 4}$ when $x \neq 4$, then $f(4) =$

(A) 1   (B) $\dfrac{8}{7}$ (C) −1   (D) 0   (E) undefined

GO ON TO THE NEXT PAGE

5. If $x^2 - 2xy + 3y^2 = 8$, then $\dfrac{dy}{dx} =$

(A) $\dfrac{8 + 2y - 2x}{6y - 2x}$

(B) $\dfrac{3y - x}{y - x}$

(C) $\dfrac{2x - 2y}{6y - 2x}$

(D) $\dfrac{1}{3}$

(E) $\dfrac{y - x}{3y - x}$

GO ON TO THE NEXT PAGE

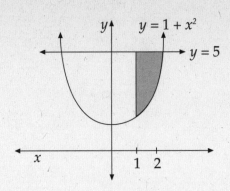

6. Which of the following integrals correctly corresponds to the area of the shaded region in the figure above ?

(A) $\int_1^2 (x^2 - 4)dx$

(B) $\int_1^2 (4 - x^2)dx$

(C) $\int_1^5 (x^2 - 4)dx$

(D) $\int_1^5 (x^2 + 4)dx$

(E) $\int_1^5 (4 - x^2)dx$

7. If $f(x) = \sec x + \csc x$, then $f'(x) =$

(A) 0

(B) $\sec^2 x + \csc^2 x$

(C) $\csc x - \sec x$

(D) $\sec x \tan x + \csc x \cot x$

(E) $\sec x \tan x - \csc x \cot x$

GO ON TO THE NEXT PAGE

8. An equation of the line normal to the graph of $y = \sqrt{(3x^2 + 2x)}$ at (2, 4) is

(A) $-4x + y + 20$ (B) $4x + 7y = 20$ (C) $-7x + 4y = 2$ (D) $7x + 4y = 30$ (E) $4x + 7y = 36$

---

9. $\displaystyle\int_{-1}^{1} \frac{4}{1 + x^2}\,dx =$

(A) 0   (B) π (C) 1   (D) 2π (E) 2

---

10. If $f(x) = \cos^2 x$, then $f''(\pi) =$

(A) -2 (B) 0   (C) 1   (D) 2   (E) 2π

---

11. If $f(x) = \dfrac{5}{x^2 + 1}$ and $g(x) = 3x$, then $g(f(2)) =$

(A) -3 (B) $\dfrac{5}{37}$ (C) 3   (D) 5   (E) $\dfrac{37}{5}$

GO ON TO THE NEXT PAGE

12. $\int x\sqrt{5x^2-4}\ dx =$

(A) $\dfrac{1}{10}\left(5x^2-4\right)^{\frac{3}{2}}+C$

(B) $\dfrac{1}{15}\left(5x^2-4\right)^{\frac{3}{2}}+C$

(C) $-\dfrac{1}{5}\left(5x^2-4\right)^{-\frac{1}{2}}+C$

(D) $\dfrac{20}{3}\left(5x^2-4\right)^{\frac{3}{2}}+C$

(E) $\dfrac{3}{20}\left(5x^2-4\right)^{\frac{3}{2}}+C$

---

13. The slope of the line tangent to the graph of $3x^2 + 5 \ln y = 12$ at $(2, 1)$ is

(A) $-\dfrac{12}{5}$ (B) $\dfrac{12}{5}$ (C) $\dfrac{5}{12}$ (D) 12   (E) −7

---

14. The equation $y = 2 - 3\sin\dfrac{\pi}{4}(x - 1)$ has a fundamental period of

(A) $\dfrac{1}{8}$ (B) $\dfrac{\pi}{4}$ (C) $\dfrac{4}{\pi}$ (D) 8   (E) 2π

GO ON TO THE NEXT PAGE

15. If $f(x) = \begin{cases} x^2 + 5 \text{ if } x < 2 \\ 7x - 5 \text{ if } x \geq 2 \end{cases}$, for all real numbers x, which of the following must be true?

    I. $f(x)$ is continuous everywhere.
    II. $f(x)$ is differentiable everywhere.
    III. $f(x)$ has a local minimum at $x = 2$.

(A) I only   (B) I and II only   (C) II and III only   (D) I and III only   (E) I, II, and III

16. For what value of x does the function $f(x) = x^3 - 9x^2 - 120x + 6$ have a local minimum?

(A) 10   (B) 4   (C) 3   (D) −4   (E) −10

17. The acceleration of a particle moving along the x-axis at time t is given by $a(t) = 4t - 12$. If the velocity is 10 when $t = 0$ and the position is 4 when $t = 0$, then the particle is changing direction at

(A) $t = 1$ (B) $t = 3$ (C) $t = 5$
(D) $t = 1$ and $t = 5$ (E) $t = 1$ and $t = 3$ and $t = 5$

GO ON TO THE NEXT PAGE

18. The average value of the function $f(x) = (x-1)^2$ on the interval from $x = 1$ to $x = 5$ is

(A) $-\dfrac{16}{3}$　　　(B) $\dfrac{16}{3}$　　　(C) $\dfrac{64}{3}$　　　(D) $\dfrac{66}{3}$　　　(E) $\dfrac{256}{3}$

19. $\displaystyle\int \left(e^{3\ln x} + e^{3x} \right)\, dx =$

(A) $3 + \dfrac{e^{3x}}{3} + C$

(B) $\dfrac{x^4}{4} + 3e^{3x} + C$

(C) $\dfrac{e^{x^4}}{4} + 3e^{3x} + C$

(D) $\dfrac{e^{x^4}}{4} + \dfrac{e^{3x}}{3} + C$

(E) $\dfrac{x^4}{4} + \dfrac{e^{3x}}{3} + C$

GO ON TO THE NEXT PAGE

20. If $f(x) = \sqrt{(x^3 + 5x + 121)} \, (x^2 + x + 11)$ then $f'(0) =$

(A) $\dfrac{5}{2}$ (B) $\dfrac{27}{2}$ (C) 22 (D) $22 + \dfrac{2}{\sqrt{5}}$ (E) $\dfrac{247}{2}$

21. If $f(x) = 5^{3x}$ then $f'(x) =$

(A) $5^{3x}(\ln 125)$ (B) $\dfrac{5^{3x}}{3\ln 5}$ (C) $3(5^{2x})$ (D) $3(5^{3x})$ (E) $3x(5^{3x-1})$

22. A solid is generated when the region in the first quadrant enclosed by the graph of $y = (x^2 + 1)^3$, the line $x = 1$, the x-axis, and the y-axis is revolved about the x-axis. Its volume is found by evaluating which of the following integrals?

(A) $\pi\displaystyle\int_1^8 \left(x^2 + 1\right)^3 dx$ (B) $\pi\displaystyle\int_1^8 \left(x^2 + 1\right)^6 dx$ (C) $\pi\displaystyle\int_0^1 \left(x^2 + 1\right)^3 dx$

(D) $\pi\displaystyle\int_0^1 \left(x^2 + 1\right)^6 dx$ (E) $2\pi\displaystyle\int_0^1 \left(x^2 + 1\right)^6 dx$

GO ON TO THE NEXT PAGE

23. $\lim\limits_{x \to 0} 4\, \dfrac{\sin x \cos x - \sin x}{x^2} =$

(A) 2   (B) $\dfrac{40}{3}$ (C) ∞ (D) 0   (E) undefined

---

24. If $\dfrac{dy}{dx} = \dfrac{(3x^2 + 2)}{y}$ and $y = 4$ when $x = 2$, then when $x = 3$, $y =$

(A) 18   (B) $\sqrt{66}$ (C) 58   (D) $\sqrt{74}$ (E) $\sqrt[3]{58}$

---

25. $\displaystyle\int \dfrac{dx}{9 + x^2} =$

(A) $\quad 3\tan^{-1}\left(\dfrac{x}{3}\right) + C$

(B) $\quad \dfrac{1}{3}\tan^{-1}\left(\dfrac{x}{3}\right) + C$

(C) $\quad \dfrac{1}{9}\tan^{-1}\left(\dfrac{x}{3}\right) + C$

(D) $\quad \dfrac{1}{3}\tan^{-1}(x) + C$

(E) $\quad \dfrac{1}{9}\tan^{-1}(x) + C$

GO ON TO THE NEXT PAGE

26. If $f(x) = \cos^3(x+1)$ then $f'(\pi) =$

(A) $-3\cos^2(\pi+1)\sin(\pi+1)$

(B) $3\cos^2(\pi+1)$

(C) $3\cos^2(\pi+1)\sin(\pi+1)$

(D) $3\pi\cos^2(\pi+1)$

(E) 0

27. $\int x\sqrt{x+3}\,dx =$

(A) $\dfrac{2}{3}(x)^{\frac{3}{2}} + 6(x)^{\frac{1}{2}} + C$

(B) $\dfrac{2(x+3)^{\frac{3}{2}}}{3} + C$

(C) $\dfrac{2}{5}(x+3)^{\frac{5}{2}} - 2(x+3)^{\frac{3}{2}} + C$

(D) $\dfrac{3(x+3)^{\frac{3}{2}}}{2} + C$

(E) $\dfrac{4x^2(x+3)^{\frac{3}{2}}}{3} + C$

GO ON TO THE NEXT PAGE

28. If $f(x) = \ln\big(\ln(1-x)\big)$, then $f'(x) =$

(A) $-\dfrac{1}{\ln(1-x)}$

(B) $\dfrac{1}{(1-x)\ln(1-x)}$

(C) $\dfrac{1}{(1-x)^2}$

(D) $-\dfrac{1}{(1-x)\ln(1-x)}$

(E) $-\dfrac{1}{\ln(1-x)^2}$

GO ON TO THE NEXT PAGE

Part B consists of 17 questions that will be answered on side 2 of the answer sheet. Following are the directions for Section I, Part B.

A GRAPHING CALCULATOR IS REQUIRED FOR SOME QUESTIONS ON THIS PART OF THE EXAMINATION.

Directions: Solve each of the following problems, using the available space for scratchwork. After examining the form of the choices, decide which is the best of the choices given and fill in the corresponding oval on the answer sheet. No credit will be given for anything written in the test book. Do not spend too much time on any one problem.

BE SURE YOU ARE USING SIDE 2 OF THE ANSWER SHEET TO RECORD YOUR ANSWERS TO QUESTIONS NUMBERED 29–45.

YOU MAY NOT RETURN TO SIDE 1 OF THE ANSWER SHEET.

In this test:

(1) The *exact* numerical value of the correct answer does not always appear among the choices given. When this happens, select from among the choices the number that best approximates the exact numerical value.

(2) Unless otherwise specified, the domain of a function f is assumed to be the set of all real numbers x for which $f(x)$ is a real number.

Note: Question numbers with an asterisk (*) indicate a graphing calculator-active question.

29. $\int_{0}^{\frac{\pi}{4}} \sin x \, dx + \int_{-\frac{\pi}{4}}^{0} \cos x \, dx =$

(A) $-\sqrt{2}$ (B) -1 (C) 0 (D) 1 (E) $\sqrt{2}$

30. Boats A and B leave the same place at the same time. Boat A heads due North at 12 km/hr. Boat B heads due East at 18 km/hr. After 2.5 hours, how fast is the distance between the boats increasing (in km/hr)?

(A) 21.63   (B) 31.20   (C) 75.00   (D) 9.84   (E) 54.08

---

31. $\lim\limits_{h\to 0} \dfrac{\tan\left(\frac{\pi}{6}+h\right)-\tan\left(\frac{\pi}{6}\right)}{h} =$

(A) $\dfrac{\sqrt{3}}{3}$ (B) $\dfrac{4}{3}$ (C) $\sqrt{3}$ (D) 0   (E) $\dfrac{3}{4}$

---

32. If $\displaystyle\int_{30}^{100} f(x)dx = A$ and $\displaystyle\int_{50}^{100} f(x)dx = B$ then $\displaystyle\int_{30}^{50} f(x)dx =$

(A) $A + B$ (B) $A - B$ (C) 0   (D) $B - A$ (E) 20

---

33. If $f(x) = 3x^2 - x$, and $g(x) = f^{-1}(x)$, then $g'(10)$ could be

(A) 59   (B) $\dfrac{1}{59}$ (C) $\dfrac{1}{10}$ (D) 11   (E) $\dfrac{1}{11}$

GO ON TO THE NEXT PAGE

34. The graph of $y = x^3 - 5x^2 + 4x + 2$ has a local minimum at

(A) (0.46, 2.87)   (B) (0.46, 0)   (C) (2.87, −4.06)   (D) (4.06, 2.87)   (E) (1.66, −0.59)

---

35. The volume generated by revolving about the y-axis the region enclosed by the graphs $y = 9 - x^2$ and $y = 9 - 3x$, for $0 \leq x \leq 2$, is

(A) −8π   (B) 4π   (C) 8π   (D) 24π   (E) 48π

---

36. The average value of the function $f(x) = \ln^2 x$ on the interval [2, 4] is

(A) −1.204   (B) 1.204   (C) 2.159   (D) 2.408   (E) 8.636

---

37. $\dfrac{d}{dx} \displaystyle\int_0^{3x} \cos(t)\,dt =$

(A) sin3x (B) −3sin3x (C) cos3x (D) 3sin3x (E) 3cos3x

GO ON TO THE NEXT PAGE

38. If the definite integral $\int_1^3 (x^2 + 1)dx$ is approximated by using the Trapezoid Rule with $n = 4$, the error is

(A) 0 (B) $\dfrac{7}{3}$ (C) $\dfrac{1}{12}$ (D) $\dfrac{65}{6}$ (E) $\dfrac{97}{3}$

---

39. The radius of a sphere is increasing at a rate proportional to its radius. If the radius is 4 initially, and the radius is 10 after two seconds, what will the radius be after three seconds?

(A) 62.50 (B) 13.00 (C) 15.81 (D) 16.00 (E) 25.00

---

40. If Newton's Method is used to estimate the real root of $x^3 + x^2 + 2 = 0$, then a first approximation of $x_1 = -2$ gives a third approximation of $x_3 =$

(A) −1.802 (B) −1.750 (C) −1.698 (D) 1.698 (E) 1.750

GO ON TO THE NEXT PAGE

41. $\int \ln 2x \ dx =$

(A) $\dfrac{\ln 2x}{x} + C$

(B) $\dfrac{\ln 2x}{2x} + C$

(C) $x \ln x - x + C$

(D) $x \ln 2x - x + C$

(E) $2x \ln 2x - 2x + C$

42. If the function $f(x)$ is continuous and differentiable $= \begin{cases} ax^3 - 6x; & \text{if } x \le 1 \\ bx^2 + 4; & x > 1 \end{cases}$ then $a =$

(A) 0   (B) 1   (C) –14   (D) –24   (E) 26

43. Two particles leave the origin at the same time and move along the y-axis with their respective positions determined by the functions $y_1 = \cos 2t$ and $y_2 = 4\sin t$ for $0 < t < 6$. For how many values of t do the particles have the same acceleration?

(A) 0   (B) 1   (C) 2   (D) 3   (E) 4

GO ON TO THE NEXT PAGE

44. Find the distance traveled (to three decimal places) in the first four seconds, for a particle whose velocity is given by $v(t) = 7e^{-t^2}$; where t stands for time.

   (A) 0.976   (B) 6.204   (C) 6.359   (D) 12.720   (E) 7.000

---

45. $\displaystyle \int \tan^6 x \sec^2 x \, dx =$

   (A) $\dfrac{\tan^7}{7} + C$

   (B) $\dfrac{\tan^7 x}{7} + \dfrac{\sec^3 x}{3} + C$

   (C) $\dfrac{\tan^7 x \sec^3 x}{21} + C$

   (D) $7\tan^7 x + C$

   (E) $\dfrac{2}{7}\tan^7 x \sec x + C$

STOP

END OF SECTION I

IF YOU FINISH BEFORE TIME IS CALLED, YOU MAY CHECK YOUR WORK ON THIS SECTION.

DO NOT GO ON TO SECTION II UNTIL YOU ARE TOLD TO DO SO.

MAKE SURE YOU HAVE PLACED YOUR AP NUMBER LABEL ON YOUR ANSWER SHEET AND HAVE WRITTEN AND GRIDDED YOUR NUMBER CORRECTLY IN SECTION C OF THE ANSWER SHEET.

THERE IS NO TEST MATERIAL ON THIS PAGE.

CALCULUS AB

SECTION II

Time—1 hour and 30 minutes

Number of problems—6

Percent of total grade—50

SHOW ALL YOUR WORK. Indicate clearly the methods you use because you will be graded on the correctness of your methods as well as on the accuracy of your final answers. If you choose to use decimal approximations, your answer should be correct to three decimal places.

A GRAPHING CALCULATOR IS REQUIRED FOR SOME QUESTIONS ON THIS PART OF THE EXAMINATION.

Note: Unless otherwise specified, the domain of a function f is assumed to be the set of all real numbers x for which $f(x)$ is a real number.

1. Let f be the function given by $f(x) = 2x^4 - 4x^2 + 1$.

   (a) Find an equation of the line tangent to the graph at (2,17).
   (b) Find the x- and y-coordinates of the relative maxima and relative minima.
   (c) Find the x- and y-coordinates of the points of inflection.
   (d) Sketch the graph of $f(x)$.

---

2. A particle moves along the x-axis so that its acceleration at any time $t > 0$ is given by $a(t) = 12t - 18$. At time $t = 1$, the velocity of the particle is $v(1) = 0$ and the position is $x(1) = 9$.

   (a) Write an expression for the velocity of the particle $v(t)$.

   (b) At what values of t does the particle change direction?

   (c) Write an expression for the position $x(t)$ of the particle.

   (d) Find the total distance traveled by the particle from $t = \dfrac{3}{2}$ to $t = 6$.

GO ON TO THE NEXT PAGE

3. Let R be the region enclosed by the graphs of $y = 2 \ln x$ and $y = \dfrac{x}{2}$, and the lines $x = 2$ and $x = 8$.

   (a) Find the area of R.

   (b) Set up, <u>but do not integrate</u>, an integral expression, in terms of a single variable, for the volume of the solid generated when R is revolved about the x-axis.

   (c) Set up, but <u>do not integrate</u>, an integral expression, in terms of a single variable, for the volume of the solid generated when R is revolved about the line $x = -1$.

---

4. Water is draining at the rate of $48\pi \text{ft}^3$/minute from a conical tank whose diameter at its base is 40 feet and whose height is 60 feet.

   (a) Find an expression for the volume of water in the tank in terms of its radius.

   (b) At what rate is the radius of the water in the tank shrinking when the radius is 16 feet?

   (c) How fast is the height of the water in the tank dropping at the instant that the radius is 16 feet?

---

5. Consider the equation $x^2 - 2xy + 4y^2 = 64$.

   (a) Write an expression for the slope of the curve at any point (x, y).

   (b) Find the equation of the tangent lines to the curve at the point $x = 2$.

   (c) Find $\dfrac{d^2y}{dx^2}$ at $(0, 4)$.

GO ON TO THE NEXT PAGE

6. Let $\int_0^x \left[\cos\left(\dfrac{t}{2}\right) + \left(\dfrac{3}{2}\right) \right] dt$ on the closed interval $[0, 4]$.

   (a) Approximate $F(2)$ using four inscribed rectangles.
   (b) Find $F'(2)$.
   (c) Find the average value of $F'(x)$ on the interval $[0, 4]$.

END OF EXAMINATION

ANSWERS TO SECTION 1
FOR THE AB EXAM

(1)	B	(11)	C	(21)	A	(31)	B	(41)	D
(2)	D	(12)	B	(22)	D	(32)	B	(42)	C
(3)	C	(13)	A	(23)	D	(33)	E	(43)	D
(4)	A	(14)	D	(24)	E	(34)	C	(44)	B
(5)	E	(15)	A	(25)	B	(35)	C	(45)	A
(6)	B	(16)	A	(26)	A	(36)	B		
(7)	E	(17)	D	(27)	C	(37)	E		
(8)	E	(18)	B	(28)	D	(38)	C		
(9)	D	(19)	E	(29)	D	(39)	C		
(10)	A	(20)	B	(30)	A	(40)	C		

23

Answers and Explanations to the AB Exam

STEP-BY-STEP SOLUTIONS TO THE PROBLEMS

PROBLEM 1. If $f(x) = 5x^{\frac{4}{3}}$, then $f'(8) =$

We need to use Basic Differentiation to solve this problem.

Step 1: $f'(x) = \dfrac{4}{3}\left(5x^{\frac{1}{3}}\right)$

Step 2: Now all we have to do is plug in 8 for x and simplify.

$$\frac{4}{3}\left(5\left(8^{\frac{1}{3}}\right)\right) = \frac{4}{3}(5(2)) = \frac{40}{3}$$

The answer is (B).

> **Note:** If you had trouble with this problem, you should review the unit on Basic Differentiation.

PROBLEM 2. $\displaystyle\lim_{x\to\infty} \dfrac{5x^2 - 3x + 1}{4x^2 + 2x + 5}$ is

Step 1: To solve this problem, you need to remember how to evaluate Limits. Always do limit problems on the first pass. Whenever we have a limit of a polynomial fraction where $x \to$, we divide the numerator and the denominator, separately, by the highest power of x in the fraction.

$$\lim_{x\to\infty} \frac{5x^2 - 3x + 1}{4x^2 + 2x + 5} = \lim_{x\to\infty} \frac{\dfrac{5x^2}{x^2} - \dfrac{3x}{x^2} + \dfrac{1}{x^2}}{\dfrac{4x^2}{x^2} + \dfrac{2x}{x^2} + \dfrac{5}{x^2}}$$

Step 2: Simplify $\displaystyle\lim_{x\to\infty} \dfrac{5 - \dfrac{3}{x} + \dfrac{1}{x^2}}{4 + \dfrac{2}{x} + \dfrac{5}{x^2}}$

Step 3: Now take the limit. Remember that the $\displaystyle\lim_{x\to\infty} \dfrac{k}{x^n} = 0$, if $n > 0$ where k is a constant. Thus we get

$$\lim_{x\to\infty} \frac{5 - \dfrac{3}{x} + \dfrac{1}{x^2}}{4 + \dfrac{2}{x} + \dfrac{5}{x^2}} = \lim_{x\to\infty} \frac{5 - 0 + 0}{4 + 0 + 0} = \frac{5}{4} \, .$$

The answer is (D).

> **Note:** If you had trouble with this problem, you should review the unit on Limits.

PROBLEM 3. If $f(x) = \dfrac{3x^2 + x}{3x^2 - x}$ then $f'(x)$ is

Step 1: We need to use the Quotient Rule to evaluate this derivative. Remember, the

derivative of $\dfrac{u}{v} = \dfrac{v\dfrac{du}{dx} - u\dfrac{dv}{dx}}{v^2}$. But before we take the derivative, we should factor an

x out of the top and bottom and cancel, simplifying the quotient.

$$f(x) = \frac{3x^2 + x}{3x^2 - x} = \frac{x(3x+1)}{x(3x-1)} = \frac{3x+1}{3x-1}$$

Step 2: Now take the derivative.

$$f'(x) = \frac{(3x-1)(3) - (3x+1)(3)}{(3x-1)^2}$$

Step 3: Simplify:

$$\frac{9x - 3 - 9x - 3}{(3x-1)^2} = \frac{-6}{(3x-1)^2}$$

The answer is (C).

Note: If you had trouble with this problem, you should review the units on Product Rule, Quotient Rule, and Chain Rule.

PROBLEM 4. If the function f is continuous for all real numbers and if $f(x) = \dfrac{x^2 - 7x + 12}{x - 4}$

when $x \neq 4$, then $f(4) =$

This problem is testing your knowledge of Continuity.

Step 1: Notice that if we plug 4 into the numerator and denominator we get $\dfrac{0}{0}$,

which is undefined. So, the first thing that we should do is factor the numerator. What we are looking for is a common factor in the numerator and denominator. If we find a common factor, we can cancel the factors and simplify the problem.

We get $f(x) = \dfrac{x^2 - 7x + 12}{x - 4} = \dfrac{(x-3)(x-4)}{x-4} = (x-3)$.

Step 2: Now we plug in 4 for x and we get 1.

The answer is (A).

Note: If you had trouble with this problem, you should review the unit on Continuity.

PROBLEM 5. If $x^2 - 2xy + 3y^2 = 8$, then $\dfrac{dy}{dx} =$

Whenever we have a polynomial where the x's and y's are not separated we need to use Implicit Differentiation to find the derivative.

Step 1: Take the derivative of everything with respect to x.

$$2x\frac{dx}{dx} - 2\left(x\frac{dy}{dx} + y\frac{dx}{dx}\right) + 6y\frac{dy}{dx} = 0 \quad \text{Remember that } \frac{dx}{dx} = 1!$$

Step 2: Simplify and then put all of the terms containing $\dfrac{dy}{dx}$ on one side, and all of the other terms on the other side.

$$2x - 2x\frac{dy}{dx} - 2y + 6y\frac{dy}{dx} = 0$$

$$-2x\frac{dy}{dx} + 6y\frac{dy}{dx} = 2y - 2x$$

Factor out the $\dfrac{dy}{dx}$, then isolate it.

$$\frac{dy}{dx}(6y - 2x) = 2y - 2x$$

$$\frac{dy}{dx} = \frac{2y - 2x}{(6y - 2x)} = \frac{y - x}{3y - x}$$

The answer is (E).

> **Note:** If you had trouble with this problem, you should review the unit on Implicit Differentiation.

PROBLEM 6.

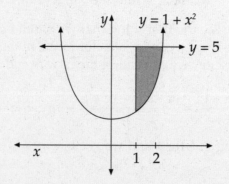

Which of the following integrals correctly corresponds to the area of the region in the figure above between the curve $y = 1 + x^2$ and the line $y = 5$ from $x = 1$ to $x = 2$?

We use integrals to find the Area Between Two Curves. If the top curve of a region is $f(x)$ and the bottom curve of a region is $g(x)$, from $x = a$ to $x = b$, then the area is found by the integral

$$\int_a^b \left[f(x) - g(x) \right] dx$$

Step 1: The top curve here is the line $y = 5$, and the bottom curve is $y = 1 + x^2$, and the region extends from the line $x = 1$ to the line $x = 2$. Thus, the integral for the area is

$$\int_1^2 \left[(5) - \left(1 + x^2 \right) \right] dx = \int_1^2 \left(4 - x^2 \right) dx$$

The answer is (B).

> **Note:** If you had trouble with this problem, you should review the unit on The Area Between Two Curves.

PROBLEM 7. If $f(x) = \sec x + \csc x$, then $f'(x) =$

This question is testing whether you know your Derivatives of Trigonometric Functions. If you do, this is an easy problem.

Step 1: The derivative of $\sec x$ is $\sec x \tan x$ and the derivative of

$\csc x$ is $-\csc x \cot x$. That makes the derivative here $\sec x \tan x - \csc x \cot x$.

The answer is (E).

> **Note:** If you had trouble with this problem, you should review the units on Derivatives of Trigonometric Functions.

PROBLEM 8. An equation of the line normal to the graph of $y = \sqrt{\left(3x^2 + 2x \right)}$ at $(2, 4)$ is

Here we do every thing that we normally do for finding the Equations of Tangent Lines, except that we use the negative reciprocal of the slope to find the normal line. This is because the normal line is perpendicular to the tangent line.

Step 1: First, find the slope of the tangent line.

$$\frac{dy}{dx} = \frac{1}{2} \left(3x^2 + 2x \right)^{-\frac{1}{2}} (6x + 2)$$

Step 2: DON'T SIMPLIFY. Immediately plug in $x = 2$. We get:

$$\frac{dy}{dx} = \frac{1}{2} \left(3x^2 + 2x \right)^{-\frac{1}{2}} (6x + 2) = \frac{1}{2} \left(3(2)^2 + 2(2) \right)^{-\frac{1}{2}} (6(2) + 2) = \frac{1}{2} (16)^{-\frac{1}{2}} (14) = \frac{7}{4}$$

This means that the slope of the tangent line at $x = 2$ is $\frac{7}{4}$, so the slope of the normal line is $-\frac{4}{7}$.

Step 3: Then the equation of the tangent line is $(y-4) = -\frac{4}{7}(x-2)$.

Step 4: Multiply through by 7 and simplify.

$$7y - 28 = -4x + 8$$
$$4x + 7y = 36$$

The answer is E.

> **Note:** If you had trouble with this problem, you should review the unit Equations of Tangent Lines.

PROBLEM 9. $\int_{-1}^{1} \frac{4}{1+x^2} dx =$

You should recognize this integral as one of the Inverse Trigonometric Integrals.

Step 1: As you should recall, $\int \frac{dx}{1+x^2} = \tan^{-1}(x) + C$. The 4 is no big deal, just multiply the integral by 4 to get $4\tan^{-1}(x)$. Then we just have to evaluate the limits of integration.

Step 2: $4\tan^{-1}(x)\Big|_{-1}^{1} = 4\tan^{-1}(1) - 4\tan^{-1}(-1) = 4\left(\frac{\pi}{4}\right) - 4\left(-\frac{\pi}{4}\right) = 2\pi$

The answer is (D).

> **Note:** If you had trouble with this problem, you should review the unit on Inverse Trigonometric Integrals. You really only need to know three of them, and you should just memorize them.

PROBLEM 10. If $f(x) = \cos^2 x$, then $f''(\pi) =$

This problem is just asking us to find a higher order Derivative of a Trigonometric Function.

Step 1: The first derivative requires the chain rule:

$f(x) = \cos^2 x$

$f'(x) = 2(\cos x)(-\sin x) = -2\cos x \sin x$

Step 2: The second derivative requires the product rule:

$$f'(x) = -2\cos x \sin x$$

$$f''(x) = -2(\cos x \cos x - \sin x \sin x) = -2\left(\cos^2 x - \sin^2 x\right)$$

Step 3: Now we plug in π for x and simplify.

$$-2\left(\cos^2(\pi) - \sin^2(\pi)\right) = -2(1-0) = -2$$

The answer is (A).

> **Note:** If you had trouble with this problem, you should review the units on Derivatives of Trigonometric Functions, The Product Rule, and The Chain Rule.

PROBLEM 11. If $f(x) = \dfrac{5}{x^2+1}$ and $g(x) = 3x$ then $g(f(2)) =$

This isn't even a calculus problem. This is a *precalculus* problem! You will get one or two of these on the AP examination for Calculus AB, but generally not for Calculus BC. They are a gift, and you should be grateful!

Step 1: To find $g(f(x))$, all you need to do is to replace all of the x's in $g(x)$ with $f(x)$'s.

$$g(f(x)) = 3f(x) = 3\left(\frac{5}{x^2+1}\right) = \frac{15}{x^2+1}$$

Step 2: Now all we have to do is plug in 2 for x.

$$g(f(2)) = \frac{15}{2^2+1} = 3$$

The answer is (C).

PROBLEM 12. $\displaystyle\int x\sqrt{5x^2-4}\,dx =$

Any time we have an integral with an x factor whose power is one less than another x factor, we can try to do the integral with u-substitution. This is our favorite technique for doing integration and the most important one to master.

Step 1: Let $u = 5x^2 - 4$ and $du = 10x\,dx$ and so $\dfrac{1}{10}du = x\,dx$.

Then we can rewrite the integral as:

$$\int x\sqrt{5x^2-4}\,dx = \frac{1}{10}\int u^{\frac{1}{2}}du$$

Step 3: Now this becomes a basic integral.

$$\frac{1}{10}\int u^{\frac{1}{2}}du = \frac{1}{10}\left(\frac{u^{\frac{3}{2}}}{\frac{3}{2}}\right)+C = \frac{1}{15}u^{\frac{3}{2}}+C$$

Step 4: Reverse the substitution and we get: $\frac{1}{15}\left(5x^2-4\right)^{\frac{3}{2}}+C$

The answer is (B).

> **Note:** If you had trouble with this problem, you should review the unit on the u-substitution.

PROBLEM 13. The slope of the line tangent to the graph of $3x^2 + 5\ln y = 12$ at (2,1) is

This is another Equation of a Tangent Line problem, combined with Implicit Differentiation. Often the AP Examination has more than one tangent line problem, so make sure that you can do these well!

By the way, do you remember the derivative of $\ln(f(x))$? It is $\frac{f'(x)}{f(x)}$.

Step 1: First, we take the derivative of the equation.

$$6x\frac{dx}{dx}+\frac{5}{y}\frac{dy}{dx}=0$$

Step 2: Next, we simplify and solve for $\frac{dy}{dx}$.

$$6x+\frac{5}{y}\frac{dy}{dx}=0$$

$$\frac{dy}{dx}=\frac{-6xy}{5}$$

Step 3: Now we plug in 2 for x and 1 for y to get the slope of the tangent line.

$$\frac{dy}{dx}=\frac{-6(2)(1)}{5}=\frac{-12}{5}$$

The answer is (A).

> **Note:** We could have plugged in directly for x and y after simplifying. On a more complicated derivative you ALWAYS want to plug in right after you differentiate. On a simple one such as this, the choice is up to you.

> **Note:** If you had trouble with this problem, you should review the units on Equation of a Tangent Line and Implicit Differentiation.

PROBLEM 14. The equation $y = 2 - 3\sin\dfrac{\pi}{4}(x - 1)$ has a fundamental period of

This is another Precalculus problem. The AP people expect you to remember a lot of your trigonometry, so if you're rusty, review the unit in the Appendix.

Step 1: In an equation of the form $f(x) = A \sin B (x \pm C) \pm D$, you should know four components. The amplitude of the equation is A, the horizontal or phase shift is $\pm C$, the vertical shift is $\pm D$, and the fundamental period is $\dfrac{2\pi}{B}$.

The same is true for $f(x) = A \cos B (x \pm C) \pm D$.

Step 2: All we have to do is plug into the formula for the period.

$$\frac{2\pi}{B} = \frac{2\pi}{\dfrac{\pi}{4}} = 8$$

The answer is (D).

PROBLEM 15. If $f(x) = \begin{cases} x^2 + 5 & \text{if } x < 2 \\ 7x - 5 & \text{if } x \geq 2 \end{cases}$, for all real numbers x, which of the following must be true?

    I. $f(x)$ is continuous everywhere.

    II. $f(x)$ is differentiable everywhere.

    III. $f(x)$ has a local minimum at $x = 2$.

This problem is testing your knowledge of the rules of continuity and differentiability. While the more formal treatment is located in the unit on Continuity, here we'll go directly to a shortcut to the right answer. This type of function is called a piecewise function because it is broken into two or more pieces, depending on the value of x that one is looking at.

Step 1: If a piecewise function is continuous at a point a, then when you plug a into each of the pieces of the function, you should get the same answer. The function consists of a pair of polynomials (Remember that all polynomials are continuous!), where the only point that might be a problem is $x = 2$. So here we'll plug 2 into both pieces of the function to see if we get the same value. If we do, then the function is continuous. If we don't, then it's discontinuous. At $x = 2$, the upper piece is equal to 9 and the lower piece is also equal to 9. So the function is continuous everywhere, and **I** is true. You should then eliminate answer choice C.

Step 2: If a piecewise function is differentiable at a point a, then when you plug a into each of the derivatives of the pieces of the function, you should get the same answer. It is the same idea as in Step 1. So here we will plug 2 into the derivatives of both pieces of the function to see if we get the same value. If we do, then the function is differentiable. If we don't, then it is non-differentiable at $x = 2$.

The derivative of the upper piece is $2x$, and at $x = 2$, the derivative is 4.

The derivative of the lower piece is 7 everywhere.

Because the two derivatives are not equal, the function is not differentiable every-where and II is false. You should then eliminate answer choices B and E.

Step 3: The slope of the function to the left of $x = 2$ is 4. The slope of the function to the right of $x = 2$ is 7. If the slope of a function has the same sign on either side of a point then the function cannot have a local minimum or maximum at that point. So III is false because of what we found in Step 2.

The answer is (A).

> **Note:** If you had trouble with this problem, you should review the unit on Continuity.

PROBLEM 16. For what value of x does the function $f(x) = x^3 - 9x^2 - 120x + 6$ have a local minimum?

This problem requires you to know how to find Maxima/Minima. This is a part of curve sketching and is one of the most important parts of Differential Calculus. A function has *critical points* where the derivative is zero or undefined (which is never a problem when the function is an ordinary polynomial). After finding the critical points we test them to determine whether they are maxima or minima or something else.

Step 1: First, as usual, take the derivative and set it equal to zero.

$$f'(x) = 3x^2 - 18x - 120$$
$$3x^2 - 18x - 120 = 0$$

Step 2: Find the values of x that make the derivative equal to zero. These are the critical points.

$$3x^2 - 18x - 120 = 0$$
$$x^2 - 6x - 40 = 0$$
$$(x - 10)(x + 4) = 0$$
$$x = \{10, -4\}$$

Step 3: In order to determine whether a critical point is a maximum or a minimum, we need to take the second derivative. $f''(x) = 6x - 18$

Step 4: Now we plug the critical points from Step 2 into the second derivative. If it yields a negative value, then the point is a maximum. If it yields a positive value, then the point is a minimum. If it yields zero, it is neither, and is most likely a point of inflection.

$$6(10) - 18 = 42$$
$$6(-4) - 18 = -42$$

Therefore 10 is a minimum.

The answer is (A).

Note: If you had trouble with this problem, you should review the unit on Maxima/Minima.

PROBLEM 17. The acceleration of a particle moving along the x-axis at time t is given by $a(t) = 4t - 12$. If the velocity is 10 when $t = 0$ and the position is 4 when $t = 0$, then the particle is changing direction at

Step 1: Because acceleration is the derivative of velocity, if we know the acceleration of a particle, we can find the velocity by integrating the acceleration with respect to t.

$$\int (4t - 12)dt = 2t^2 - 12t + C$$

Next, because the velocity is 10 at $t = 0$, we can plug in 0 for t and solve for the constant.

$$2(0)^2 - 12(0) + C = 10.$$

Therefore $C = 10$ and the velocity, $v(t)$, is $2t^2 - 12t + 10$

Step 2: In order to find when the particle is changing direction we need to know when the velocity is equal to zero, so we set $v(t) = 0$ and solve for t.

$$2t^2 - 12t + 10 = 0$$
$$t^2 - 6t + 5 = 0$$
$$(t - 5)(t - 1) = 0$$
$$t = \{1, 5\}$$

Now, provided that the acceleration is not also zero at $t = \{1, 5\}$, the particle will be changing direction at those times. The acceleration is found by differentiating the equation for velocity with respect to time: $a(t) = 4t - 12$. This is not zero at either $t = 1$ or $t = 5$. Therefore, the particle is changing direction when $t = 1$ and $t = 5$.

The answer is (D).

Note: Did you notice that you didn't even need the information about the position of the particle. Sometimes the AP gives you more information than you need. Don't let that make you suspicious!

Note: If you had trouble with this problem, you should review the unit on Position, Velocity, and Acceleration.

PROBLEM 18. The average value of the function $f(x) = (x - 1)^2$ on the interval from $x = 1$ to $x = 5$ is

Step 1: If you want to find the average value of $f(x)$ on an interval $[a, b,]$ you need to evaluate the integral $\frac{1}{b-a}\int_a^b f(x)dx$.

So here we would evaluate the integral $\dfrac{1}{5-1}\displaystyle\int_1^5 (x-1)^2\,dx$.

Step 2: $\dfrac{1}{5-1}\displaystyle\int_1^5 (x-1)^2\,dx = \dfrac{1}{4}\int_1^5 \left(x^2 - 2x + 1\right)dx$

$$= \dfrac{1}{4}\left(\dfrac{x^3}{3} - x^2 + x\right)\Bigg|_1^5 = \dfrac{1}{4}\left[\left(\dfrac{5^3}{3} - 5^2 + 5\right) - \left(\dfrac{1}{3} - 1 + 1\right)\right]$$

$$= \dfrac{1}{4}\left(\dfrac{125}{3} - 20 - \dfrac{1}{3}\right) = \dfrac{64}{12} = \dfrac{16}{3}$$

The answer is (B).

> **Note:** If you had trouble with this problem, you should review the unit on Mean Value Theorem for Integrals.

PROBLEM 19. $\displaystyle\int\left(e^{3\ln x} + e^{3x}\right)dx =$

This problem requires that you know your rules of Exponential Functions.

Step 1: First of all, $e^{3\ln x} = e^{\ln x^3} = x^3$. So we can rewrite the integral as

$$\int\left(e^{3\ln x} + e^{3x}\right)dx = \int\left(x^3 + e^{3x}\right)dx.$$

Step 2: The rule for the integral of an exponential function is $\displaystyle\int e^k\,dx = \dfrac{1}{k}e^{kx} + C$.

Now we can do this integral. $\displaystyle\int\left(x^3 + e^{3x}\right)dx = \dfrac{x^4}{4} + \dfrac{1}{3}e^{3x} + C$.

The answer is (E).

> **Note:** If you had trouble with this problem, you should review the unit on Exponential Functions.

PROBLEM 20. If $f(x) = \sqrt{\left(x^3 + 5x + 121\right)}\left(x^2 + x + 11\right)$ then $f'(0) =$

This problem is just a complicated derivative, requiring you to be familiar with the Chain Rule and the Product Rule.

Step 1: $f'(x) = \dfrac{1}{2}\left(x^3 + 5x + 121\right)^{-\frac{1}{2}}\left(3x^2 + 5\right)\left(x^2 + x + 11\right) + \left(x^3 + 5x + 121\right)^{\frac{1}{2}}(2x + 1)$

Step 2: Whenever a problem asks you to find the value of a complicated derivative at a particular point, NEVER simplify the derivative. Immediately plug in the value for x and do arithmetic instead of algebra.

$$f'(x) = \frac{1}{2}\left(0^3 + 5(0) + 121\right)^{-\frac{1}{2}}\left(3(0)^2 + 5\right)\left(0^2 + (0) + 11\right) + \left((0)^3 + 5(0) + 121\right)^{\frac{1}{2}}(2(0) + 1)$$

$$= \frac{1}{2}(121)^{-\frac{1}{2}}(5)(11) + (121)^{\frac{1}{2}}(1) = \frac{5}{2} + 11 = \frac{27}{2}$$

The answer is (B).

> **Note:** If you had trouble with this problem, you should review the units on The Chain Rule and The Product Rule.

PROBLEM 21. If $f(x) = 5^{3x}$ then $f'(x) =$

This problem requires you to know how to find the Derivative of an Exponential Function. The rule is: If a function is of the form $a^{f(x)}$, its derivative is $a^{f(x)}(\ln a) f'(x)$. Now all we have to do is follow the rule!

Step 1: $f(x) = 5^{3x}$

$$f'(x) = 5^{3x}(\ln 5)(3)$$

Step 2: If you remember your rules of logarithms, $3\ln 5 = \ln(5^3) = \ln 125$.

So we can rewrite the answer to $f'(x) = 5^{3x}(\ln 5)(3) = 5^{3x}\ln 125$.

The answer is (A).

> **Note:** If you had trouble with this problem, you should review the unit on Exponential Functions.

PROBLEM 22. A solid is generated when the region in the first quadrant enclosed by the graph of $y = (x^2 + 1)^3$, the line $x = 1$, and the x-axis, is revolved about the x-axis. Its volume is found by evaluating which of the following integrals?

This problem requires you to know how to find the Volume of a Solid of Revolution.

If you have a region between two curves, from $x = a$ to $x = b$, then the volume generated when the region is revolved around the x-axis is: $\pi\int_a^b \left[f(x)\right]^2 - \left[g(x)\right]^2 dx$, if $f(x)$ is above $g(x)$ throughout the region.

Step 1: First, we have to determine what the region looks like. The curve looks like this:

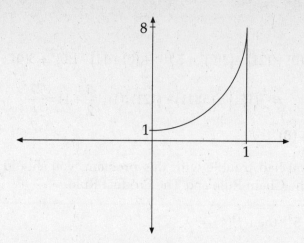

The shaded region is the part that we are interested in. Notice that the curve is always above the x-axis (which is $g(x)$). Now we just follow the formula:

$$\pi\int_0^1\left[\left(x^2+1\right)^3\right]^2-[0]^2\,dx=\pi\int_0^1\left(x^2+1\right)^6\,dx$$

The answer is (D).

> **Note:** If you had trouble with this problem, you should review the unit on Finding the Volume of a Solid of Revolution.

PROBLEM 23. $\displaystyle\lim_{x\to0}4\frac{\sin x\cos x-\sin x}{x^2}=$

This problem requires us to evaluate the Limit of a Trigonometric Function.

There are two important trigonometric limits to memorize:

$$\lim_{x\to0}\frac{\sin x}{x}=1\text{ and }\lim_{x\to0}\frac{1-\cos x}{x}=0$$

Step 1: The first step that we always take when evaluating the limit of a trigonometric function is to rearrange the function so that it looks like some combination of the limits above. We can do this by factoring a $\sin x$ out of the numerator.

Now we can break this into limits that we can easily evaluate.

$$\lim_{x\to0}4\frac{\sin x\cos x-\sin x}{x^2}=4\lim_{x\to0}\left(\frac{\sin x}{x}\right)\left(\frac{\cos x-1}{x}\right)$$

$$\left(\text{Note that }\lim_{x\to0}\frac{1-\cos x}{x}=\lim_{x\to0}\frac{\cos x-1}{x}=0.\right)$$

Step 2: Now if we take the limit as $x\to0$ we get: $4(1)(0)=0$

The answer is (D).

Note: If you had trouble with this problem, you should review the unit on Limits of Trigonometric Functions.

PROBLEM 24. If $\dfrac{dy}{dx} = \dfrac{(3x^2 + 2)}{y}$ and $y = 4$ when $x = 2$, then when $x = 3$, $y =$

This is a very basic Differential Equation. As with many of the more difficult topics in Calculus, the AP examination only tends to ask us to solve very straightforward differential equations. In fact, on the AB examination, you are only going to need to know one technique for getting these right. It is called *Separation of Variables*.

Step 1: First, take all of the terms with a y in them and put them on the left side of the equal sign. Take all of the terms with an x in them and put them on the right side of the equal sign. Then we get:

$$y\,dy = (3x^2 + 2)\,dx$$

Step 2: Now integrate both sides.

$$\int y\,dy = \int \left(3x^2 + 2\right)dx$$

$$\frac{y^2}{2} = x^3 + 2x + c$$

Notice how we only use one constant. All we have to do now is solve for C. We do this by plugging in 2 for x and 4 for y.

$$\frac{16}{2} = 2^3 + 4 + C$$
$$C = -4$$

So we can rewrite the equation as $\dfrac{y^2}{2} = x^3 + 2x - 4$.

Step 3: Now if we plug in 3 for x, we will get y.

$$\frac{y^2}{2} = 27 + 6 - 4$$
$$y^2 = 58$$
$$y = \pm\sqrt{58}$$

The answer is (E).

Note: If you had trouble with this problem, you should review the unit on Differential Equations.

PROBLEM 25. $\displaystyle\int \frac{dx}{9+x^2} =$

This is another Inverse Trigonometric Integral.

Step 1: We know that. $\displaystyle\int \frac{dx}{1+x^2} = \tan^{-1}(x)+C$. (See problem 9 if you're not sure of this.) The trick here is to get the denominator of the fraction in the integrand to be of the correct form. If we factor 9 out of the denominator we get:

$$\int \frac{dx}{9+x^2} = \int \frac{dx}{9\left(1+\dfrac{x^2}{9}\right)} = \frac{1}{9}\int \frac{dx}{1+\dfrac{x^2}{9}} = \frac{1}{9}\int \frac{dx}{1+\left(\dfrac{x}{3}\right)^2}$$

Step 2: Now if we use u-substitution we will be able to evaluate this integral.

Let $u = \dfrac{x}{3}$ and $du = \dfrac{1}{3}dx$ or $3du = dx$. Then we have:

$$\frac{1}{9}\int \frac{dx}{1+\left(\dfrac{x}{3}\right)^2} = \frac{1}{9}\int \frac{3du}{1+u^2} = \frac{1}{3}\int \frac{du}{1+u^2} = \frac{1}{3}\tan^{-1}(u)+C$$

Step 3: Now all we have to do is reverse the u-substitution and we're done.

$$\frac{1}{3}\tan^{-1}(u)+C = \frac{1}{3}\tan^{-1}\left(\frac{x}{3}\right)+C$$

The answer is (B).

> **Note:** If you had trouble with this problem, you should review the unit on Inverse Trigonometric Integrals.

PROBLEM 26: If $f(x) = \cos^3(x+1)$ then $f'(\pi) =$

Think of $\cos^3(x+1)$ as $\left[\cos(x+1)\right]^3$.

Step 1: First, we take the derivative of the outside function and ignore the inside functions. The derivative of u^3 is $3u^2$

We get: $\dfrac{d}{dx}[\ \ ]^3 = 3[\ \ ]^2$.

Step 2: Next, we take the derivative of the cosine term and multiply. The derivative of $\cos u$ is $-\sin u$.

$$\frac{d}{dx}\left[\cos(\ )\right]^3 = -3\left[\cos(\ )\right]^2 \sin(\ )$$

Step 3: Finally, we take the derivative of $x+1$ and multiply. The derivative of $x+1$ is 1.

$$\frac{d}{dx}\left[\cos(x+1)\right]^3 = -3\left[\cos(x+1)\right]^2 \sin(x+1)$$

The answer is (A).

Note: If you had trouble with this problem, you should review the section on **The Chain Rule.**

Problem 27: $\displaystyle\int x\sqrt{x+3}\, dx =$

We can do this integral with u-substitution.

Step 1: Let $u = x+3$. Then $du = dx$ and $u-3 = x$.

Step 2: Substituting, we get:

$$\int x\sqrt{x+3}\, dx = \int (u-3)u^{\frac{1}{2}}\, du.$$

Why is this better than the original integral, you might ask? Because now we can distribute and the integral becomes easy.

Step 3: When we distribute, we get:

$$\int (u-3)u^{\frac{1}{2}}\, du = \int \left(u^{\frac{3}{2}} - 3u^{\frac{1}{2}} \right) du$$

Step 4: Now we can integrate:

$$\int \left(u^{\frac{3}{2}} - 3u^{\frac{1}{2}} \right) du = \frac{2}{5}u^{\frac{5}{2}} - 3\cdot\frac{2}{3}u^{\frac{3}{2}} + C$$

Step 5: Substituting back, we get:

$$\frac{2}{5}u^{\frac{5}{2}} - 3\cdot\frac{2}{3}u^{\frac{3}{2}} + C = \frac{2}{5}(x+3)^{\frac{5}{2}} - 3\cdot\frac{2}{3}(x+3)^{\frac{3}{2}} + C.$$

The answer is (C).

Note: If you had trouble with this problem, review the section on **u-substitution.**

PROBLEM 28: If $f(x) = \ln(\ln(1-x))$, then $f'(x) =$

Here, we use the chain rule.

Step 1: First, take the derivative of the outside function.

The derivative of $\ln u$ is $\dfrac{du}{u}$.

We get:

$$\frac{d}{dx}\ln(\ln(\ )) = \frac{1}{\ln(\ )}.$$

Step 2: Now we take the derivative of the function in the denominator. Once again, the function is $\ln u$.

We get:

$$\frac{d}{dx}\ln(\ln(1-x)) = \frac{1}{\ln(1-x)} \cdot \frac{-1}{1-x} = -\frac{1}{(1-x)\ln(1-x)}.$$

The answer is (D).

Note: If you had trouble with this problem, you should review the section on **The Chain Rule**.

PROBLEM 29. $\displaystyle\int_0^{\frac{\pi}{4}} \sin x \ dx + \int_{\frac{\pi}{4}}^0 \cos x \ dx =$

This is a pair of basic Trigonometric Integrals. You should have memorized several trigonometric integrals, particularly $\displaystyle\int \sin x \ dx = -\cos x + C$ and. $\displaystyle\int \cos x \ dx = -\sin x + C$

Step 1: $\displaystyle\int_0^{\frac{\pi}{4}} \sin x \ dx + \int_{\frac{\pi}{4}}^0 \cos x \ dx = -\cos x\Big|_0^{\frac{\pi}{4}} + |\sin x\Big|_{\frac{\pi}{4}}^0$

Step 2: Now we evaluate the limits of integration, and we're done

$$-\cos x\Big|_0^{\frac{\pi}{4}} + |\sin x\Big|_{-\frac{\pi}{4}}^0 = \left(-\cos\frac{\pi}{4}\right) - (-\cos(0)) + (\sin(0)) - \left(\sin\left(-\frac{\pi}{4}\right)\right) = -\frac{1}{\sqrt{2}} + 1 + 0 + \frac{1}{\sqrt{2}} = 1$$

The answer is (D).

Note: If you had trouble with this problem, you should review the unit on Trigonometric Integrals.

PROBLEM 30. Boats A and B leave the same place at the same time. Boat A heads due north at 12 km/hr. Boat B heads due east at 18 km/hr. After 2.5 hours, how fast is the distance between the boats increasing (in km/hr)?

Step 1: The boats are moving at right angles to each other and are thus forming a right triangle with the distance between them forming the hypotenuse. Whenever we see right triangles in related rates problems, we look to use the Pythagorean Theorem. Call the distance that Boat A travels y, and the distance that Boat B travels x. Then the rate at which Boat A goes north is $\frac{dy}{dt}$, and the rate at which Boat B travels is $\frac{dx}{dt}$. The distance between the two boats is z, and we are looking for how fast z is growing, which is $\frac{dz}{dt}$. Now we can use the Pythagorean Theorem to set up the relationships: $x^2 + y^2 = z^2$

Step 2: Differentiating both sides we obtain:

$$2x\frac{dx}{dt} + 2y\frac{dy}{dt} = 2z\frac{dz}{dt} \text{ or } x\frac{dx}{dt} + y\frac{dy}{dt} = z\frac{dz}{dt}$$

Step 3: After 2.5 hours, Boat A has traveled 30 km and Boat B has traveled 45 km. Because of the Pythagorean Theorem, we also know that, when $y = 30$ and $x = 45$, $z = 54.08$.

Step 4: Now we plug everything into the equation from Step 2 and solve for $\frac{dz}{dt}$.

$$(45)(18) + (30)(12) = (54.08)\frac{dz}{dt}$$

$$1170 = (54.08)\frac{dz}{dt}$$

$$21.63 = \frac{dz}{dt}$$

The answer is (A).

Note: If you had trouble with this problem, you should review the unit on Related Rates.

PROBLEM 31. $\lim\limits_{h \to 0} \dfrac{\tan\left(\dfrac{\pi}{6} + h\right) - \tan\left(\dfrac{\pi}{6}\right)}{h} =$

This may *appear* to be a limit problem, but it is *actually* testing to see whether you know The Definition of the Derivative.

Step 1: You should recall that the Definition of the Derivative says

$$\lim_{h \to 0} \frac{f(x+h) - f(x)}{h} = f'(x).$$

Thus, if we replace $f(x)$ with $\tan(x)$, we can rewrite the problem as:

$$\lim_{h \to 0} \frac{\tan(x+h) - \tan(x)}{h} = [\tan(x)]'$$

Step 2: The derivative of $\tan x$ is $\sec^2 x$. Thus

$$\lim_{h \to 0} \frac{\tan\left(\frac{\pi}{6} + h\right) - \tan\left(\frac{\pi}{6}\right)}{h} = \sec^2\left(\frac{\pi}{6}\right)$$

Step 3: Because $\sec\left(\frac{\pi}{6}\right) = \frac{2}{\sqrt{3}}, \sec^2\left(\frac{\pi}{6}\right) = \frac{4}{3}$

The answer is (B).

> **Note:** If you had trouble with this problem, you should review the units on The Definition of the Derivative and Derivatives of Trigonometric Functions.

PROBLEM 32. If $\int_{30}^{100} f(x)dx = A$ and $\int_{50}^{100} f(x)dx = B$ then $\int_{30}^{50} f(x)dx =$

This question is testing your knowledge of the rules of Definite Integrals.

Step 1: Generally speaking, $\int_a^b f(x)dx + \int_b^c f(x)dx = \int_a^c f(x)dx$

So here, $\int_{30}^{50} f(x)dx + \int_{50}^{100} f(x)dx = \int_{30}^{100} f(x)dx$

If we substitute $\int_{30}^{100} f(x)dx = A$ and $\int_{50}^{100} f(x)dx = B$, we get $\int_{30}^{50} f(x)dx + B = A$

The answer is (B).

> **Note:** If you had trouble with this problem, you should review the unit on Definite Integrals.

PROBLEM 33. If $f(x) = 3x^2 - x$, and $g(x) = f^{-1}(x)$, then $g'(10)$ could be

This problem requires you to know how to find the Derivative of an Inverse Function.

Step 1: The rule for finding the derivative of an inverse function is:

$$\text{If } y = f(x) \text{ and if } g(x) = f^{-1}(x) \text{ than } g'(x) = \frac{1}{f'(y)}$$

Step 2: In order to use the formula, we need to find the derivative of f and the value of x that corresponds to $y = 10$.

First, $f'(x) = 6x - 1$. Second, when $y = 10$ we get $10 = 3x^2 - x$.

If we solve this for x we get $x = 2$ (and $x = -\frac{5}{3}$ but we'll use 2. It's easier.)

Step 3: Plugging into the formula, we get $\dfrac{1}{f'(y)} = \dfrac{1}{(6)(2)-1} = \dfrac{1}{11}$.

The answer is (E).

> **Note:** There was another possible answer using $x = -\dfrac{5}{3}$, but that doesn't give us one of the answer choices. Generally, the AP examination sticks to the easier answer. They are testing whether you know what to do and usually NOT trying to trick you.
>
> **Note:** If you had trouble with this problem, you should review the unit on the Derivative of an Inverse Function.

PROBLEM 34. The graph of $y = x^3 - 5x^2 + 4x + 2$ has a local minimum at

This is another Maxima/Minima question. Fortunately, we can now use a calculator to solve this problem.

If you are comfortable with a graphing calculator, do the following:

Step 1: Graph the function in a standard window.

Step 2: Go to 2nd CALC Minimum and follow the directions. If you are unfamiliar with this, see the unit on Using the TI-82 Calculator. You should get –4.06. Wasn't that easy?

The answer is (C).

If you are not comfortable with a graphing calculator, you can do the following:

Step 1: Take the derivative of the function and set it equal to zero.

$$f'(x) = 3x^2 - 10x + 4 = 0$$

Step 2: Use the quadratic formula to solve for x. You should get $x = \{287, 0.46\}$.

Step 3: Now take the second derivative of the function.

$$f''(x) = 6x - 10$$

Step 4: Plug each of the critical values from Step 2 into the second derivative. If you get a positive value, the point is a minimum. If you get a negative value, the point is a maximum. If you get zero, the point is probably a point of inflection (don't worry about that here).

$$f''(2.87) = 7.21$$
$$f''(.46) = -7.21$$

So 2.87 is the x-coordinate of the minimum. To find the y-coordinate, just plug 2.87 into $f(x)$ and you get -4.06.

The answer is (C).

> **Note:** If you had trouble with this problem, you should review the unit on Maxima/Minima. *Any time that you are asked to find a maximum or a minimum of an equation in the calculator section of the multiple choice part of the AP Exam, you should use the 2nd CALC functions.*

PROBLEM 35. The volume generated by revolving about the y-axis the region enclosed by the graphs $y = 9 - 2x$ and $y = 9 - 3x$, for $0 \le x \le 2$, is

This is another Volume of a Solid of Revolution problem. As you should have noticed by now, these are very popular on the AP Examination and show up in both the multiple-choice section and in the Long Problem section. If you are not good at these, go back and review the unit carefully. You cannot afford to get these wrong on the AP! The good thing about *this* volume problem is that it is in the calculator part of the multiple choice section, so you can use a graphing calculator to assist you.

Step 1: First, graph the two curves on the same set of axes. The graph should look like this:

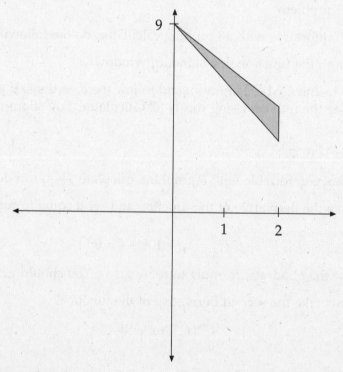

Step 2: We are being asked to rotate this region around the y-axis, and both of the functions are in terms of x, so we should use the method of shells. We use this method whenever we take a vertical slice of a region and rotate it around an axis parallel to the slice (review the unit if you are not sure what it means). This will give us a region that looks like this:

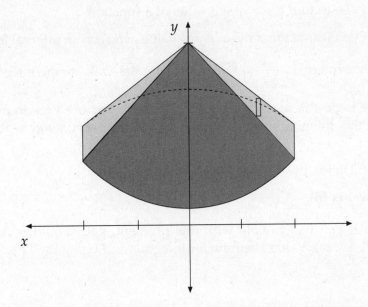

Step 3: The formula for the method of shells says that if you have a region between two curves, $f(x)$ and $g(x)$ from $x = a$ to $x = b$, then the volume generated when the region is revolved around the y - axis is: $2\pi \int_a^b x\big[f(x) - g(x)\big]dx$; if $f(x)$ is above $g(x)$ throughout the region. Thus our integral is.

$$2\pi \int_0^2 x\big[(9 - x^2) - (9 - 3x)\big]dx$$

We can simplify this integral to $2\pi \int_0^2 x(3x - x^2)dx = 2\pi \int_0^2 (3x^2 - x^3)dx$.

Step 4: If you are comfortable with your TI-82 calculator you can use the **fnint** function under the MATH menu. You should enter: **fnint** $(3x^2 - x^3)$, $(x, 0, 2)$ and multiply the answer by 2 . You should get: 25.1327. If you then try each of the answer choices you will find that this equals 8 .

You can always evaluate the integral by hand:

$$2\pi \int_0^2 (3x^2 - x^3)dx = 2\pi \left(x^3 - \frac{x^4}{4} \right)\Big|_0^2 = 8\pi$$

The answer is (C).

Note: If you had trouble with this problem, you should review the unit on Finding the Volume of a Solid of Revolution. You also should familiarize yourself with the calculator by reviewing the unit on Using the TI-82 Calculator.

Problem 36. The average value of the function $f(x) = \ln^2 x$ on the interval $[2,4]$ is

This problem requires you to be familiar with the Mean Value Theorem for Integrals which we use to find the average value of a function.

Step 1: If you want to find the average value of $f(x)$ on an interval $[a, b]$, you need to evaluate the integral. $\dfrac{1}{b-a}\displaystyle\int_a^b f(x)dx$ So here we evaluate the integral $\dfrac{1}{2}\displaystyle\int_2^4 \ln^2 x \ dx$.

You have to do this integral on your calculator because you do not know how to evaluate this integral analytically unless you are very good with integration by parts!

Use **fnint**. Divide this by 2 and you will get 1.204.

The answer is (B).

> **Note:** If you had trouble with this problem, you should review the units on Mean Value Theorem for Integrals and Using the TI-82 Calculator.

PROBLEM 37. $\dfrac{d}{dx}\displaystyle\int_0^{3x} \cos(t)dt =$

This problem is testing your knowledge of the Second Fundamental Theorem of Calculus. The theorem states that $\dfrac{d}{dx}\displaystyle\int_a^u f(t)dt = f(u)\dfrac{du}{dx}$, where a is a constant and u is a function of x. So all we have to do is follow the theorem. $\dfrac{d}{dx}\displaystyle\int_0^{3x} \cos(t)dt = 3\cos 3x$

The answer is (E).

> **Note:** If you had trouble with this problem, you should review the unit on The Fundamental Theorem of Calculus.

PROBLEM 38. If the definite integral $\displaystyle\int_1^3 \left(x^2 + 1\right)dx$ is approximated by using the Trapezoid Rule with $n = 4$, the error is

This problem will require you to be familiar with the Trapezoid Rule. This is very easy to do on the calculator, and some of you may even have written programs to evaluate this. Even if you haven't, the formula is easy. The area under a curve from $x = a$ to $x = b$, divided into n intervals is approximated by the Trapezoid Rule and is

$$\left(\frac{1}{2}\right)\left(\frac{b-a}{2}\right)\left[y_0 + 2y_1 + 2y_2 + 2y_3 \ldots + 2y_{n-2} + 2y_{n-1} + y_n\right]$$

This formula may look scary, but it actually is quite simple, and the AP Examination never uses a very large value for n anyway.

Step 1: $\dfrac{b-a}{n} = \dfrac{3-1}{4} = \dfrac{1}{2}$. Plugging into the formula, we get:

$$\frac{1}{4}\left[\left(1^2+1\right)+2\left(1.5^2+1\right)+2\left(2^2+1\right)+2\left(2.5^2+1\right)+\left(3^2+1\right)\right]$$

This is easy to plug into your calculator and you will get 10.75 or $\dfrac{43}{4}$

Step 2: In order to find the error, we now need to know the actual value of the integral. You could do this easily on your calculator using **fnint** or if you want to do it analytically, you get:

$$\int_1^3 x^2+1\,dx = \frac{x^3}{3}+x\Big|_1^3 = \frac{32}{2} \text{ or } 10.666$$

Step 3: The error is $\dfrac{43}{4}-\dfrac{32}{3}=\dfrac{1}{12}$

The answer is (C).

> **Note:** If you had trouble with this problem, you should review the unit on Rectangular, Trapezoid, and Simpson's Rules.

PROBLEM 39. The radius of a sphere is increasing at a rate proportional to itself. If the radius is 4 initially, and the radius is 10 after two seconds, then what will the radius be after three seconds?

This is not a Related Rate problem, this is a Differential Equation! It just happens to involve a rate.

Step 1: If we translate the first sentence into an equation we get: $\dfrac{dR}{dt}=kR$

Put all of the terms that contain an R on the left of the equals sign, and all of the terms that contain a t on the right hand side. $\dfrac{dR}{R}=k\,dt$

Step 2: Integrate both sides: $\displaystyle\int \frac{dR}{R} = k\int dt$

Step 3: If we solve this for R we get $R=Ce^{kt}$ (see the Unit on Differential Equations).

Now we need to solve for C and k. First we solve for C by plugging in the information that the radius is 4 initially. This means that $R=4$ when $t=0$.

$$4=Ce^0 \text{ then } C=4.$$

Next we solve for k by plugging in the information that $R = 10$ when $t = 2$.

$$10 = 4e^{2k}$$

$$\frac{5}{2} = e^{2k}$$

$$\ln \frac{5}{2} = 2k$$

$$\frac{1}{2} \ln \frac{5}{2} = k$$

Step 4: Now we have our final equation: $R = 4e^{\left(\frac{1}{2}\ln\frac{5}{2}\right)t}$

If we plug in $t = 3$ we get: $R = 4e^{\left(\frac{1}{2}\ln\frac{5}{2}\right)(3)} = 15.811$

The answer is (C).

> **Note:** If you had trouble with this problem, you should review the unit on Differential Equations.

PROBLEM 40. If Newton's method is used to estimate the real root of $x^3 + x^2 + 2$, then a first approximation of $x_1 = -2$ gives a third approximation of $x_3 =$

> Newton's Method is a formula for approximating the roots of a polynomial. One puts a guess, called x_1, into the formula and gets out a value that is an approximate root, called x_2. One then puts x_2 into the formula and gets a *better* approximation of the root, called x_3, and so on. One can continue in this fashion and get a more and more accurate approximation. If you don't have a calculator, it is difficult to go more than a couple of iterations into the formula because the decimals start to get ridiculous. Fortunately, we *have* a calculator, so this will be easy. Furthermore, there is a program that you can put into your calculator that will do this trivially. It is in the unit on Using the TI-82 Calculator. But assuming that you don't have the program . . .

Step 1: The formula is: $x_2 = x_1 - \frac{f(x_1)}{f'(x_1)}$

First, we need to find the derivative: $f'(x) = 3x^2 + 2x$.

Now we plug in –2 for x_1 and we get: $x_2 = -2 - \frac{(-2)^3 + (-2)^2 + 2}{3(-2)^2 + 2(-2)} = -1.75$

Now we plug –1.75 into the formula and we get:

$$x^3 = -1.75 - \frac{(-1.75)^3 + (-1.75)^2 + 2}{3(-1.75)^2 + 2(-1.75)} = -1.698$$

The answer is (C).

> **Note:** If you had trouble with this problem, you should review the unit on Newton's Method.

PROBLEM 41. $\displaystyle\int \ln 2x \ dx =$

This is a simple integral that we do using Integration By Parts. The AB Examination only has the simplest of these types of integrals, although the BC Examination has harder ones. Furthermore, you should memorize that $\displaystyle\int \ln(ax) \ dx = x\ln(ax) - x + C$, which makes this integral easy.

Step 1: The formula for Integration By Parts is: $\displaystyle\int u\,dv = uv - \int v\,du$

The trick is that we have to let $dv = dx$.

$$\text{Let } u = \ln 2x \qquad \text{and } dv = dx$$

$$du = \frac{2}{2x}du = \frac{1}{x}dx \text{ and } v = x$$

Plugging in to the formula we get:

$$\int \ln 2x\,dx = x\ln 2x - \int dx = x\ln 2x - x + C$$

The answer is (D).

> **Note:** If you had trouble with this problem, you should review the unit on Integration By Parts.

PROBLEM 42. For the function $f(x) = \begin{cases} ax^3 - 6x; \text{if } x \le 1 \\ bx^2 + 4; x > 1 \end{cases}$ to be continuous and differentiable, a must be

This question is testing your knowledge of the rules of Continuity, where we also discuss differentiability.

Step 1: If the function is continuous, then if we plug 1 into th e top and bottom pieces of the function we should get the same answer.

$$a(1^3) - 6(1) = b(1^2) + 4$$

$$a - 6 = b + 4$$

Step 2: If the function is differentiable, then if we plug 1 into the derivatives of the top and bottom pieces of the function we should get the same answer.

$$3a(1^2) - 6 = 2b(1)$$
$$3a - 6 = 2b$$

Step 3: Now we have a pair of simultaneous equations. If we solve them, we get $a = -14$

The answer is (C).

> **Note:** If you had trouble with this problem, you should review the unit on Continuity.

PROBLEM 43. Two particles leave the origin at the same time and move along the y-axis with their respective positions determined by the functions $y_1 = \cos 2t$ and $y_2 = 4 \sin t$ for $0 < t < 6$. For how many values of t do the particles have the same acceleration?

If you want to find acceleration, all you have to do is take the second derivative of the position functions.

Step 1: $\qquad \dfrac{dy_1}{dx} = -2 \sin 2t \quad$ and $\quad \dfrac{dy_2}{dx} = 4 \cos t$

$$\dfrac{d^2 y_1}{dx^2} = -4 \cos 2t \quad \text{and} \quad \dfrac{d^2 y_2}{dx^2} = -4 \sin t$$

Step 2: Now all we have to do is to graph both of these equations on the same set of axes on a calculator. You should make the window from $x = 0$ to $x = 7$ (leave yourself a little room so that you can see the whole range that you need). You should get a picture that looks like this:

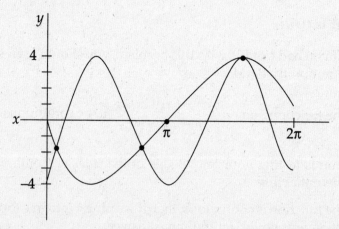

Where the graphs intersect, the acceleration is the same. There are three points of intersection.

The answer is (D).

> **Note:** If you had trouble with this problem, you should review the unit on Position, Velocity, and Acceleration.

PROBLEM 44: Find the distance traveled (to three decimal places) in the first four seconds, for a particle whose velocity is given by $v(t) = 7e^{-t^2}$; where t stands for time.

Step 1: If we want to find the distance traveled, we take the integral of velocity from the starting time to the finishing time. Therefore, we need to evaluate $\int_0^4 7e^{-t^2} dt$.

Step 2: But we have a problem! We can't take the integral of e^{-t^2}. This means that the AP want you to find the answer using your calculator.

Step 3: On the TI 82/83, you enter: **fnint(7e^(-x^2),x,0,4)**. You should get 6.203588383.

Rounded to three decimal places, the answer is 6.204 .

The answer is (B).

PROBLEM 45: $\int \tan^6 x \sec^2 x \, dx =$

We can do this integral with u-substitution.

Step 1: Let $u = \tan x$. Then $du = \sec^2 x \, dx$.

Step 2: Substituting, we get: $\int \tan^6 x \sec^2 x \, dx = \int u^6 du$.

Step 3: This is an easy integral: $\int u^6 du = \frac{u^7}{7} + C$.

Step 4: Substituting back, we get: $\frac{\tan^7 x}{7} + C$.

The answer is (A).

Note: If you had trouble with this problem, review the section on **u-substitution**.

ANSWERS AND EXPLANATIONS TO SECTION II

PROBLEM 1. Let f be the function given by $y = f(x) = 2x^4 - 4x^2 + 1$.

(a) Find an equation of the line tangent to the graph at $(-2, 17)$.

In order to find the equation of a Tangent Line at a particular point we need to take the derivative of the function and plug in the x and y values at that point to give us the slope of the line.

Step 1: The derivative is: $f'(x) = 8x^3 - 8x$. If we plug in $x = -2$, we get:

$$f'(-2) = 8(-2)^3 - 8(2) = -48$$

This is the slope m.

Step 2: Now we use the slope-intercept form of the equation of a line, $y - y_1 = m(x - x_1)$, and plug in the appropriate values of x, y, and m.

$$y - 17 = -48(x + 2).$$

If we simplify this we get $y = -48x - 79$.

This was worth 2 points—1 for finding the slope and 1 for coming up with the correct equation in any form.

(b) Find the x and y-coordinates of the relative maxima and relative minima.

If we want to find the maxima/minima, we need to take the derivative and set it equal to zero. The values that we get are called critical points. We will then test each point to see if it is a maximum or a minimum.

Step 1: We already have the first derivative from part (a), so we can just set it equal to zero:

$$8x^3 - 8x = 0$$

If we now solve this for x we get:

$$8x(x^2 - 1) = 0 \quad 8x(x + 1)(x - 1) = 0 \quad x = 0, 1, -1$$

These are our critical points. In order to test if a point is a maximum or a minimum, we usually use the *second derivative test*. We plug each of the critical points into the second derivative. If we get a positive value, the point is a relative minimum. If we get a negative value, the point is a relative maximum. If we get zero, the point is a point of inflection.

Step 2: The second derivative is $f''(x) = 24x^2 - 8$. If we plug in the critical points we get:

$$f''(0) = 24(0)^2 - 8 = -8$$

$$f''(1) = 24(1)^2 - 8 = 16$$

$$f''(-1) = 24(-1)^2 - 8 = 16$$

So $x = 0$ is a relative maximum, and $x = 1, -1$ are relative minima.

Step 3: In order to find the y-coordinates, we plug the x values back into the original equation, and solve.

$$f(0) = 1$$

$$f(1) = -1$$

$$f(-1) = -1$$

and our points are

$(0,1)$ is a relative maximum

$(1,-1)$ is a relative minimum

$(-1,-1)$ is a relative minimum

This was worth 3 points—1 for correctly identifying each critical point.

(c) Find the x-coordinates of the points of inflection.

If we want to find the points of inflection, we set the second derivative equal to zero. The values that we get are the x-coordinates of the points of inflection.

Step 1: We already have the second derivative from part (b), so all we have to do is set it equal to zero and solve for x:

$$24x^2 - 8 = 0 \quad x^2 = \frac{1}{3} \quad x = \pm\sqrt{\frac{1}{3}}$$

Step 2: In order to find the y-coordinates, we plug the x values back into the original equation, and solve.

$$f\left(\sqrt{\frac{1}{3}}\right) = 2\left(\sqrt{\frac{1}{3}}\right)^4 - 4\left(\sqrt{\frac{1}{3}}\right)^2 + 1 = \frac{2}{9} - \frac{4}{3} + 1 = -\frac{1}{9}$$

$$f\left(-\sqrt{\frac{1}{3}}\right) = 2\left(-\sqrt{\frac{1}{3}}\right)^4 - 4\left(-\sqrt{\frac{1}{3}}\right)^2 + 1 = \frac{2}{9} - \frac{4}{3} + 1 = -\frac{1}{9}$$

So the points of inflection are: $\left(\sqrt{\frac{1}{3}}, -\frac{1}{9}\right)$ and $\left(-\sqrt{\frac{1}{3}}, -\frac{1}{9}\right)$.

This was worth 2 points: 1 for correctly identifying each point of inflection.

(d) Sketch the graph of $f(x)$.

You can graph this function easily using your calculator. Just put the equation into $y=$ and set the window to $x\min = -4$, $x\max = 4$, $y\min = -6$, and $y\max=6$. Then hit GRAPH.

You should get this:

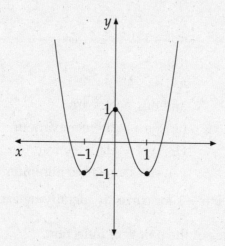

Viewing Window

[-4 x 4] by [-6 x 6]

This was worth 2 points for graphing correctly.

PROBLEM 2. A particle moves along the x-axis so that its acceleration at any time $t > 0$ is given by $a(t) = 12t - 18$. At time $t = 1$, the velocity of the particle is $v(1) = 0$ and the position is $x(1) = 9$.

(a) Write an expression for the velocity of the particle $v(t)$.

Step 1: We know that the derivative of velocity with respect to time is acceleration, so the integral of acceleration with respect to time is velocity.

$$\int a(t)\,dt = v(t). \int 12t - 18\,dt = 6t^2 - 18t + C = v(t)$$

If we plug in the information that at time $t = 1$, $v(1) = 0$, we can solve for C.

$$6(1)^2 - 18(1) + C = 0$$
$$-12 + C = 0$$
$$C = 12$$

This means that the velocity of the particle is $6t^2 - 18t + 12$.

This was worth 2 points—1 for integrating correctly and 1 for solving for the constant.

(b) At what values of t does the particle change direction?

When a particle is in motion, it changes direction at the time when its velocity is zero. So all we have to do is set velocity equal to zero and solve for t.

$$6t^2 - 18t + 12 = 0$$
$$t^2 - 3t + 2 = 0$$
$$(t-2)(t-1) = 0$$
$$t = 1, 2$$

This was worth 2 points—1 for realizing that you should set velocity equal to zero and 1 for getting the correct answers.

(c) Write an expression for the position x (t) of the particle.

We know that the derivative of position with respect to time is velocity, so the integral of velocity with respect to time is position. $\int v(t)\,dt = x(t)$.

$$\int 6t^2 - 18t + 12\,dt = 2t^3 - 9t^2 + 12t + C = x(t)$$

If we plug in the information that at time $t = 1$, x $(1) = 9$, we can solve for C.

$$2(1)^3 - 9(1)^2 + 12(1) + C = 9$$
$$5 + C = 9$$
$$C = 4 \text{ so } x(t) = 2t^3 - 9t^2 + 12t + 4$$

This was worth 2 points—1 for integrating correctly and 1 for solving for the constant.

(d) Find the total distance traveled by the particle from $t = \dfrac{3}{2}$ to $t = 6$?

Step 1: Normally, all that we have to do to find the distance traveled is to integrate the velocity equation from the starting time to the ending time. But we have to watch out for whether the particle changes direction. If so, we have to break the integration into two parts—a positive integral for when it is traveling to the right, and a negative integral for when it is traveling to the left.

We know that the particle changes direction at $t = 1$ and at $t = 2$. We need to know which direction the particle is moving at time $t = \dfrac{3}{2}$. We can do this by plugging $\dfrac{3}{2}$ into the velocity and looking at its sign. $6\left(\dfrac{3}{2}\right)^2 - 18\left(\dfrac{3}{2}\right) + 12 = -\dfrac{3}{2}$.

This is negative, so the particle is moving to the left from $t = \frac{3}{2}$ to $t = 2$. And, because we know that the particle changes direction at $t = 2$, it must be moving to the right after that. Therefore we are going to need two integrals, one for $t = \frac{3}{2}$ to $t = 2$ and one for $t = 2$ to $t = 6$. So we need to evaluate:

$$\int_{\frac{3}{2}}^{2} -\left(6t^2 - 18t + 12\right)dt + \int_{2}^{6}\left(6t^2 - 18t + 12\right)dt$$

We already integrated these in part (c), $\int\left(6t^2 - 18t + 12\right)dt = 2t^3 - 9t^2 + 12t + C$, but now, instead of solving for the constant, we evaluate the equation at the limits of integration:

$$-\left(2(2)^3 - 9(2)^2 + 12(2)\right) + \left(2\left(\frac{3}{2}\right)^3 - 9\left(\frac{3}{2}\right)^2 + 12\left(\frac{3}{2}\right)\right) +$$

$$\left(2(6)^3 - 9(6)^2 + 12(6)\right) - \left(2(2)^3 - 9(2)^2 + 12(2)\right)$$

$$-\left(2t^3 - 9t^2 + 12t\right)\Big|_{\frac{3}{2}}^{2} + \left(2t^3 - 9t^2 + 12t\right)\Big|_{2}^{6} = -4 + 4.5 + 180 - 4 = 176.5 \text{ or } \frac{353}{2}$$

This was worth 3 points—1 apiece for setting up each of the two integrals and 1 for the correct answer.

PROBLEM 3. Let R be the region enclosed by the graphs of $y = 2 \ln x$ and $y = \frac{x}{2}$, and the lines $x = 2$ and $x = 8$.

(a) Find the area of R.

Step 1: If there are two curves, $f(x)$ and $g(x)$, where $f(x)$ is always above $g(x)$, on the interval $[a, b]$, then the area of the region between the two curves is found by:

$$\int_{a}^{b}\left(f(x) - g(x)\right)dx$$

In order to determine whether one of the curves is above the other, we can graph them on the calculator. Put the equations $y_1 = 2 \ln x$ and $y_2 = \frac{x}{2}$ into **y=** and set the window to **xmin** = 2, **xmax** = 10, **ymin** = 0, and **ymax**=5. Then hit **GRAPH**. The graph looks like this:

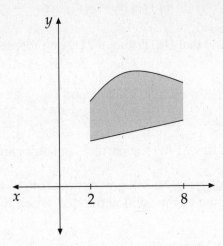

As we can see, the graph of $y = 2 \ln x$ is above $y = \dfrac{x}{2}$ on the entire interval, so all we

have to do is evaluate the integral $\displaystyle\int_2^8 \left(2 \ln x - \dfrac{x}{2} \right) dx =$.

Step 2: We can do the integration one of two ways—on the calculator or analytically.

Calculator: Evaluate **fnint**$\big((2 \ln x - (x/2)), x, 2, 8 \big) = 3.498$

Analytically: $\displaystyle\int_2^8 \left(2 \ln x - \dfrac{x}{2} \right) dx = 2 \int_2^8 \ln x \, dx - \dfrac{1}{2} \int_2^8 x \, dx =$

$$= 2(x \ln x - x)\Big|_2^8 - \dfrac{1}{2}\left(\dfrac{x^2}{2} \right)\Bigg|_2^8 = 18.498 - 15 = 3.498$$

By the way, you should have memorized $\displaystyle\int \ln x \, dx = x \ln x - x$, or you can do it as one
of the basic Integration By Parts integrals.

This was worth 3 points—1 for setting up the integral, 1 for antidifferentiating
correctly, and 1 for evaluating the limits correctly.

(b) Set up, but *do not integrate,* an integral expression, in terms of a single
variable, for the volume of the solid generated when R is revolved about
the *x-axis.*

Step 1: If there are two curves, $f(x)$ and $g(x)$, where $f(x)$ is always above $g(x)$, on the
interval $[a, b]$, then the volume of the solid generated when the region is revolved
about the *x*-axis is found by using the method of washers:

$$\pi \int_a^b [f(x)]^2 - [g(x)]^2 \, dx$$

Here, we already know that $f(x)$ is above $g(x)$ on the interval, so the integral we need to evaluate is:

$$\pi \int_2^8 [2 \ln x]^2 - \left[\frac{x}{2}\right]^2 \, dx$$

This was worth 2 points—1 for the correct constants and 1 for the correct integral.

(c) Set up, but <u>do not integrate</u>, an integral expression, in terms of a single variable, for the volume of the solid generated when R is revolved about the line $x = -1$.

Step 1: Now we have to revolve the area around a <u>vertical</u> axis. If there are two curves, $f(x)$ and $g(x)$, where $f(x)$ is always above $g(x)$, on the interval $[a, b]$, then the volume of the solid generated when the region is revolved about the y-axis is found by using the method of shells:

$$2\pi \int_a^b x[f(x) - g(x)] \, dx$$

When we are rotating around a vertical axis, we use the same formula as when we rotate around the y-axis, but we have to account for the shift away from $x = 0$. Here we have a curve that is 1 unit farther away from the line $x = -1$ than it is from the y-axis, so we add 1 to the radius of the shell (For a more detailed explanation of shifting axes, see the unit on Finding the Volume of a Solid of Revolution). This gives us the equation:

$$2\pi \int_2^8 (x+1)\left[2 \ln x - \frac{x}{2}\right] dx$$

This was worth 3 points—1 for the correct constants, 1 for using the shells method, and 1 for getting the shift correct.

Problem 4. Water is draining at the rate of 48 ft^3/sec from a conical tank whose diameter at its base is 40 feet and whose height is 60 feet.

(a) Find an expression for the volume of water (in ft^3/sec) in the tank in terms of its radius.

The formula for the volume of a cone is: $V = \frac{1}{3}\pi R^2 H$, where R is the radius of the cone, and H is the height. The ratio of the height of a cone to its radius is constant at any point on the edge of the cone, so we also know that $\frac{H}{R} = \frac{60}{20} = 3$ (Remember that the radius is half the diameter.). If we solve this for H and substitute, we get:

$$H = 3R$$

$$V = \frac{1}{3}\pi R^2(3R) = \pi R^3$$

This was worth 3 points—1 for the formula of a cone, 1 for the ratio of height to radius, and 1 for the correct final formula.

(b) At what rate (in ft/sec) is the radius of the water in the tank shrinking when the radius is 16 feet?

Step 1: This is a Related Rates question. We now have a formula for the volume of the cone in terms of its radius, so if we differentiate it in terms of t we should be able to solve for the rate of change of the radius $\frac{dR}{dt}$.

We are given that the rate of change of the volume and the radius are, respectively:

$$\frac{dV}{dt} = 48\pi \text{ and } R = 16.$$

Differentiating the formula for the volume, we get: $\frac{dV}{dt} = 3\pi R^2 \frac{dR}{dt}$.

Now we plug in and get: $48\pi = 3\pi 16^2 \frac{dR}{dt}$. Finally, if we solve for $\frac{dR}{dt}$, we get:

$$\frac{dR}{dt} = \frac{1}{16} ft/\sec.$$

This was worth 2 points—1 for differentiating correctly and 1 for the correct answer.

(c) How fast (in ft/sec) is the height of the water in the tank dropping at the instant that the radius is 16 feet?

Step 1: This is the same idea as the previous problem, except that we want to solve for $\frac{dH}{dt}$. In order to do this, we need to go back to our ratio of height to radius and solve it for the radius:

$$\frac{H}{R} = 3 \quad \text{or} \quad \frac{H}{3} = R.$$

Substituting for R in the original equation, we get: $V = \frac{1}{3}\pi\left(\frac{H}{3}\right)^2 H = \frac{\pi H^3}{27}$.

Step 2: Now we need to know what H is when R is 16. Using our ratio:

$$H = 3(16) = 48.$$

Step 3: Now if we differentiate we get:

$$\frac{dV}{dt} = \frac{\pi H^2}{9} \frac{dH}{dt}.$$

Now we plug in and solve:

$$48\pi = \frac{\pi(48)^2}{9} \frac{dH}{dt}$$

$$\frac{dH}{dt} = \frac{3}{16}$$

One should also note that, because $H = 3R$, $\dfrac{dH}{dt} = 3\dfrac{dR}{dt}$. Thus, after we found $\dfrac{dR}{dt}$ in part 2, we merely had to multiply it by 3 to find the answer for part 3.

This was worth 3 points—1 for the new volume formula, 1 for differentiating correctly, and 1 for the correct answer.

PROBLEM 5. Consider the equation $x^2 - 2xy + 4y^2$.

(a) Write an expression for the slope of the curve at any point (x, y).

Step 1: The slope of the curve is just the derivative. But, here, we have to use Implicit Differentiation to find the derivative. If we take the derivative of each term with respect to x, we get:

$$2x\frac{dx}{dx} - 2\left(x\frac{dy}{dx} + y\frac{dx}{dx} \right) + 8y\frac{dy}{dx} = 0.$$

Remember that $\dfrac{dx}{dx} = 1$, which gives us:

$$2x - 2\left(x\frac{dy}{dx} + y \right) + 8y\frac{dy}{dx} = 0.$$

Step 2: Now just simplify and solve for $\dfrac{dy}{dx}$.

$$2x - 2x\frac{dy}{dx} - 2y + 8y\frac{dy}{dx} = 0$$

$$x - x\frac{dy}{dx} - y + 4y\frac{dy}{dx} = 0$$

$$-x\frac{dy}{dx} + 4y\frac{dy}{dx} = y - x$$

$$\left(4y - x\right)\frac{dy}{dx} = y - x$$

$$\frac{dy}{dx} = \frac{y - x}{4y - x}$$

This was worth 2 points—1 for the differentiation and 1 for simplifying correctly.

(b) Find the equation of the tangent lines to the curve at the point $x = 2$.

Step 1: We are going to use the point-slope form of a line, $y - y_1 = m(x - x_1)$, where (x_1, y_1) is a point on the curve and the derivative at that point is the slope m. First, we need to know the value of y when $x = 2$. If we plug 2 for x into the original equation, we get:

$$4 - 4y + 4y^2 = 64$$

$$4y^2 - 4y - 60 = 0$$

Using the quadratic formula, we get:

$$y = \frac{1 \pm \sqrt{61}}{2} = 4.41, -3.41$$

Notice that there are two values of y when $x = 2$, which is why there are two tangent lines.

Step 2: Now that we have our points, we need the slope of the tangent line at $x = 2$.

$$\frac{dy}{dx} = \frac{y - x}{4y - x}.$$

At $y = 4.41$, $\dfrac{dy}{dx} = \dfrac{4.41 - 2}{4(4.41) - 2} = 0.15$.

At $y = -3.41$, $\dfrac{dy}{dx} = \dfrac{-3.41 - 2}{4(-3.41) - 2} = 0.35$

Step 3: Plugging into our equation for the tangent line, we get:

$$y - 4.41 = 0.15(x - 2)$$
$$y + 3.41 = 0.35(x - 2)$$

It is not necessary to simplify these equations.

These were worth 3 points—1 for finding the y-coordinates, 1 for finding the correct slopes, and 1 for the correct lines.

(c) Find $\dfrac{d^2y}{dx^2}$ at $(0, 4)$.

Step 1: Once we have the first derivative, we have to differentiate again to find $\dfrac{d^2y}{dx^2}$.

But, we have to use implicit differentiation again.

$$\frac{dy}{dx} = \frac{y - x}{4y - x}.$$

Using the quotient rule $\dfrac{d^2y}{dx^2} = \dfrac{(4y - x)\left(\dfrac{dy}{dx} - \dfrac{dx}{dx}\right) - (y - x)\left(4\dfrac{dy}{dx} - \dfrac{dx}{dx}\right)}{(4y - x)^2}.$

Simplifying we get: $\dfrac{d^2y}{dx^2} = \dfrac{(4y - x)\left(\dfrac{dy}{dx} - 1\right) - (y - x)\left(4\dfrac{dy}{dx} - 1\right)}{(4y - x)^2}$

Now, we plug in $\dfrac{y - x}{4y - x}$ for $\dfrac{dy}{dx}$, which gives us:

$$\frac{d^2y}{dx^2} = \frac{(4y - x)\left(\dfrac{y - x}{4y - x} - 1\right) - (y - x)\left(4\dfrac{y - x}{4y - x} - 1\right)}{(4y - x)^2}.$$

Now we would have to use a lot of algebra to simplify this but, fortunately, we can just plug $(0, 4)$ in immediately for x and y, and solve from there.

$$\frac{d^2y}{dx^2} = \frac{(16)\left(\dfrac{4}{16} - 1\right) - (4)\left(4\dfrac{4}{16} - 1\right)}{(16)^2} = \frac{-3}{64}$$

This was worth 3 points—1 for differentiating correctly, 1 for the substitution for $\dfrac{dy}{dx}$, and 1 for the correct answer.

PROBLEM 6. Let $F(x) = \displaystyle\int_0^x \left[\cos\left(\dfrac{t}{2}\right) + \left(\dfrac{3}{2}\right) \right] dt$ on the closed interval $[0, 4\pi]$.

(a) Approximate $F(2\pi)$ using four inscribed rectangles.

Step 1: This means that we need to find $\displaystyle\int_0^{2\pi} \left[\cos\left(\dfrac{t}{2}\right) + \left(\dfrac{3}{2}\right) \right] dt$.

The graph of $\cos\left(\dfrac{t}{2}\right) + \left(\dfrac{3}{2}\right)$ from 0 to 2π, using four inscribed rectangles looks like:

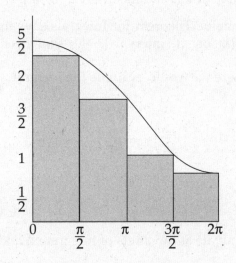

If we are cutting the interval $[0, 2\pi]$ into 4 rectangles, the width of each rectangle is $\dfrac{\pi}{2}$.

The height of each rectangle depends on the x-coordinate.

Step 2: We can now set up the calculation for the area of the rectangles:

$$\text{Area} = \frac{\pi}{2}\left[\left(\cos\frac{\pi}{4} + \frac{3}{2} \right) + \left(\cos\frac{\pi}{2} + \frac{3}{2} \right) + \left(\cos\frac{3\pi}{4} + \frac{3}{2} \right) + \left(\cos\pi + \frac{3}{2} \right) \right]$$

$$= \frac{\pi}{2}\left[\left(\frac{3}{2} + \frac{1}{\sqrt{2}} \right) + \left(\frac{3}{2} \right) + \left(\frac{3}{2} - \frac{1}{\sqrt{2}} \right) + \left(\frac{1}{2} \right) \right] = \frac{5\pi}{2} \approx 7.854$$

This was worth 3 points—1 for setting up the correct integral, 1 for setting up the correct rectangle calculation, and 1 for the correct answer.

(b) Find $F'(2\pi)$

Step 1: The Second Fundamental Theorem of Calculus says that if $f(x)$ is a continuous function, and a is a constant, then $\dfrac{d}{dx}\displaystyle\int_a^x f(t)\,dt = f(x)$.

So here we have: $\dfrac{d}{dx}\displaystyle\int_0^x \cos\left(\dfrac{t}{2}\right)+\dfrac{3}{2}\,dt = \cos\left(\dfrac{x}{2}\right)+\dfrac{3}{2}$

Step 2: Now we plug in 2π for x and we get $\cos\pi + \dfrac{3}{2} = \dfrac{1}{2}$.

This was worth 2 points—1 for the correct derivative and 1 for the correct answer.

(c) Find the average value of $F'(x)$ on the interval $[0, 4\pi]$.

Step 1: The Mean Value Theorem for Integrals says that if you want to find the average value of $f(x)$ on an interval $[a, b]$, you need to evaluate the integral $\dfrac{1}{b-a}\displaystyle\int_a^b f(x)\,dx$. So here we would evaluate the integral $\dfrac{1}{4\pi-0}\displaystyle\int_0^{4\pi}\left(\cos\left(\dfrac{x}{2}\right)+\dfrac{3}{2}\right)dx$.

$$\dfrac{1}{4\pi}\int_0^{4\pi}\left(\cos\left(\dfrac{x}{2}\right)+\dfrac{3}{2}\right)dx = \dfrac{1}{4\pi}\int_0^{4\pi}\cos\dfrac{x}{2}\,dx + \dfrac{3}{8\pi}\int_0^{4\pi}dx =$$

$$\dfrac{1}{4\pi}2\sin\dfrac{x}{2}\Big|_0^{4\pi} + \dfrac{3}{8\pi}x\Big|_0^{4\pi} = \dfrac{1}{2\pi}\sin\dfrac{x}{2}\Big|_0^{4\pi} + \dfrac{3}{8\pi}x\Big|_0^{4\pi}$$

Step 2: Now we evaluate at the limits of integration and we get:

$$\dfrac{1}{2\pi}\sin\dfrac{x}{2}\Big|_0^{4\pi} + \dfrac{3}{8\pi}x\Big|_0^{4\pi} = \dfrac{1}{2\pi}(\sin 2\pi - \sin 0) + \dfrac{3}{8\pi}(4\pi) = \dfrac{3}{2}$$

This was worth 3 points—1 for setting up the integral correctly, 1 for integrating correctly, and 1 for the correct answer.

24

The Princeton Review AP Calculus BC Diagnostic Exam

CALCULUS BC

Time—1 hour and 30 minutes.

Number of questions—40

Percent of total grade—50

Part A consists of 28 questions that will be answered on side 1 of the answer sheet. Following are the directions for Section I, Part A.

<u>Directions:</u> Solve each of the following problems, using the available space for scratchwork. After examining the form of the choices, decide which is the best of the choices given and fill in the corresponding oval on the answer sheet. No credit will be given for anything written in the test book. Do not spend too much time on any one problem.

In this test:

(1) Unless otherwise specified, the domain of a function f is assumed to be the set of all real numbers x for which $f(x)$ is a real number.

1. If $7 = xy - e^{xy}$, then $\dfrac{dy}{dx} =$

(A) $x - e^y$

(B) $y - e^x$

(C) $\dfrac{ye^{xy} + y}{x - xe^{xy}}$

(D) $\dfrac{-y}{x}$

(E) $\dfrac{ye^{xy} + y}{x + xe^{xy}}$

2. The volume of the solid that results when the area between the curve $y = e^x$ and the line $y = 0$, from $x = 1$ to $x = 2$, is revolved around the x-axis is

(A) $2\pi(e^4 - e^2)$

(B) $\dfrac{\pi}{2}(e^4 - e^2)$

(C) $\dfrac{\pi}{2}(e^2 - e)$

(D) $2\pi(e^2 - e)$

(E) $2\pi e^2$

GO ON TO THE NEXT PAGE

3. $\int \dfrac{x-18}{(x+3)(x-4)}\,dx =$

(A) $\int \dfrac{5dx}{(x+3)(x-4)}$

(B) $\int \dfrac{dx}{(x+3)(x-4)}$

(C) $\int \dfrac{3dx}{x+3} + \int \dfrac{2dx}{x-4}$

(D) $\int \dfrac{15dx}{x+3} - \int \dfrac{14dx}{x-4}$

(E) $\int \dfrac{3dx}{x+3} - \int \dfrac{2dx}{x-4}$

4. If $y = 5x^2 + 4x$ and $x = \ln t$ then $\dfrac{dy}{dt} =$

(A) $\dfrac{10}{t} + 4$ (B) $10t\ln t + 4t$ (C) $\dfrac{10\ln t + 4}{t}$ (D) $\dfrac{5}{t^2} + \dfrac{4}{t}$ (E) $10\ln t + \dfrac{4}{t}$

GO ON TO THE NEXT PAGE

5. $\int_0^{\frac{\pi}{2}} \sin^5 x \cos x \, dx =$

(A) $\dfrac{1}{6}$ (B) $-\dfrac{1}{6}$ (C) 0 (D) –6 (E) 6

6. The tangent line to the curve $y = x^3 - 4x + 8$ at the point $(2, 8)$ has an x-intercept at

(A) $(-1, 0)$ (B) $(1, 0)$ (C) $(0, -8)$ (D) $(0, 8)$ (E) $(8, 0)$

7. The graph in the xy–plane represented by $x = 3\sin(t)$ and $y = 2\cos(t)$ is

(A) a circle (B) an ellipse (C) a hyperbola (D) a parabola (E) a line

GO ON TO THE NEXT PAGE

8. $\displaystyle\int \frac{dx}{\sqrt{4-9x^2}} =$

(A) $\dfrac{1}{6}\sin^{-1}\left(\dfrac{3x}{2}\right)+C$

(B) $\dfrac{1}{2}\sin^{-1}\left(\dfrac{3x}{2}\right)+C$

(C) $6\sin^{-1}\left(\dfrac{3x}{2}\right)+C$

(D) $3\sin^{-1}\left(\dfrac{3x}{2}\right)+C$

(E) $\dfrac{1}{3}\sin^{-1}\left(\dfrac{3x}{2}\right)+C$

---

9. $\displaystyle\lim_{x\to\infty} 4x\sin\left(\dfrac{1}{x}\right)$ is

(A) 0   (B) 2   (C) 4   (D) 4π (E) nonexistent

GO ON TO THE NEXT PAGE →

10. The position of a particle moving along the x-axis at time t is given by $x(t) = e^{\cos(2t)}$. For which of the following values of t will $x'(t) = 0$?

   I. $t = 0$

   II. $t = \dfrac{\pi}{2}$

   III. $t = \pi$

(A) I only   (B) II only   (C) I and III only   (D) I and II only   (E) I, II, and III

---

11. $\displaystyle\lim_{h \to 0} \dfrac{\sec(\pi + h) - \sec(\pi)}{h} =$

(A) -1 (B) 0 (C) $\dfrac{1}{\sqrt{2}}$ (D) 1 (E) $\sqrt{2}$

GO ON TO THE NEXT PAGE

12. Let f be a function defined for all real x. If, for each real number L, there exists a $\delta > 0$ and an $\varepsilon > 0$, such that $L - \varepsilon < f(x) < L + \varepsilon$ for $0 < |x| < \delta$, then

(A) $\lim_{x \to 0} f(x) = L$

(B) $\lim_{x \to 0} f(x) = 0$

(C) $\lim_{x \to \infty} f(x) = L$

(D) $\lim_{x \to \infty} f(x) = 0$

(E) $\lim_{x \to L} f(x) = 0$

13. The radius of convergence of $\sum_{n=1}^{\infty} \dfrac{a^n}{(x+2)^n}$; $a > 0$ is

(A) $(a - 2) \leq x \leq (a + 2)$

(B) $(a - 2) < x < (a + 2)$

(C) $(-a - 2) > x > (a - 2)$

(D) $(a - 2) > x > (-a - 2)$

(E) $(a - 2) \leq x \leq (-a - 2)$

GO ON TO THE NEXT PAGE

14. $\int_0^1 \sin^{-1}(x)dx =$

(A) 0
(B) $\dfrac{\pi+2}{2}$
(C) $\dfrac{\pi-2}{2}$
(D) $\dfrac{\pi}{2}$
(E) $\dfrac{-\pi}{2}$

15. The equation of the line *normal* to $y = \sqrt{\dfrac{5-x^2}{5+x^2}}$ at $x = 2$ is

(A) $81x - 60y = 142$
(B) $81x + 60y = 182$
(C) $20x + 27y = 49$
(D) $20x + 27y = 31$
(E) $81x - 60y = 182$

16. If c satisfies the conclusion of the Mean Value Theorem for Derivatives for $f(x) = 2\sin x$ on the interval $[0, \pi]$ then c could be

(A) 0
(B) $\dfrac{\pi}{4}$
(C) $\dfrac{\pi}{2}$

(D) π
(E) There is no value of c on $[0, \pi]$

GO ON TO THE NEXT PAGE

17. The average value of $f(x) = x \ln x$ on the interval $[1, \ e]$ is

(A) $\dfrac{e^2 + 1}{4}$ (B) $\dfrac{e^2 + 1}{4(e+1)}$ (C) $\dfrac{e+1}{4}$ (D) $\dfrac{e^2 + 1}{4(e-1)}$ (E) $\dfrac{3e^2 + 1}{4(e-1)}$

---

18. A 17-foot ladder is sliding down a wall at a rate of –5 feet/sec. When the top of the ladder is 8 feet from the ground, how fast is the foot of the ladder sliding away from the wall (in feet/sec.)?

(A) $\dfrac{75}{8}$ (B) $\dfrac{8}{3}$ (C) $\dfrac{3}{8}$ (D) -16 (E) $\dfrac{-75}{3}$

---

19. If $\dfrac{dy}{dx} = 3y \cos x$, and $y = 8$ when $x = 0$, then $y =$

(A) $8e^{3\sin x}$ (B) $8e^{3\cos x}$ (C) $8e^{3\sin x} + 3$ (D) $3\dfrac{y^2}{2}\cos x + 8$ (E) $3\dfrac{y^2}{2}\sin x + 8$

GO ON TO THE NEXT PAGE

20. The length of the curve determined by $x = 3t$ and $y = 2t^2$ from $t = 0$ to $t = 3$ is

(A) $\int_0^9 \sqrt{9t^2 + 4t^4}\, dt$

(B) $\int_0^{162} \sqrt{9 - 16t^2}\, dt$

(C) $\int_0^{162} \sqrt{9 + 16t^2}\, dt$

(D) $\int_0^3 \sqrt{9 - 16t^2}\, dt$

(E) $\int_0^9 \sqrt{9 + 16t^2}\, dt$

---

21. If a particle moves in the xy-plane so that at time $t > 0$ its position vector is $\left(e^{t^2}, e^{-t^3}\right)$ then its velocity at time $t = 3$ is

(A) $(\ln 6, \ln(-27))$

(B) $(\ln 9, \ln(-27))$

(C) (e^9, e^{-27})

(D) $(6e^9, -27e^{-27})$

(E) $(9e^9, -27e^{-27})$

---

22. The graph of $f(x) = \sqrt{11 + x^2}$ has a point of inflection at

(A) $\left(0, \sqrt{11}\right)$

(B) $\left(-\sqrt{11}, 0\right)$

(C) $\left(0, -\sqrt{11}\right)$

(D) $\left(\sqrt{\dfrac{11}{2}}, \sqrt{\dfrac{33}{2}}\right)$

(E) There is no point of inflection.

GO ON TO THE NEXT PAGE

23. What is the volume of the solid generated by rotating about the y-axis the region enclosed by $y = \sin x$ and the x-axis, from $x = 0$ to $x = \pi$?

(A) π^2　　　　(B) $2\pi^2$　　　　(C) $4\pi^2$　　　　(D) 2　　　　(E) 4

24. $\displaystyle\int_{\frac{2}{\pi}}^{\infty} \frac{\sin\left(\frac{1}{t}\right)}{t^2}\,dt =$

(A) 1　　　　(B) 0　　　　(C) –1　　　　(D) 2　　　　(E) Undefined

25. A rectangle is to be inscribed between the parabola $y = 4 - x^2$ and the x-axis, with its base on the x-axis. A value of x that maximizes the area of the rectangle is

(A) 0　　　(B) $\dfrac{2}{\sqrt{3}}$　　　(C) $\dfrac{2}{3}$　　　(D) $\dfrac{4}{3}$　　　(E) $\dfrac{\sqrt{3}}{2}$

GO ON TO THE NEXT PAGE

26. $\displaystyle\int \frac{dx}{\sqrt{9-x^2}} =$

(A) $\sin^{-1} 3x + C$

(B) $\ln\left|x + \sqrt{9-x^2}\right| + C$

(C) $\dfrac{1}{3}\sin^{-1} x + C$

(D) $\sin^{-1}\dfrac{x}{3} + C$

(E) $\dfrac{1}{3}\ln\left|x + \sqrt{9-x^2}\right| + C$

---

27. Find $\displaystyle\lim_{x\to\infty} x^{\frac{1}{x}}$

(A) 0   (B) 1   (C) ∞ (D) -1 (E) $-\infty$

GO ON TO THE NEXT PAGE

28. What is the sum of the Maclaurin series $\pi - \dfrac{\pi^3}{3!} + \dfrac{\pi^5}{5!} - \dfrac{\pi^7}{7!} + \ldots + (-1)^n \dfrac{\pi^{2n+1}}{(2n+1)!} + \ldots$?

(A) 1   (B) 0   (C) −1   (D) e (E) There is no sum.

GO ON TO THE NEXT PAGE

Part B consists of 17 questions that will be answered on side 2 of the answer sheet. Following are the directions for Section I, Part B.

A GRAPHING CALCULATOR IS REQUIRED FOR SOME QUESTIONS ON THIS PART OF THE EXAMINATION.

Directions: Solve each of the following problems, using the available space for scratchwork. After examining the form of the choices, decide which is the best of the choices given and fill in the corresponding oval on the answer sheet. No credit will be given for anything written in the test book. Do not spend too much time on any one problem.

BE SURE YOU ARE USING SIDE 2 OF THE ANSWER SHEET TO RECORD YOUR ANSWERS TO QUESTIONS NUMBERED 29–45.

YOU MAY NOT RETURN TO SIDE 1 OF THE ANSWER SHEET.

In this test:

(1) The *exact* numerical value of the correct answer does not always appear among the choices given. When this happens, select from among the choices the number that best approximates the exact numerical value.

(2) Unless otherwise specified, the domain of a function f is assumed to be the set of all real numbers x for which $f(x)$ is a real number.

Note: Question numbers with an asterisk (*) indicate a graphing calculator-active question.

29. The first three non-zero terms in the Taylor series about $x = 0$ for $f(x) = \cos x$

(A) $x + \dfrac{x^3}{3!} + \dfrac{x^5}{5!}$

(B) $x - \dfrac{x^3}{3!} + \dfrac{x^5}{5!}$

(C) $1 - \dfrac{x^2}{2!} - \dfrac{x^4}{4!}$

(D) $1 - \dfrac{x^2}{2!} + \dfrac{x^4}{4!}$

(E) $1 + \dfrac{x^2}{2!} + \dfrac{x^4}{4!}$

GO ON TO THE NEXT PAGE

30. $\int \cos^3 x \ dx =$

(A) $\dfrac{\cos^4 x}{4} + C$

(B) $\dfrac{\sin^4 x}{4} + C$

(C) $\sin x - \dfrac{\sin^3 x}{3} + C$

(D) $\sin x + \dfrac{\sin^3 x}{3} + C$

(E) $\sin^3 x + C$

---

31. If $f(x) = (3x)^{(3x)}$ then $f'(x) =$

(A) $(3x)^{(3x)}(3\ln(3x) + 3)$

(B) $(3x)^{(3x)}(3\ln(3x) + 3x)$

(C) $(9x)^{(3x)}(\ln(3x) + 1)$

(D) $(3x)^{(3x-1)}(3x)$

(E) $(3x)^{(3x-1)}(9x)$

---

32. To what limit does the sequence $S_n = \dfrac{3+n}{3^n}$ converge as n approaches infinity?

(A) 1

(B) $\dfrac{1}{3}$

(C) 0

(D) ∞

(E) 3

GO ON TO THE NEXT PAGE

33. $\int \dfrac{18x-17}{(2x-3)(x+1)}\,dx =$

(A) $8\ln(2x-3) + 7\ln(x+1) + C$

(B) $2\ln(2x-3) + 7\ln(x+1) + C$

(C) $4\ln(2x-3) + 7\ln(x+1) + C$

(D) $7\ln(2x-3) + 2\ln(x+1) + C$

(E) $\dfrac{7}{2}\ln(2x-3) + 4\ln(x+1) + C$

34. A particle moves along a path described by $x = \cos^3 t$ and $y = \sin^3 t$. The distance that the particle travels along the path from $t = 0$ to $t = \dfrac{\pi}{2}$ is

(A) 0.75　　　(B) 1.50　　　(C) 0　　　(D) −3.50　　　(E) −0.75

35. The sale price of an item is $800 - 35x$ dollars and the total manufacturing cost is $2x^3 - 140x^2 + 2{,}600x + 10{,}000$ dollars, where x is the number of items. What number of items should be manufactured in order to optimize the manufacturer's total profit?

(A) 35　　　(B) 25　　　(C) 10　　　(D) 15　　　(E) 20

GO ON TO THE NEXT PAGE

36. The area enclosed by the polar equation $r = 4 + \cos\theta$, for $0 \le \theta \le 2\pi$, is

(A) 0   (B) $\dfrac{9\pi}{2}$ (C) 18π (D) $\dfrac{33\pi}{2}$ (E) $\dfrac{33\pi}{4}$

---

37. Use the trapezoid rule with $n = 4$ to approximate the area between the curve $y = (x^3 - x^2)$ and the x-axis from $x = 3$ to $x = 4$.

(A) 35.266   (B) 27.766   (C) 63.031   (D) 31.516   (E) 25.125

---

38. If $f(x) = \sum\limits_{k=1}^{\infty} (\cos^2 x)^k$, then $f\left(\dfrac{\pi}{4}\right)$ is

(A) -2 (B) -1 (C) 0   (D) 1   (E) 2

GO ON TO THE NEXT PAGE

39. The volume of the solid that results when the area between the graph of $y = x^2 + 2$ and the graph of $y = 10 - x^2$ from $x = 0$ to $x = 2$ is rotated around the x-axis is

(A) $2\pi \int_0^2 y\left(\sqrt{y-2}\right)dy + 2\pi \int_0^2 y\left(\sqrt{10-y}\right)dy$

(B) $2\pi \int_2^6 y\left(\sqrt{y-2}\right)dy + 2\pi \int_6^{10} y\left(\sqrt{10-y}\right)dy$

(C) $2\pi \int_2^6 y\left(\sqrt{y-2}\right)dy - 2\pi \int_6^{10} y\left(\sqrt{10-y}\right)dy$

(D) $2\pi \int_0^2 y\left(\sqrt{y-2}\right)dy - 2\pi \int_0^2 y\left(\sqrt{10-y}\right)dy$

(E) $2\pi \int_0^2 y\left(\sqrt{10-y}\right)dy - 2\pi \int_0^2 y\left(\sqrt{y-2}\right)dy$

---

40. $\displaystyle\int_0^4 \frac{dx}{\sqrt{9+x^2}} =$

(A) ln3 (B) ln4 (C) –ln2 (D) –ln4 (E) Undefined

---

41. The rate that an object cools is directly proportional to the difference between its temperature (in Kelvins) at that time and the surrounding temperature (in Kelvins). If an object is initially at $35K$, and the surrounding temperature remains constant at $10K$, it takes 5 minutes for the object to cool to $25K$. How long will it take for the object to cool to $20K$?

(A) 6.66 min. (B) 7.50 min. (C) 7.52 min. (D) 8.97 min. (E) 10.00 min.

GO ON TO THE NEXT PAGE

42. $\int e^x \cos x \, dx =$

   (A) $\dfrac{e^x}{2}(\sin x + \cos x) + C$

   (B) $\dfrac{e^x}{2}(\sin x - \cos x) + C$

   (C) $\dfrac{e^x}{2}(\cos x - \sin x) + C$

   (D) $2e^x(\sin x + \cos x) + C$

   (E) $e^x(\sin x + \cos x) + C$

---

43. Two particles leave the origin at the same time and move along the y-axis with their respective positions determined by the functions $y_1 = \cos 2t$ and $y_2 = 4 \sin t$ for $0 < t < 6$. For how many values of t do the particles have the same acceleration?

   (A) 0   (B) 1   (C) 2   (D) 3   (E) 4

---

44. The minimum value of the function $y = x^3 - 7x + 11$; $x \geq 0$ is approx.

   (A) 18.128   (B) 9.283   (C) 6.698   (D) 5.513   (E) 3.872

GO ON TO THE NEXT PAGE

45. Use Euler's method with $h = 0.2$ to estimate $y(1)$, $y' = y$ if and $y(0) = 1$.

(A) 1.200   (B) 2.0746   (C) 2.488   (D) 4.838   (E) 9.677

STOP

END OF SECTION I

IF YOU FINISH BEFORE TIME IS CALLED, YOU MAY CHECK YOUR WORK ON THIS SECTION.

DO NOT GO ON TO SECTION II UNTIL YOU ARE TOLD TO DO SO.

MAKE SURE YOU HAVE PLACED YOUR AP NUMBER LABEL ON YOUR ANSWER SHEET AND HAVE WRITTEN AND GRIDDED YOUR NUMBER CORRECTLY IN SECTION C OF THE ANSWER SHEET.

CALCULUS BC

SECTION II

Time—1 hour and 30 minutes

Number of problems—6

Percent of total grade—50

SHOW ALL YOUR WORK. Indicate clearly the methods you use because you will be graded on the correctness of your methods as well as on the accuracy of your final answers. If you choose to use decimal approximations, your answer should be correct to three decimal places.

A GRAPHING CALCULATOR IS REQUIRED FOR SOME QUESTIONS ON THIS PART OF THE EXAMINATION.

<u>Note:</u> Unless otherwise specified, the domain of a function f is assumed to be the set of all real numbers x for which $f(x)$ is a real number.

1. Let R be the region enclosed by the graphs of $y = 2\ln x$ and $y = \dfrac{x}{2}$, and the lines $x = 2$ and $x = 8$.

   (a) Find the area of R.

   (b) Set up, but <u>do not integrate</u>, an integral expression, in terms of a single variable, for the volume of the solid generated when R is revolved about the x-axis.

   (c) Set up, but <u>do not integrate</u>, an integral expression, in terms of a single variable, for the volume of the solid generated when R is revolved about the line $x = -1$.

2. Let f be the function given by $f(x) = 2x^4 - 4x^2 + 1$.

   (a) Find an equation of the line tangent to the graph at $(2, 17)$.

   (b) Find the x and y-coordinates of the relative maxima and relative minima.

   (c) Find the x-coordinates of the points of inflection.

   (d) Sketch the graph of $f(x)$.

GO ON TO THE NEXT PAGE

3. Water is draining at the rate of $48\pi\,\text{ft}^3/\text{sec}$ from a conical tank whose diameter at its base is 40 feet and whose height is 60 feet.

   (a) Find an expression for the volume of water in the tank in terms of its radius.

   (b) At what rate is the radius of the water in the tank shrinking when the radius is 16 feet?

   (c) How fast is the height of the water in the tank dropping at the instant that the radius is 16 feet?

---

4. Let f be the function given by $f(x) = e^{-4x^2}$.

   (a) Find the first four nonzero terms and the general term of the power series for $f(x)$ about $x = 0$.

   (b) Find the interval of convergence of the power series for $f(x)$ about $x = 0$. Show the analysis that leads to your conclusion.

   (c) Use term-by-term differentiation to show that $f'(x) = -8xe^{-4x^2}$.

---

5. Two particles travel in the xy-plane. For time $t \geq 0$, the position of particle A is given by $x = t + 1$ and $y = (t + 1)^2 - 2t - 2$, and the position of particle B is given by $x = 4t - 2$ and $y = -2t + 2$.

   (a) Find the velocity vector for each particle at time $t = 2$.

   (b) Set up an integral expression for the distance traveled by particle A from time $t = 1$ to $t = 3$. Do not evaluate the integral.

   (c) At what time do the two particles collide? Justify your answer.

   (d) Sketch the path of both particles from time $t = 0$ to $t = 4$. Indicate the direction of each particle along its path.

GO ON TO THE NEXT PAGE

6. Let f and g be functions that are differentiable throughout their domains and that have the following properties:

(i) $f(x + y) = f(x)g(y) + g(x)f(y)$

(ii) $\lim\limits_{a \to 0} f(a) = 0$

(iii) $\lim\limits_{h \to 0} \dfrac{g(h) - 1}{h} = 0$

(iv) $f'(0) = 1$

(a) Use L'Hopital's rule to show that $\lim\limits_{a \to 0} \dfrac{f(a)}{a} = 1$.

(b) Use the definition of the derivative to show that $f'(x) = g(x)$.

(c) Find $\displaystyle\int \dfrac{g(x)}{f(x)} dx$.

END OF EXAMINATION

THERE IS NO TEST MATERIAL ON THIS PAGE.

ANSWERS TO SECTION 1
FOR THE BC EXAM

(1)	D	(11)	B	(21)	D	(31)	A	(41)	D
(2)	B	(12)	A	(22)	E	(32)	C	(42)	A
(3)	E	(13)	D	(23)	B	(33)	B	(43)	D
(4)	C	(14)	C	(24)	A	(34)	B	(44)	E
(5)	A	(15)	A	(25)	B	(35)	E	(45)	C
(6)	B	(16)	C	(26)	D	(36)	D		
(7)	B	(17)	D	(27)	B	(37)	D		
(8)	E	(18)	B	(28)	B	(38)	D		
(9)	C	(19)	A	(29)	C	(39)	B		
(10)	E	(20)	E	(30)	C	(40)	A		

25

Answers and Explanations to the BC Exam

STEP-BY-STEP SOLUTIONS OF THE PROBLEMS

PROBLEM 1. If $7 = xy - e^{xy}$, then $\dfrac{dy}{dx} =$

We need to use Implicit Differentiation to solve this problem.

Step 1: $0 = \left(x\dfrac{dy}{dx} + y\dfrac{dx}{dx} \right) - \left[\left(x\dfrac{dy}{dx} + y\dfrac{dx}{dx} \right)e^{xy} \right]$ $\qquad$ *Remember:* $\dfrac{dx}{dx} = 1$

Step 2: $0 = \left(x\dfrac{dy}{dx} + y \right) - \left[\left(x\dfrac{dy}{dx} + y \right)e^{xy} \right]$

Step 3: $0 = \left(x\dfrac{dy}{dx} + y \right) - xe^{xy}\dfrac{dy}{dx} - ye^{xy}$

Step 4: $ye^{xy} - y = x\dfrac{dy}{dx} - xe^{xy}\dfrac{dy}{dx}$

Step 5: $ye^{xy} - y = \dfrac{dy}{dx}\left(x - xe^{xy} \right)$

Step 6: $\dfrac{\left(ye^{xy} - y \right)}{\left(x - xe^{xy} \right)} = \dfrac{dy}{dx}$

Step 7: $\dfrac{\left(ye^{xy} - y \right)}{\left(x - xe^{xy} \right)} = \dfrac{y\left(e^{xy} - 1 \right)}{x\left(1 - e^{xy} \right)} = \dfrac{-y}{x}$

The answer is (D).

> **Note:** If you had trouble with this problem, you should review the units on The Product Rule, Differentiation of Exponential Functions, and Implicit Differentiation.

PROBLEM 2. The volume of the solid that results when the area between the curve $y = e^x$ and the line $y = 0$, from $x = 1$ to $x = 2$, is revolved around the x-axis is

Step 1: We need to use the formula for finding the Volume of a Solid of Revolution. The equations in the problems are given to us in terms of x, and we are rotating around the x-axis, so we can use the method of washers. The curve $y = e^x$ is always above the curve $y = 0$ (which is the x-axis), so we don't have to break this up into two integrals.

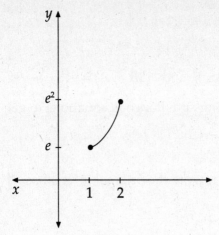

Therefore, the integral will be: $\pi\int_1^2\left(\left(e^x\right)^2-0^2\right)dx$

Step 2: $\pi\int_1^2\left(\left(e^x\right)^2-0^2\right)dx=\pi\int_1^2 e^{2x}\,dx$

Step 3: $\pi\int_1^2 e^{2x}\,dx=\pi\left(\dfrac{1}{2}e^{2x}\right)\Big|_1^2$

Step 4: $\pi\left(\dfrac{1}{2}e^{2x}\right)\Big|_1^2=\pi\left(\dfrac{1}{2}e^4-\dfrac{1}{2}e^2\right)$

Step 5: $\pi\left(\dfrac{1}{2}e^4-\dfrac{1}{2}e^2\right)=\dfrac{\pi}{2}\left(e^4-e^2\right)$

The answer is (B).

Note: If you had trouble with this problem, you should review the unit on Finding the Volume of a Solid of Revolution.

PROBLEM 3. $\displaystyle\int\frac{x-18}{(x+3)(x-4)}\,dx=$

Step 1: We need to use partial fractions to evaluate this integral.

First, we write the integrand as: $\quad\dfrac{x-18}{(x+3)(x-4)}=\dfrac{A}{(x+3)}+\dfrac{B}{(x-4)}$

Step 2: Multiply both sides by: $(x+3)\,(x-4)$

Step 3: Now we have: $x-18=A(x-4)+B(x+3)$

Step 4: $x-18=Ax-4A+Bx+3B$

Step 5: $x - 18 = (Ax + Bx) + (3B - 4A)$

Step 6: $x - 18 = x(A + B) + (3B - 4A)$

Step 7: Thus: $(A + B) = 1$ and $(3B - 4A) = -18$

Step 8: Solve this using simultaneous equations to get: $A = 3$ and $B = -3$

Step 9: We can now rewrite the original integral as: $\int \left(\dfrac{3}{x+3} + \dfrac{-2}{x-4} \right) dx$

Step 10: Which is the same as: $\int \dfrac{3dx}{x+3} - \int \dfrac{2dx}{x-4}$

The answer is (E).

> **Note:** If you had trouble with this problem, you should review the unit on Method of Partial Fractions.

PROBLEM 4. If $y = 5x^2 + 4x$ and $x = \ln t$ then $\dfrac{dy}{dt} =$

We can solve this problem with the Chain Rule.

Step 1: $\dfrac{dy}{dx} = 10x + 4$

Step 2: $\dfrac{dx}{dt} = \dfrac{1}{t}$

Step 3: $\dfrac{dy}{dt} = \dfrac{dy}{dx}\dfrac{dx}{dt}$ so $\dfrac{dy}{dt} = (10x + 4)\left(\dfrac{1}{t} \right)$

Step 4: Substitute for x so that $\dfrac{dy}{dt} = (10\ln t + 4)\left(\dfrac{1}{t} \right) = \dfrac{10\ln t + 4}{t}$

The answer is (C).

You could also have solved this by first substituting for x in the original equation and getting y in terms of t, and then differentiating with the chain rule.

> **Note:** If you had trouble with this problem, you should review the unit on The Chain Rule.

PROBLEM 5. $\int_0^{\frac{\pi}{2}} \sin^5 x \cos x \, dx =$

This Trigonometric integral is solved by using u-substitution.

Step 1: Let $\begin{array}{l} u = \sin x \\ du = \cos x \, dx \end{array}$.

Step 2: The integral is now written as: $\int_0^1 u^5 du =$

Note: The limits of integration change because

$$u = \sin \frac{\pi}{2} = 1 \text{ and } u = \sin 0 = 0$$

Step 3: $\int_0^1 u^5 du = \frac{u^6}{6} \Big|_0^1$

Step 4: $\frac{u^6}{6} \Big|_0^1 = \frac{1}{6} - 0 = \frac{1}{6}$

The answer is (A).

Note: If you had trouble with this problem, you should review the unit on Integrals of Trigonometric Functions.

PROBLEM 6. The equation of the tangent line to the curve $y = x^3 - 4x + 8$ at the point (2, 8) has an x-intercept at

Step 1: First find the slope of the tangent line. $\frac{dy}{dx} = 3x^2 - 4$

Step 2: Plug x into $\frac{dy}{dx} = 3(2)^2 - 4 = 12 - 4 = 8$.

This means that the slope of the tangent line at $x=2$ is 8.

Step 3: Then the equation of the tangent line is $(y - 8) = 8(x - 2)$.

Step 4: The x-intercept is found by plugging in $y=0$ and solving for x.

$$0 - 8 = 8(x - 2) \text{ Therefore } x = 1.$$

The answer is (B).

Note: If you had trouble with this problem, you should review the unit on Equations of Tangent Lines.

PROBLEM 7. The graph in the xy- plane represented by $x = 3\sin(t)$ and $y = 2\cos(t)$ is

This is a Parametric Equation and is solved by eliminating t from the equations and finding a direct relationship between y and x. These problems can often be quite difficult, but, fortunately, on the AP, they only give very easy versions of parametric equations.

Step 1: $\dfrac{x}{3} = \sin(t)$ and $\dfrac{y}{2} = \cos(t)$

Step 2: Because $\sin^2(t) + \cos^2(t) = 1$, we can substitute and we get $\left(\dfrac{x}{3}\right)^2 + \left(\dfrac{y}{2}\right)^2 = 1$.

This is an ellipse.

The answer is (B).

Note: If you had trouble with this problem, you should review the units on Parametric Equations, Trigonometry, and Analytic Geometry. The last two sections are contained in the Appendix on Prerequisite Mathematics.

PROBLEM 8. $\displaystyle\int \dfrac{dx}{\sqrt{4 - 9x^2}} =$

You should recognize this as an Inverse Trigonometric Integral of the form $\dfrac{du}{\sqrt{1 - u^2}}$, which is $\sin^{-1}(u)$. You also should know this by looking at the answer choices. One of the difficulties of the AP is that you are required to recognize many different types of integrals by sight, and to know which techniques to use to solve them.

Step 1: First, we need to use a little algebra to convert the integrand to the form $\dfrac{du}{\sqrt{1 - u^2}}$.

Rewrite $\sqrt{(4 - 9x^2)} = \sqrt{4\left(1 - \dfrac{9x^2}{4}\right)} = 2\sqrt{\left(1 - \dfrac{9x^2}{4}\right)} = 2\sqrt{1 - \left(\dfrac{3x}{2}\right)^2}$.

The integral then becomes: $\displaystyle\int \dfrac{dx}{\sqrt{4 - 9x^2}} = \int \dfrac{dx}{\left(2\sqrt{1 - \left(\dfrac{3x}{2}\right)^2}\right)} = \dfrac{1}{2}\int \dfrac{dx}{\sqrt{1 - \left(\dfrac{3x}{2}\right)^2}}$

Step 2: Now we can use **u-substitution**. Let $u = \dfrac{3x}{2}$ and $du = \dfrac{3}{2}dx$ and $\dfrac{2}{3}du = dx$

We can now rewrite the integral as: $\displaystyle\int \dfrac{dx}{\sqrt{4-9x^2}} = \left(\dfrac{1}{2}\right)\left(\dfrac{2}{3}\right)\displaystyle\int \dfrac{du}{\sqrt{1-u^2}} = \dfrac{1}{2}\displaystyle\int \dfrac{du}{\sqrt{1-u^2}}$

Step 3: You should have memorized this last integral. It is $\left(\dfrac{1}{3}\right)\sin^{-1}(u) + C$.

Step 4: Now reverse the u-substitution and we have:

$$\left(\dfrac{1}{3}\right)\sin^{-1}(u) + C = \left(\dfrac{1}{3}\right)\sin^{-1}\left(\dfrac{3x}{2}\right) + C$$

The answer is (E).

> **Note:** If you had trouble with this problem, you should review the units on Inverse Trigonometric Integrals and u-substitution. It is important that you memorize the three common inverse trigonometric integrals.

PROBLEM 9. $\displaystyle\lim_{x\to\infty} 4x\sin\left(\dfrac{1}{x}\right) =$

To solve this problem, you need to remember how to evaluate Limits, particularly of trigonometric functions. There is a program for evaluating limits that you can put in your TI-82 calculator. See the unit on limits for the program.

Step 1: As you should recall, the $\displaystyle\lim_{x\to0} \dfrac{\sin x}{x} = 1$. (This can be shown with L'Hopital's Rule. Differentiate the top and bottom of the limit to obtain $\displaystyle\lim_{x\to0} \dfrac{\cos x}{1}$, which just

equals 1.) Thus, we need to find a way to convert this integral into one that looks like $\dfrac{\sin x}{x}$.

We can do this with a simple substitution. Let $y = \dfrac{1}{x}$. Now we can change this limit

from $\displaystyle\lim_{x\to\infty} 4x\sin\left(\dfrac{1}{x}\right)$ to $\displaystyle\lim_{y\to0} 4\left(\dfrac{\sin y}{y}\right)$.

Step 2: $\displaystyle\lim_{y\to0} 4\left(\dfrac{\sin y}{y}\right) = 4(1) = 4$.

The answer is (C).

Note: If you had trouble with this problem, you should review the unit on Limits. You should pay particular attention to L'Hopital's Rule, and to limits of trigonometric functions.

PROBLEM 10. The position of a particle moving along the x-axis at time t is given by $x(t) = e^{\cos(2t)}, 0 \leq t \leq \pi$. For which of the following values of t will $x'(t) = 0$?

This problem requires you to know Derivatives of Exponential Functions, and Derivatives of Trigonometric Functions. Also, whenever you are given restrictions on the domain of a function, pay careful attention to the restrictions.

Step 1: $x'(t) = e^{\cos(2t)}(-2\sin(2t))$

Step 2: Now that we have the derivative, set it equal to zero.

$$e^{\cos(2t)}(-2\sin(2t)) = 0$$

Step 3: Because $e^{\cos(2t)}$ can never equal zero (Did you know this? Make sure that you do!), we only have to set $-2\sin(2t)$ equal to zero. This will be true wherever $\sin(2t) = 0$. Because sin (t) = 0 at 0, π, 2π, 3π,... we know thatAlthough there are an infinite number of solutions to this problem, because of the domain restriction we are only concerned with the first three solutions. Thus all three Roman Numeral answer choices work.

The answer is (E).

Note: If you had trouble with this problem, you should review the units on: Derivative of Exponential Functions and Derivatives of Trigonometric Functions.

PROBLEM 11. $\lim\limits_{h \to 0} \dfrac{\sec(\pi + h) - \sec(\pi)}{h} =$

This may *appear* to be a limit problem, but it is *actually* testing to see whether you know The Definition of the Derivative.

Step 1: You should recall that $\lim\limits_{h \to 0} \dfrac{f(x+h) - f(x)}{h} = f'(x)$. Thus, if we replace $f(x)$ with sec (x), we can rewrite the problem as

$$\lim\limits_{h \to 0} \dfrac{\sec(x+h) - \sec(x)}{h} = \left[\sec(x)\right]'.$$

Step 2: The derivative of sec(x) is sec(x)tan(x). Thus

$$\lim\limits_{h \to 0} \dfrac{\sec(\pi + h) - \sec(\pi)}{h} = \sec(\pi)\tan(\pi).$$

Step 3: Because sec(π) = 1 and tan(π) = 0, this is equal to 0.

The answer is (B).

Note: If you had trouble with this problem, you should review the units on: The Definition of the Derivative and Derivatives of Trigonometric Functions.

PROBLEM 12. Let f be a function defined for all real x. If, for each real number L, there exists a $\delta > 0$ and an $\varepsilon > 0$, such that $L - \varepsilon < f(x) < L + \varepsilon$ for $0 < |x| < \delta$, then

This is what is known as either the Strict Definition of a Limit, or the "Delta-Epsilon Method". This is something that you either have studied, in which case this is a very easy question, or you have not studied, in which case, you have no idea what this question is asking! The strict definition of a Limit says that this problem should give the result that $\lim\limits_{x \to 0} f(x) = L$.

The answer is (A).

Note: If you had trouble with this problem, you should review the unit on the Strict Definition of a Limit.

PROBLEM 13. The radius of convergence of $\sum\limits_{n=1}^{\infty} \dfrac{a^n}{(x+2)^n}$ is

Step 1:
An infinite series of the form $\sum\limits_{n=0}^{\infty} r^n$ will converge if $|r| < 1$. So all we have to do is to set $\left| \dfrac{a}{x+2} \right| < 1.!$

Step 2: $\left| \dfrac{a}{x+2} \right| < 1$ means that $-1 < \dfrac{a}{x+2} < 1$. We can rewrite this as $-1 > \dfrac{x+2}{a} > 1$.

Step 3: Now we have $-a > x + 2 > a$ or $-a - 2 > x > a - 2$.

The answer is (D).

Note: If you had trouble with this problem, you should review the unit on Infinite Series.

PROBLEM 14. $\int_0^1 \sin^{-1}(x)\,dx =$

Step 1: Your first reaction to this integral may very well be "I don't know how to find the integral of an inverse trigonometric function. I only know how to find the derivative of an inverse trigonometric function!" That's okay. This is actually an Integration By Parts problem. First of all, we are going to ignore the limits of integration until the end of this problem, and just focus on finding the integral itself. As you should recall, the formula for integration by parts is:

$$\int u\,dv = uv - \int v\,du.$$

Step 2: Let $u = \sin^{-1} x$ and $dv = dx$

then

$$du = \frac{dx}{\sqrt{1-x^2}} \quad \text{and} \quad v = x$$

Now, using integration by parts, we have : $\int \sin^{-1}(x)\,dx = x\sin^{-1}(x) - \int \frac{x\,dx}{\sqrt{1-x^2}}$

Step 3: We can now solve this latter integral with u-substitution.

Let $u = 1 - x^2$ and $du = -2x\,dx$.

$$\frac{-1}{2}\,du = dx$$

Then we have: $\int \frac{dx}{\sqrt{1-x^2}} = \frac{-1}{2} \int u^{\frac{-1}{2}}\,du = -u^{\frac{1}{2}}$

Step 4: Substituting back for u gives us: $\int \sin^{-1}(x)\,dx = x\sin^{-1}(x) + \sqrt{\left(1-x^2\right)}$

Step 5: Now we evaluate at the limits of integration:

$$\left(x\sin^{-1} x + \sqrt{1-x^2} \right)\Big|_0^1 = \left((1)\sin^{-1}(1) + \sqrt{0} \right) - \left(0\sin^{-1} 0 + \sqrt{1} \right) = \left(\frac{\pi}{2} \right) - (1) = \frac{\pi-2}{2}$$

The answer is (C).

> **Note:** If you had trouble with this problem, you should review the unit on Integration By Parts.

PROBLEM 15. The equation of the line *normal* to $y = \sqrt{\dfrac{5-x^2}{5+x^2}}$ at $x = 2$ is

Step 1: This problem requires you to know how to find Equations of Tangent Lines. We will use the slope-intercept formula of a line: $(y - y_1) = m(x - x_1)$

Step 2: Where $x = 2$, $y = \sqrt{\dfrac{5-(2^2)}{5+(2^2)}} = \sqrt{\dfrac{5-4}{5+4}} = \sqrt{\dfrac{1}{9}} = \dfrac{1}{3}$, so

$$x_1 = 2 \quad \text{and} \quad y_1 = \frac{1}{3}$$

Step 3: $\dfrac{dy}{dx} = \dfrac{1}{2}\left(\dfrac{5-x^2}{5+x^2}\right)^{\frac{-1}{2}}\left(\dfrac{(5+x^2)(-2x)-(5-x^2)(2x)}{(5+x^2)^2}\right)$ This would now require some

messy algebra to simplify, but fortunately we don't have to. We can plug in 2 for x

right now and solve for $\dfrac{dy}{dx} = \dfrac{1}{2}\left(\dfrac{5-2^2}{5+2^2}\right)^{\frac{-1}{2}}\left(\dfrac{(5+2^2)(-4)-(5-2^2)(4)}{(5+2^2)^2}\right),$

which simplifies to $\dfrac{1}{2}\left(\dfrac{1}{9}\right)^{\frac{-1}{2}}\left(\dfrac{(9)(-4)-(1)(4)}{(9)^2}\right) = \dfrac{3}{2}\left(\dfrac{-36-4}{81}\right) = \dfrac{-20}{27}.$

Step 4: If we were finding the equation of a *tangent line*, we would use $\dfrac{-20}{27}$ for m in

the equation, but, as you should recall, because we are finding the equation of the

normal line, we use the *negative reciprocal* of $\dfrac{-20}{27}$ for m, which is $\dfrac{27}{20}$ and plug it into

the equation of the line.

Step 5: Now we have $(y - y_1) = m(x - x_1)$ which becomes $y - \dfrac{1}{3} = \dfrac{27}{20}(x-2).$

Multiply through by 60 to get: $60y - 20 = 81x - 162$ or $81x - 60y = 142$.

The answer is (A).

> **Note:** If you had trouble with this problem, you should review the unit
> on Equations of Tangent Lines.

PROBLEM 16. If c satisfies the conclusion of the Mean Value Theorem for Derivatives for
$f(x) = 2\sin x$ on the interval $[0, \pi]$ then c is

Step 1: The Mean Value Theorem for Derivatives states that if a function is continuous on an interval $[a, b]$, then there exists some value c in that interval where

$$\frac{f(b) - f(a)}{b - a} = f'(c).$$

Step 2: $\dfrac{f(b) - f(a)}{b - a} = \dfrac{2\sin\pi - 2\sin 0}{\pi - 0} = \dfrac{0}{\pi} = 0$

Step 3: Thus $f'(c) = 0$. Because $f'(c) = 2\cos(c)$, we need to know what value of c makes $2\cos(c) = 0$. The value is $\dfrac{\pi}{2}$

The answer is (C).

> **Note:** If you had trouble with this problem, you should review the units on Mean Value Theorem for Derivatives and Continuity.

PROBLEM 17. The average value of $x \ln x$ on the interval $[1, e]$ *is*

This problem requires you to be familiar with the Mean Value Theorem for Integrals which we use to find the average value of a function.

Step 1: If you want to find the average value of $f(x)$ on an interval $[a, b]$, you need to evaluate the integral $\dfrac{1}{b-a}\displaystyle\int_a^b f(x)dx$. So here we would evaluate the integral

$\dfrac{1}{e-1}\displaystyle\int_1^e x \ln x \, dx$.

Step 2: We are going to need to do integration by parts to evaluate this integral. Lets ignore the limits of integration for now and just do the integration.

$$\text{Let } u = \ln x \quad \text{and} \quad dv = xdx$$
$$\text{then}$$
$$du = \frac{1}{x}dx \quad \text{and} \quad v = \frac{x^2}{2}$$

Now the integral becomes: $\displaystyle\int x \ln x \, dx = \dfrac{x^2}{2}\ln x - \int\left(\dfrac{x^2}{2}\right)\left(\dfrac{1}{x}\right)dx = \dfrac{x^2}{2}\ln x - \dfrac{1}{2}\int x \, dx$

Thus we have: $\displaystyle\int_1^e x \ln x \, dx = \left(\dfrac{x^2}{2}\ln x - \dfrac{x^2}{4}\right)\Bigg|_1^e$

Step 3: $\left(\dfrac{x^2}{2}\ln x - \dfrac{x^2}{4}\right)\Bigg|_1^e = \left(\dfrac{e^2}{2}\ln e - \dfrac{e^2}{4}\right) - \left(\dfrac{1}{2}\ln 1 - \dfrac{1}{4}\right) = \left(\dfrac{e^2}{2} - \dfrac{e^2}{4} + \dfrac{1}{4}\right)$

This can be simplified to: $\dfrac{e^2 + 1}{4}$

Step 4: Don't forget to multiply by $\dfrac{1}{e-1}$!! This gives the final result of: $\dfrac{e^2 + 1}{4(e-1)}$.

The answer is (D).

Note: If you had trouble with this problem, you should review the units on Mean Value Theorem for Integrals and on Integration By Parts.

PROBLEM 18. A 17 foot ladder is sliding down a wall at a rate of –5 *feet/sec*. When the top of the ladder is 8 feet from the ground, how fast is the foot of the ladder sliding away from the wall (in *feet/sec.*)?

Step 1: The ladder forms a right triangle with the wall, with the ladder itself as the hypotenuse. Whenever we see right triangles in related rates problems, we look to use the Pythagorean Theorem.

Call the distance from the top of the ladder to the ground y, and the distance from the foot of the ladder to the wall x. Then the rate at which the top of the ladder is sliding down the wall is $\dfrac{dy}{dt}$, and the rate at which the foot of the ladder is sliding away from the wall is $\dfrac{dx}{dt}$, which is what we need to find. Now we use the Pythagorean Theorem to set up the relationships: $x^2 + y^2 = 17^2$

Step 2: Differentiating both sides we obtain: $2x\dfrac{dx}{dt} + 2y\dfrac{dy}{dt} = 0$.

Step 3: Because of the Pythagorean Theorem, we also know that, when $y = 8$, $x = 15$.

Step 4: Now we plug everything into the equation from Step 2 and solve for $\dfrac{dx}{dt}$.

$$2(15)\frac{dx}{dt} + 2(8)(-5) = 0 \quad \text{and} \quad \frac{dx}{dt} = \frac{8}{3}.$$

The answer is (B).

> **Note:** If you had trouble with this problem, you should review the unit on Related Rates.

PROBLEM 19. If $\dfrac{dy}{dx} = 3y\cos x$, and $y = 8$ when $x = 0$, then $y =$

Step 1: Separate the variables, by putting all of the terms containing y on the left hand side of the equals sign, and all of the terms containing x on the right hand side.

$$\frac{dy}{y} = 3\cos x \, dx$$

Step 2: Integrate both sides. $\displaystyle\int \frac{dy}{y} = 3\int \cos x \, dx \quad \ln y = 3\sin x + C$

Whenever we have a differential equation where the solution is in terms of $\ln y$, we always solve the equation for y. This involves raising e to the power of each side. This gives us: $y = Ce^{3\sin x}$; $e^{\ln y} = e^{3\sin x + C}$; $y = e^{3\sin x} e^C$. Now you should notice that e^C is just a constant, so we call that C (Confusing, isn't it?!), and write the equation as: $y = Ce^{3\sin x}$.

Step 3: Now plug in $y = 8$ and $x = 0$ in order to solve for C.

$8 = Ce^{3\sin 0} = Ce^0 = C$
$8 = C$

Step 4: This gives the final equation of $y = 8e^{3\sin x}$.

The answer is (A).

> **Note:** If you had trouble with this problem, you should review the unit on Differential Equations.

PROBLEM 20. The length of the curve determined by $x = 3t$ and $y = 2t^2$ from $t = 0$ to $t = 9$ is

This problem requires you to find an Arc Length. This is a simple integral formula.

Step 1: The formula for the arc length of a curve given in parametric form on the interval $[a,b]$ is $\int_a^b \sqrt{\left(\dfrac{dx}{dt}\right)^2 + \left(\dfrac{dy}{dt}\right)^2}\, dt$.

Step 2: $\dfrac{dx}{dt} = 3$ and $\dfrac{dy}{dt} = 4t$, so $\int_a^b \sqrt{\left(\dfrac{dx}{dt}\right)^2 + \left(\dfrac{dy}{dt}\right)^2}\, dt = \int_0^9 \sqrt{3^2 + (4t)^2}\, dt$

Step 3: $\int_0^9 \sqrt{3^2 + (4t)^2}\, dt = \int_0^9 \sqrt{9 + 16t^2}\, dt$

The answer is (E).

> **Note:** If you had trouble with this problem, you should review the unit on Arc Length.

PROBLEM 21. If a particle moves in the xy-plane so that at time $t > 0$ its position vector is $\left(e^{t^2}, e^{-t^3}\right)$ then its velocity at time $t = 3$ is

This problem requires you to know how to find the velocity of a moving object.

Step 1: All you need to do to find the velocity of a moving object at a particular instant in time is to take the derivative of its position function at that time. The derivative of $e^{t^2} = 2te^{t^2}$ and the derivative of $e^{-t^3} = -3t^2e^{-t^3}$.

Step 2: Now we plug in 3 for t and we get $2te^{t^2} = 6e^9$ and $-3t^2e^{-t^3} = -27e^{-27}$.

The answer is (D).

> **Note:** If you had trouble with this problem, you should review the unit on Position, Velocity, and Acceleration.

PROBLEM 22. The graph of $f(x) = \sqrt{11+x^2}$ has a point of inflection at

This problem requires you to know how to find the critical points on a graph, which is a crucial part of Graphing Functions.

Step 1: The points of inflection on a graph are generally at points where the second derivative is zero, but not necessarily at *all* points where the second derivative is zero. The good thing about the AP exam is that, in the multiple choice part of the test, you do not have to worry about exceptions to the second derivative rule. Thus, all we have to do here is to take the second derivative of the function and set it equal to zero.

$$f'(x) = \frac{1}{2}\left(11+x^2\right)^{\frac{-1}{2}}(2x) = \frac{x}{\left(11+x^2\right)^{\frac{1}{2}}}$$

and

$$f''(x) = \frac{\left(11+x^2\right)^{\frac{1}{2}}(1) - (x)\frac{1}{2}\left(11+x^2\right)^{\frac{-1}{2}}(2x)}{\left(11+x^2\right)}$$

$$= \frac{\left(11+x^2\right)^{\frac{1}{2}} - \dfrac{x^2}{\left(11+x^2\right)^{\frac{1}{2}}}}{\left(11+x^2\right)^{\frac{1}{2}}}$$

$$= \frac{\left(11+x^2\right) - x^2}{\left(11+x^2\right)^{\frac{3}{2}}}$$

$$= \frac{11}{\left(11+x^2\right)^{\frac{3}{2}}}$$

Step 2: When we are setting a rational function equal to zero, all we have to do is set the numerator equal to zero, and that value will be our point of inflection. Always double check that the value that makes the numerator equal to zero does not also make the DENOMINATOR equal to zero. If it does, this is NOT a point of inflection. Here, there is no point where 11 equals zero, so there is no point of inflection. All of that work for nothing!

The answer is (E).

> **Note:** If you had trouble with this problem, you should review the unit on Graphing Functions.

PROBLEM 23. What is the volume of the solid generated by rotating about the y-axis the region enclosed by $y = \sin x$ and the x-axis, from $x = 0$ to $x = \pi$?

This problem requires us to find the Volume of a Solid of Revolution.

Step 1: Whenever you find the volume of a solid of revolution, you should first draw the graph of the equation so that you are sure of exactly what the curve looks like. You can graph this easily on your calculator, and should get something that looks like this:

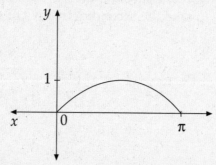

Step 2: Because we are revolving this curve around the y-axis, it will be easier to use the "shells" formula than the "washers" formula.

Using this formula, we get: $2\pi \int_0^\pi x \sin x \, dx$. This is one of our basic Integration By Parts integrals.

Let $u = x$ and $dv = \sin x \, dx$
            then
$du = dx$ and $v = -\cos x \, dx$

then $\int x \sin x \, dx = -x \cos x + \int \cos x \, dx = -x \cos x + \sin x$

Step 3: Now we evaluate at the limits of integration and we get:

$$2\pi (-x\cos x + \sin x)\Big|_0^\pi = 2\pi\big[(-\pi\cos\pi + \sin\pi) - (0 + \sin 0)\big] = 2\pi^2$$

The answer is (B).

Note: If you had trouble with this problem, you should review the unit on Finding the Volume of a Solid of Revolution.

PROBLEM 24. $\displaystyle\int_{\frac{2}{\pi}}^{\infty} \frac{\sin\left(\frac{1}{t}\right)}{t^2}\,dt =$

This is an Improper Integral. As with many of the more difficult topics in Calculus, the AP examination only tends to ask us to solve very straightforward Improper Integrals. The trick is to change the improper integral into a proper one.

Step 1: First, we need to rewrite the integral as a limit. Let $a = \infty$, and evaluate:

$\displaystyle\lim_{a\to\infty}\int_{\frac{2}{\pi}}^{a} \frac{\sin\left(\frac{1}{t}\right)}{t^2}\,dt$. This gets rid of the "infinity problem".

At this point, we Ignore the limits of integration while we figure out how to do the integral.

Using u-substitution, let $u = \dfrac{1}{t}$ and $du = \dfrac{-1}{t^2}\,dt$

$\displaystyle\int \frac{\sin\left(\frac{1}{t}\right)}{t^2}\,dt = -\int \sin u\,du = \cos u$. Substituting back, we get $\displaystyle\cos\frac{1}{t}\Big|_{\frac{2}{\pi}}^{a}$

Step 2: Now we evaluate the integral at the limits of integration and get:

$$\cos\frac{1}{t}\Big|_{\frac{2}{\pi}}^{a} = \cos\frac{1}{a} - \cos\frac{\pi}{2} = \cos\frac{1}{a}.$$

Step 3: Finally, we take the limit and we get: $\displaystyle\lim_{a\to\infty}\cos\frac{1}{t} = \cos 0 = 1$

The answer is (A).

Note: If you had trouble with this problem, you should review the units on Improper Integrals and u-substitution.

PROBLEM 25. A rectangle is to be inscribed between the parabola $y = 4 - x^2$ and the x-axis, with its base on the x-axis. The value of x that maximizes the area of the rectangle is

This is a Maximum/Minimum problem. It is usually helpful to draw a picture first, so we know what we are looking for.

Step 1: If we graph $y = 4 - x^2$ and sketch in a rectangle, we get something like this:

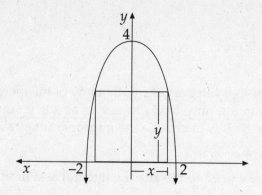

Step 2: Notice that the length of the base of the rectangle is $2x$ and the height of the rectangle is y. This means that the area of the rectangle is: $A = 2xy$. If we substitute $4 - x^2$ for y, we get: $A = x(4 - x^2) = 4x - x^3$.

Step 3: If we want to find the maximum area, all we have to do is take the derivative of A and set it equal to zero.

$$\frac{dA}{dx} = 4 - 3x^2 = 0$$

$$4 = 3x^2$$

$$x = \pm\sqrt{\frac{4}{3}} = \pm\frac{2}{\sqrt{3}}$$

Because the answer is looking for a length, we are satisfied with the positive answer. Besides, $-\dfrac{2}{\sqrt{3}}$ isn't an answer choice.

The answer is (B).

> **Note:** If you had trouble with this problem, you should review the unit on Maximum/Minimum Problems.

PROBLEM 26: $\displaystyle\int \frac{dx}{\sqrt{9 - x^2}} =$

If you look at this integral, you should notice that it is similar to the integral

$$\int \frac{dx}{\sqrt{1 - x^2}} = \sin^{-1} x + C.$$

Step 1: If we factor 9 out of the radicand, we can convert our integral into the one above.

$$\int \frac{dx}{\sqrt{9-x^2}} = \int \frac{dx}{\sqrt{9\left(1-\frac{x^2}{9}\right)}} = \int \frac{dx}{3\sqrt{\left(1-\frac{x^2}{9}\right)}} = \frac{1}{3}\int \frac{dx}{\sqrt{\left(1-\frac{x^2}{9}\right)}}.$$

Step 2: Now we can use u-substitution. Let $u = \frac{x}{3}$. Then $du = \frac{1}{3}dx$ and $3du = dx$. We get:

$$\frac{1}{3}\int \frac{dx}{\sqrt{\left(1-\frac{x^2}{9}\right)}} = \frac{1}{3}\int \frac{3du}{\sqrt{1-u^2}} = \int \frac{du}{\sqrt{1-u^2}} = \sin^{-1}u + C$$

Step 3: Substituting back, we get:

$$\sin^{-1}\frac{x}{3} + C.$$

Note: If you had trouble with this problem, you should review the sections on **Trig Functions** and on ***u*-substitution**.

The answer is (D).

PROBLEM 27: Find $\lim\limits_{x \to \infty} x^{\frac{1}{x}}$

This is a limit of an *indeterminate form*. We use L'Hopital's rule to evaluate these limits, but we need to get this into the form $\frac{f(x)}{g(x)}$ so that we can take the derivative of the top and bottom. We do this using logarithms.

Step 1: First, let $y = x^{\frac{1}{x}}$.

Step 2: Now, take the log of both sides (remember, when we say log, we mean *natural log*, not *common log*).

We get: $\ln y = \ln x^{\frac{1}{x}}$, which we can rewrite, using log rules, as $\frac{1}{x}\ln x$ or $\frac{\ln x}{x}$.

Step 3: Now we can use L'Hopital's rule. Take the derivative of the top and bottom.

$$\lim_{x\to\infty}\frac{\ln x}{x} = \lim_{x\to\infty}\frac{\frac{1}{x}}{1} = 0 \cdot$$

Be careful! This is NOT the answer! We just found that that the limit of $\ln y \to 0$, therefore, $y \to e^0 = 1$. Whenever you use this technique, remember to reverse the logarithm at the end.

The answer is (B).

Note: If you had trouble with this problem, review the section on *L'Hopital's Rule*.

PROBLEM 28: What is the sum of the Maclaurin series $\pi - \frac{\pi^3}{3!} + \frac{\pi^5}{5!} - \frac{\pi^7}{7!} + \ldots$

$+(-1)^n\frac{\pi^{2n+1}}{(2n+1)!} + \ldots$?

This series is of the form $x - \frac{x^3}{3!} + \frac{x^5}{5!} - \frac{x^7}{7!} + \ldots + (-1)^n\frac{x^{2n+1}}{(2n+1)!} + \cdots$. This, as we know, is the Maclaurin Series expansion of $\sin x$. In the problem, we simply have π instead of x. Therefore, this is equal to $\sin \pi = 0$.

The answer is (B).

Note: If you had trouble with this problem, you should review the section on **The Chain Rule**.

PROBLEM 29. The first three non-zero terms in the Taylor series about $x = 0$ of $\cos x$ are

This is a Taylor Series problem. There are four Taylor Series that you should memorize. *This is one of them.* But just in case you didn't memorize it . . .

Step 1: The formula for a Taylor Series about the point $x = a$ is:

$$f(a) + f'(a)(x-a) + f''(a)\frac{(x-a)^2}{2!} + f'''(a)\frac{(x-a)^3}{3!} + \ldots + f^{(n)}(a)\frac{(x-a)^n}{n!} + \ldots$$

Step 2: First, let's take the first few derivatives of $\cos x$:

$$f(x) = \cos x$$
$$f'(x) = -\sin x$$
$$f''(x) = -\cos x$$
$$f'''(x) = \sin x$$
$$f^{(4)}(x) = \cos x$$

Step 3: Next, we evaluate each of these at $a = 0$:

$$f(0) = \cos 0 = 1$$
$$f'(0) = -\sin 0 = 0$$
$$f''(0) = -\cos 0 = -1$$
$$f'''(0) = \sin 0 = 0$$
$$f^{(4)}(0) = \cos 0 =$$

Step 4: Now if we plug in 0 for a throughout the formula we get:

$$f(0) + f'(0)(x) + f''(0)\frac{(x)^2}{2!} + f'''(0)\frac{(x)^3}{3!} + .f^{(4)}(0)\frac{(x)^4}{4!} + ...$$

$$= 1 + (0)(x) + (-1)\left(\frac{x^2}{2!}\right) + (0)\left(\frac{x^3}{3!}\right) + (1)\left(\frac{x^4}{4!}\right) + ...$$

$$= 1 + 0 - \frac{x^2}{2!} + 0 + \frac{x^4}{4!} = 1 - \frac{x^2}{2!} + \frac{x^4}{4!}$$

The answer is (C).

You should make sure to memorize the four Taylor Series in the unit. One of them almost always shows up on the AP Examination!

> **Note:** If you had trouble with this problem, you should review the unit on Taylor Series.

PROBLEM 30. $\int \cos^3 x \, dx =$

This is a Trigonometric Integral. Generally, these are solved by using the trigonometric substitutions that you learned in precalculus. If you are unfamiliar with these, you should go back and review them.

Step 1: First, rewrite $\cos^3 x$ as $\cos x (\cos^2 x)$. Then, because $\cos^2 x = (1 - \sin^2 x)$, we can rewrite the integral as

$$\int \cos^3 x \, dx = \int \cos x (1 - \sin^2 x) dx = \int \cos x \, dx - \int \cos x \sin^2 x \, dx$$

Step 2: The first integral is easy. $\int \cos x \, dx = \sin x$. We do the second integral with u-substitution.

Let $u = \sin x$ and $du = \cos x dx$. Then $\int \cos x \sin^2 x \, dx = \int u^2 du = \dfrac{u^3}{3} + C$.

Substituting back for u and combining, gives us

$$\int \cos x \, dx - \int \cos x \sin^2 x \, dx = \sin x - \frac{\sin^3 x}{3} + C$$

The answer is (C).

> **Note:** If you had trouble with this problem, you should review the unit on Trigonometric Integrals. You may also need to review the trigonometric formulas in the Appendix on Prerequisite Mathematics.

PROBLEM 31. If $f(x) = (3x)^{(3x)}$ then $f'(x) =$

This problem requires us to find the Derivative of an Exponential Function. Any power of x can be written as a power of e, which is what we will use to do this derivative.

Step 1: Rewrite $(3x)^{(3x)}$ as $e^{(3x)\ln(3x)}$. Now, we can take the derivative of this using the Chain Rule and the Product Rule.

Step 2: $f'(x) = \left[(3x)\left(\dfrac{3}{3x}\right) + 3\ln(3x)\right]e^{(3x)\ln(3x)} = (3 + 3\ln(3x))e^{(3x)\ln(3x)}$

Step 3: Now, replacing $e^{(3x)\ln(3x)}$ with $(3x)^{(3x)}$, gives us $(3x)^{(3x)}(3 + 3\ln(3x))$.

The answer is (A).

> **Note:** If you had trouble with this problem, you should review the unit on Derivatives of Exponential Functions.

PROBLEM 32. To what limit does the sequence $s_n = \dfrac{3+n}{3^n}$ converge as n approaches infinity?

This will require us to use one of the convergence tests for Infinite Series. If we use the Ratio Test, we will be able to tell if the sequence converges, and, if so, to what value?

Step 1: $s_n = \dfrac{3+n}{3^n}$ and $s_{n+1} = \dfrac{4+n}{3^{n+1}}$. So the Ratio Test says that, if $\displaystyle\lim_{n\to\infty}\left|\dfrac{s_{n+1}}{s_n}\right| < 1$, then the

sequence will converge, and if $\displaystyle\lim_{n\to\infty}\left|\dfrac{s_{n+1}}{s_n}\right| > 1$, then the series diverges–in other words,

it has no limit.

Step 2: $\dfrac{s_{n+1}}{s_n} = \dfrac{\dfrac{4+n}{3^{n+1}}}{\dfrac{3+n}{3^n}} = \dfrac{4+n}{3+n}\cdot\dfrac{3^n}{3^{n+1}} = \dfrac{1}{3}\left(\dfrac{4+n}{3+n}\right).$

If we take the limit, $\displaystyle\lim_{n\to\infty}\dfrac{1}{3}\left(\dfrac{4+n}{3+n}\right) = \dfrac{1}{3}.$

Because this is less than one, the sequence converges to zero.

The answer is (C).

> **Note:** If you had trouble with this problem, you should review the unit on Infinite Series.

PROBLEM 33. $\displaystyle\int\dfrac{18x-17}{(2x-3)(x+1)}dx =$

This is another Partial Fractions integral.

Step 1: Write the integrand as: $\dfrac{18x-17}{(2x-3)(x+1)} = \dfrac{A}{(2x-3)} + \dfrac{B}{(x+1)}$

Step 2: Multiply both sides by: $(2x-3)(x+1)$

Step 3: Now we have: $18x - 17 = A(x+1) + B(2x-3)$

Step 4: $18x - 17 = Ax + A + 2Bx - 3B$

Step 5: $18x - 17 = (Ax + 2Bx) + (A + 3B)$

Step 6: $18x - 17 = x(A + 2B) + (A + 3B)$

Step 7: Thus: $(A + 2B) = 18$ and $(A - 3B) = -17$

Step 8: Solve this using simultaneous equations to get: $A = 4$ and $B = 7$

Step 9: We can now rewrite the original integral as:

$$\int\left(\dfrac{4}{2x-3} + \dfrac{7}{x+1}\right)dx = 4\int\dfrac{dx}{2x-3} + 7\int\dfrac{dx}{x+1}$$

Step 10: These are both basic *ln* integrals, and we get:

$$4\int \frac{dx}{2x-3} + 7\int \frac{dx}{x+1} = 2\ln|2x-3| + 7\ln|x+1| + C$$

The answer is (B).

Note: If you had trouble with this problem, you should review the unit on the Method of Partial Fractions.

PROBLEM 34. A particle moves along a path described by $x = \cos^3 t$ and $y = \sin^3 t$. Find the distance that the particle travels along the path from $t = 0$ to $t = \dfrac{\pi}{2}$.

This is another Arc Length problem.

Step 1: The formula for the arc length of a curve given in parametric form on the interval $[a,b]$ is $\displaystyle\int_a^b \sqrt{\left(\frac{dx}{dt}\right)^2 + \left(\frac{dy}{dt}\right)^2}\, dt$.

Step 2: $\dfrac{dx}{dt} = -3\cos^2 t \sin t$ and $\dfrac{dy}{dt} = 3\sin^2 t \cos t$, so

$$\int_a^b \sqrt{\left(\frac{dx}{dt}\right)^2 + \left(\frac{dy}{dt}\right)^2}\, dt = \int_0^{\frac{\pi}{2}} \sqrt{\left(-3\cos^2 t \sin t\right)^2 + \left(3\sin^2 t \cos t\right)^2}\, dt$$

$$= \int_0^{\frac{\pi}{2}} \sqrt{\left(9\cos^4 t \sin^2 t\right) + \left(9\sin^4 t \cos^2 t\right)}\, dt$$

Step 3: Now you have a choice.

If you are good with your calculator, you can calculate this integral using the integral function under the MATH menu of the TI-82. You want to input:

fnint $\sqrt{\left(9(\cos x)^4(\sin x)^2\right) + \left(9(\sin x)^4(\cos x)^2\right)}$, x, 0, $(\pi/2))$. This will give you 1.5.

If you are not comfortable with this on the calculator, or if you prefer to do the integration, you need to do the following. Reduce the integrand by factoring out $\sin^2 t \cos^2 t$ and we get:

$$\int_0^{\frac{\pi}{2}} \sqrt{9\left(\sin^2 t \cos^2 t\right)\left(\sin^2 t \cos^2 t\right)}\, dt = \int_0^{\frac{\pi}{2}} \sqrt{9\left(\sin^2 t \cos^2 t\right)(1)}\, dt$$

$$\int_0^{\frac{\pi}{2}} \sqrt{9\sin^2 t \cos^2 t}\, dt = \int_0^{\frac{\pi}{2}} 3\sin t \cos t\, dt$$

Step 4: Using u-substitution, let $u = \sin t$ and $du = \cos t$. Then we get:

$$\int_0^{\frac{\pi}{2}} 3\sin t \cos t \, dt = 3\int_0^1 u \, du = 3\left(\frac{u^2}{2}\right)\Big|_0^1 = \frac{3}{2}.$$

The answer is (B).

> **Note:** If you had trouble with this problem, you should review the unit on Arc Length. Any time that you are asked to evaluate a definite integral in the calculator section of the multiple choice part of the AP Examination, you should use the **fnint** function.

PROBLEM 35. The sales price of an item is $800 - 35x$ dollars and the total manufacturing cost is $2x^3 - 140x^2 + 2600x + 10000$ dollars, where x is the number of items. What number of items should be manufactured in order to optimize the manufacturer's total profit?

> Anytime that you see the word "optimize", you will be doing a Maximum/Minimum problem. The profit is the number of items sold times the difference between the sales price of each object and its cost. The number of items is x, so the sales price is $800x - 35x^2$.

Step 1: Let P equal profit.

$$P = 800x - 35x^2 - (2x^3 - 140x^2 + 2600x + 10000) = -2x^3 + 105x^2 - 1800x - 10000$$

Step 2: $\dfrac{dP}{dx} = -6x^2 + 210x - 1800$. Setting it equal to 0 we get:

$$-6x^2 + 210x - 1800 = 0$$

$$x^2 - 35x + 300 = 0$$

$$(x - 20)(x - 15) = 0$$

$$x = 15, 20$$

Step 3: In order to determine which of these is the maximum and which is the minimum, use the second derivative test. The second derivative of sales is

$\dfrac{d^2P}{dx^2} = -12x + 210$. At $x = 15$, we get $\dfrac{d^2P}{dx^2} = 30$, so this is a minimum. At $x = 20$, we

get $\dfrac{d^2P}{dx^2} = -30$, so this is a maximum. Therefore, the optimum number of units is 20.

The answer is (E).

> **Note:** If you had trouble with this problem, you should review the unit on Maximum/Minimum Problems.

PROBLEM 36. Find the area enclosed by the polar equation $r = 4 + \cos\theta$ for $0 \le \theta \le 2\pi$.

This problem requires you to know the polar formula for finding the area of a region. The formula is $Area = \dfrac{1}{2}\displaystyle\int_{\theta=a}^{\theta=b} r^2 d\theta$.

Step 1: $\dfrac{1}{2}\displaystyle\int_{\theta=a}^{\theta=b} r^2 d\theta = \dfrac{1}{2}\displaystyle\int_{0}^{2\pi}(4+\cos\theta)^2 d\theta$. At this point, you should switch your calcu-

lator into radian mode, and evaluate this integral using $\mathbf{fnint}\left((4+\cos x)^2, x, 0, 2\pi\right)$. Divide this by 2 and you will get 51.8363. Now evaluate each of the five answer choices to see which one gives you 51.8363.

The answer is (D).

If you don't want to do this integral with the calculator, do the following:

If we expand the integrand, we will get three integrals:

$$\frac{1}{2}\int_{0}^{2\pi}(4+\cos\theta)^2 d\theta = \frac{1}{2}\int_{0}^{2\pi}16 d\theta + \frac{1}{2}\int_{0}^{2\pi}8\cos\theta d\theta + \frac{1}{2}\int_{0}^{2\pi}\cos^2\theta d\theta.$$

The first two integrals are easy and give us:

$$\frac{1}{2}(16\theta)\Big|_{0}^{2\pi} + \frac{1}{2}(8\sin\theta)\Big|_{0}^{2\pi} + \frac{1}{2}\int_{0}^{2\pi}\cos^2\theta d\theta.$$

Evaluating the first two integrals, we get $(8\theta)\Big|_{0}^{2\pi} + (4\sin\theta)\Big|_{0}^{2\pi} = 16\pi$.

Step 2: We do the second integral using a trigonometric substitution.

$$\frac{1}{2}\int_{0}^{2\pi}\cos^2\theta d\theta = \frac{1}{2}\int_{0}^{2\pi}\frac{1+\cos 2\theta}{2}d\theta$$

If we simplify this integral, we get

$$\frac{1}{2}\int_{0}^{2\pi}\frac{1}{2}d\theta + \frac{1}{2}\int_{0}^{2\pi}\frac{\cos 2\theta}{2} = \frac{1}{4}\theta\Big|_{0}^{2\pi} + \frac{1}{8}\sin^2\theta\Big|_{0}^{2\pi} = \frac{\pi}{2}$$

If we add $16\pi + \dfrac{\pi}{2}$, we get $\dfrac{33\pi}{2}$.

The answer is (D).

> **Note:** If you had trouble with this problem, you should review the units on Trigonometric Integrals and Polar Equations.

PROBLEM 37. Use the trapezoid rule with $n = 4$ to approximate the area between the curve $f(x) = (x^3 - x^2)$ and the x-axis from $x = 3$ to $x = 4$.

This problem will require you to be familiar with the Trapezoid Rule. This is very easy to do on the calculator, and some of you may even have written programs to evaluate this. Even if you haven't, the formula is easy. The area under a curve from $x = a$ to $x = b$, divided into n intervals is approximated by the Trapezoid Rule and is

$$\left(\frac{1}{2}\right)\left(\frac{b-a}{n}\right)\left[y_0 + 2y_1 + 2y_2 + 2y_3 \cdots + 2y_{n-2} + 2y_{n-1} + y_n\right]$$

This formula may look scary, but it actually is quite simple, and the AP Examination never uses a very large value for n anyway.

Step 1: $\dfrac{b-a}{n} = \dfrac{4-3}{4} = \dfrac{1}{4}$ Plugging into the formula, we get:

$$\frac{1}{8}\left[\left(3^3 - 3^2\right) + 2\left(3.25^3 - 3.25^2\right) + 2\left(3.5^3 - 3.5^2\right) + 2\left(3.75^3 - 3.75^2\right) + \left(4^3 - 4^2\right)\right]$$

This is easy to plug into your calculator and you will get 31.516.

The answer is (D).

Note: If you had trouble with this problem, you should review the unit on Rectangular, Trapezoid, and Simpson's Rules.

PROBLEM 38. If $f(x) = \displaystyle\sum_{k=0}^{\infty} \left(\cos^2 x\right)^k$ then $f\left(\dfrac{\pi}{4}\right)$ is

Step 1: If a series is of the form $\displaystyle\sum_{n=0}^{\infty} ar^n$, and $|r| < 1$ then the sum of the series is $(a)\left(\dfrac{1}{1-r}\right)$, where a is the first term. (Notice that if the series went from one to infinity, instead of zero to infinity, then the formula would be $(a)\left(\dfrac{r}{1-r}\right)$. Be careful that you memorize the correct formula. You could avoid this confusion by deriving the formula as you do the problem. To learn how to do that, refer to the unit on Infinite Series.) Here r is $\cos^2 x$ and a is $\dfrac{1}{2}$, and $\left|\cos^2 x\right| < 1$. This is very important.

If $|r| \geq 1$ the formula doesn't work. So the sum is $\left(\dfrac{1}{2}\right)\left(\dfrac{1}{1-\cos^2 x}\right)$.

Using trigonometric substitution, this sum can be simplified to $\left(\dfrac{1}{2}\right)\dfrac{1}{\sin^2 x}$.

Now if we plug in $\dfrac{\pi}{4}$ for x, we get $\left(\dfrac{1}{2}\right)\dfrac{1}{\sin^2\left(\dfrac{\pi}{4}\right)} = 1$.

The answer is D.

> **Note:** If you had trouble with this problem, you should review the unit on Infinite Series.

PROBLEM 39. The volume of the solid that results when the area between the graph of $y = x^2 + 2$ and the graph of $y = 10 - x^2$ from $x = 0$ to $x = 2$ is rotated around the x-axis is

This is another Volume of a Solid of Revolution problem. As you should have noticed by now, these are very popular on the AP Examination and show up in both the multiple choice section and in the Long Problem section. If you are not good at these, go back and review the unit carefully. You cannot afford to get these wrong on the AP! The good thing about this volume problem is that it is in the calculator part of the multiple choice section, so you can use a program and your graphing calculator to assist you with this problem.

Step 1: First, graph the two curves on the same set of axes. The graph should look like this:

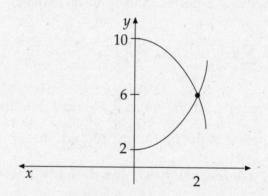

Step 2: Now look at the answer choices. Notice that each answer choice is the sum of two integrals, and all of the functions are in terms of y. This means that you are going to have to use the method of shells, and that you will have to convert the functions from being in terms of x to being in terms of y. Also, when you make a horizontal slice in this region, you are using different curves above and below the intersection point, so you will have to do two integrals. Okay. One thing at a time.

Step 3: First let's find where the two curves intersect. We can do this either with the calculator or algebraically.

Using the calculator, go to 2ᴺᴰ CALC (Above TRACE) and pick choice 5 for intersect. Follow the instructions and you will get the point of intersection. Find both the x- and y-coordinates.

Algebraically, we set the two equations equal to each other.

$$x^2 + 2 = 10 - x^2$$

$$2x^2 = 8$$
$$x^2 = 4$$
$$x = \pm 2$$

We will use $x = 2$, and plugging in 2 for x, we get $y = 6$.

Now we have to find what y is when $x = 0$. On the lower curve, $y = 2$. On the higher curve, $y = 10$. These then are our limits of integration. We will have two integrals, one from 2 to 6, the other from 6 to 10. By process of elimination, this means that the answer can only be B or C.

Step 3: Now we have to convert each of the equations to being in terms of y.

$$\begin{array}{ll} y = x^2 + 2 & y = 10 - x^2 \\ x^2 = y - 2 & x^2 = 10 - y \\ x = \pm\sqrt{y - 2} & x = \pm\sqrt{10 - y} \end{array}$$

Step 4: We are only concerned with the positive roots for this region, so using the Shells formula, we get that the volume is $2\pi\int_2^6 y\left(\sqrt{y-2}\right)dy + 2\pi\int_6^{10} y\left(\sqrt{10-y}\right)dy$.

The answer is (B).

> **Note:** If you had trouble with this problem, you should review the unit on Finding the Volume of a Solid of Revolution.

PROBLEM 40. $\int_0^4 \dfrac{dx}{\sqrt{9 + x^2}} =$

Remember, this is the calculator part of the test. You can simply evaluate this on your calculator using **fnint**. If you are not comfortable with this, we will show you how to do this algebraically. This integral requires a Trigonometric Substitution. These types of integrals require a lot of algebra, so you should leave them for the second pass.

Step 1: Using your calculator, enter **fnint** $((1/(\sqrt{(9+x^2)})),x,0,4)$. You should get 1.0986. Evaluate each of the answer choices to see which has the same value, or which is closest.

The answer is (A).

Step 2: If you are not using your calculator, you first have to do a substitution. Let's ignore the limits of integration for now, and just evaluate the integral. Whenever we have an integral of the form $\sqrt{(a^2+x^2)}$, we do the substitution $x = a\tan\theta$. So here we let $x = 3\tan\theta$ and $dx = 3\sec^2\theta\,d\theta$

Substituting into the integrand, we get: $\displaystyle\int \frac{dx}{\sqrt{9+x^2}} = \int \frac{3\sec^2\theta}{\sqrt{9+9\tan^2\theta}}\,d\theta$.

If we factor 9 out of the radical in the denominator, we get :

$$\int \frac{3\sec^2\theta}{\sqrt{9+9\tan^2\theta}}\,d\theta = \int \frac{3\sec^2\theta}{3\sqrt{1+\tan^2\theta}}\,d\theta = \int \frac{\sec^2\theta}{\sec\theta}\,d\theta = \int \sec\theta\,d\theta.$$

You should have memorized this integral. It is $\ln|\sec\theta + \tan\theta|$.

Step 3: Now we want to evaluate the limits of integration, but in order to do that, we should substitute back for x. If $x = 3\tan\theta$ then $\dfrac{x}{3} = \tan\theta$. Using the Pythagorean Theorem, $\dfrac{\sqrt{9+x^2}}{3} = \sec\theta$

So now we have $\ln|\sec\theta + \tan\theta| = \ln\left|\dfrac{\sqrt{9+x^2}}{3} + \dfrac{x}{3}\right|$. If we evaluate the limits of integration, we get:

$$\ln\left|\frac{\sqrt{9+x^2}}{3} + \frac{x}{3}\right|_0^4 = \ln\left|\frac{5}{3} + \frac{4}{3}\right| - \ln|1| = \ln 3.$$

The answer is (A).

> **Note:** If you had trouble with this problem, you should review the unit on Trigonometric Substitutions.

PROBLEM 41. The rate that an object cools is directly proportional to the difference between its temperature (in Kelvin) at that time and the surrounding temperature (in Kelvin). If an object is initially at $35K$, and the surrounding temperature remains constant at $10K$, it takes 5 minutes for the object to cool to $25K$. How long will it take for the object to cool to $20K$?

This is a Differential Equation. Each AP Examination tends to contain one differential equation word problem. They usually give you the same type of equation and are actually not terribly difficult, once you understand the question.

Step 1: The first sentence tells us what the equation is going to be. Let T stand for temperature at a particular time, and S stand for the surrounding temperature. Then our equation is: $\dfrac{dT}{dt} = k(T - S)$. Time is always represented by t, and k is a constant.

This equation is solvable using separation of variables. Put everything that contains a T on the left side, and everything that contains a t on the right side:

$$\frac{dT}{(T-S)} = kdt.$$

If we integrate both sides we get: $\displaystyle\int \frac{dT}{(T-S)} = k\int dt$

and performing the integration gives us: $\ln|T - S| = kt + C$.

Step 2: Whenever we have an equation of this form, we then exponentiate both sides, giving us:

$|T - S| = e^{kt+C}$ or $|T - S| = Ce^{kt}$

If we plug in the rest of the information from the problem, we can solve for the constants k and C.

The initial temperature tells us that at time $t = 0$, $T = 35$, and $S = 10$. So $|35 - 10| = Ce^{k(0)}$ and $25 = C$

Then at time $t = 5$, $T = 25$. So, $|25 - 10| = 25e^{k(5)}$

$$15 = 25e^{k(5)}$$

$$\frac{3}{5} = e^{5k}$$

$$\frac{1}{5}\ln\left(\frac{3}{5}\right) = k$$

This gives us the final equation $|T - 10| = 25e^{\left(\frac{1}{5}\ln\left(\frac{3}{5}\right)\right)(t)}$

Step 3: Finally, we plug in the last bit of information, that $T = 20$ to solve for t.

$$|20 - 10| = 25e^{\left(\frac{1}{5}\ln\left(\frac{3}{5}\right)\right)(t)}$$

$$\frac{2}{5} = e^{\left(\frac{1}{5}\ln\left(\frac{3}{5}\right)\right)(t)}$$

$$\ln\frac{2}{5} = \frac{1}{5}\ln\left(\frac{3}{5}\right)(t)$$

$$5\frac{\ln(2/5)}{\ln(3/5)} = t$$

$$t = 8.97$$

The answer is (D).

Note: If you had trouble with this problem, you should review the unit on Differential Equations.

PROBLEM 42. $\displaystyle\int e^x \cos x\,dx =$

You should recognize this integral as an elementary **Integration By Parts** integral. If so, this won't be very hard to do.

Step 1: Let $u = e^x$ and $dv = \cos x\,dx$

$du = e^x dx \qquad v = \sin x$

Then we have: $\displaystyle\int e^x \cos x\,dx = e^x \sin x - \int e^x \sin x\,dx$

Step 2: We need to do integration by parts *a second time* to evaluate the second integral.

$$\text{Let } u = e^x \text{ and } dv = \sin x\,dx$$

$$du = e^x dx \qquad v = -\cos x$$

$$\int e^x \cos x\,dx = e^x \sin x - \int e^x \sin x\,dx$$

Now we have $\displaystyle\int e^x \cos x\,dx = e^x \sin x + e^x \cos x - \int e^x \cos x\,dx$

Step 3: Although this looks as if we are back where we started, and will have to do a *third* integration, if we add $\displaystyle\int e^x \cos x\,dx$ to both sides we get:

$$2\int e^x \cos x\,dx = e^x \sin x + e^x \cos x.$$

Now if we divide both sides by 2 we get:

$$\int e^x \cos x \, dx = \frac{e^x \sin x + e^x \cos x}{2}$$

The answer is (A).

Note: If you had trouble with this problem, you should review the unit on Integration by Parts.

PROBLEM 43. Two particles leave the origin at the same time and move along the y-axis with their respective positions determined by the functions $y_1 = \cos 2t$ and $y_2 = 4 \sin t$ for $0 < t < 6$. For how many values of t do the particles have the same acceleration?

If you want to find acceleration, all you have to do is take the second derivative of the position functions.

Step 1: $\dfrac{dy_1}{dx} = -2 \sin 2t$ and $\dfrac{dy_2}{dx} = 4 \cos t$

$\dfrac{d^2 y_1}{dx^2} = -4 \cos 2t$ and $\dfrac{d^2 y_2}{dx^2} = -4 \sin t$

Step 2: Now all we have to do is to graph both of these equations on the same set of axes on a calculator. You should make the window from $x = 0$ to $x = 7$ (leave yourself a little room so that you can see the whole range that you need). You should get a picture that looks like this:

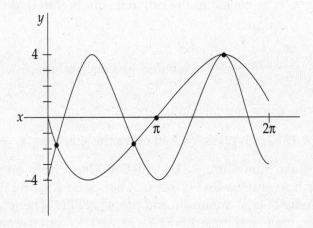

Where the graphs intersect, the acceleration is the same. There are three points of intersection.

The answer is (D).

Note: If you had trouble with this problem, you should review the unit on Position, Velocity, and Acceleration.

PROBLEM 44: The minimum value of the function $y = x^3 - 7x + 11$; $x \geq 0$ is approx.

You have two options. First, let's do the problem without using the calculator.

If we want to find the minimum, we take the derivative and set it equal to zero. We get: $\frac{dy}{dx} = 3x^2 - 7 = 0$. Now we solve for x:

$$3x^2 = 7$$

$$x^2 = \frac{7}{3}$$

$$x = \pm\sqrt{\frac{7}{3}} \, .$$

We are only concerned with positive values of x (note the restriction in the problem), so let's look at $x = \sqrt{\frac{7}{3}}$. First, we have to determine whether this is a minimum or a maximum. The simplest way to do this is with the second derivative test. Take the second derivative and we get: $\frac{d^2y}{dx^2} = 6x$. Now, plug in $x = \sqrt{\frac{7}{3}}$. The second derivative is positive there, so the point is a minimum.

Now, we plug $x = \sqrt{\frac{7}{3}}$ back into the original equation to find the y value.

$$y = \left(\sqrt{\frac{7}{3}}\right)^3 - 7\left(\sqrt{\frac{7}{3}}\right) + 11 \approx 3.872 \text{ (rounded to 3 decimal places)}.$$

Now let's do this using the calculator.

First, using the TI 82/83, press **Y=** and enter the graph as $Y_1 = x^{\wedge}3 - 7x + 11$.

In order to find the minimum, we use the **CALC** functions on the calculator. Press **2nd TRACE** to bring up the **CALC** menu. Then press **3** to find the minimum.. Move the cursor to the left of the minimum and press **ENTER**. Then move the cursor to the right of the minimum and press **ENTER** again. Then put the cursor on the approximate location of the minimum and press **ENTER** one last time. The calculator should display that the minimum is at $x = 1.5275237$ $y = 3.8715489$. We only need 3 decimal places, so the answer is $y = 3.872$.

The answer is (E).

> Note: If you had trouble with this problem, you should review the sections on **Curve Sketching** and on **Maxima and Minima**.

PROBLEM 45: Use Euler's method with $h = 0.2$ to estimate $y(1)$, if $y' = y$ and $y(0) = 1$.

We are given that the curve goes through the point $(0,1)$. We will call the coordinates of this point $x_0 = 0$ and $y_0 = 1$. The slope is found by plugging $y_0 = 1$ into $y' = y$, so we have an initial slope of $y'_0 = 1$.

Now we need to find the next set of points.

Step 1: Increase x_0 by h to get x_1.

$$x_1 = 0.2$$

Step 2: Multiply h by y'_0 and add to y_0 to get y_1.

$$y_1 = 1 + 0.2(1) = 1.2$$

Step 3: Find y'_1 by plugging y_1 into the equation for y'

$$y'_1 = 1.2$$

Repeat until you get to $x = 1$.

Step 1: Increase x_1 by h to get x_2.

$$x_2 = 0.4$$

Step 2: Multiply h by y'_1 and add to y_1 to get y_2.

$$y_2 = 1.2 + 0.2(1.2) = 1.44$$

Step 3: Find y'_2 by plugging y_2 into the equation for y'

$$y'_2 = 1.44$$

Step 1: $x_3 = x_2 + h$.

$$x_3 = 0.6$$

Step 2: $y_3 = y_2 + h(y'_2)$

$$y_3 = 1.44 + 0.2(1.44) = 1.728$$

Step 3: $y'_3 = y_3$

$$y'_3 = 1.728$$

Step 1: $x_4 = x_3 + h$.

$$x_4 = 0.8$$

Step 2: $y_4 = y_3 + h(y_3')$

$$y_4 = 1.728 + 0.2(1.728) = 2.0736$$

Step 3: $y_4' = y_4$

$$y_4' = 2.0736$$

Step 1: $x_5 = x_4 + h.$

$$x_5 = 1.0$$

Step 2: $y_5 = y_4 + h(y_4')$

$$y_5 = 2.0736 + 0.2(2.0736) = 2.48832$$

We don't need to go any farther because we are asked for the value of y when $x = 1$.

The answer is $y = 2.48832$.

The answer is (C).

Note: If you had trouble with this problem, review the section on *Differential Equations*.

ANSWERS AND EXPLANATIONS TO SECTION II

PROBLEM 1. Let R be the region enclosed by the graphs of $y = 2\ln x$ and $y = \dfrac{x}{2}$, and the lines $x = 2$ and $x = 8$.

(a) Find the area of R.

Step 1: If there are two curves, $f(x)$ and $g(x)$, where $f(x)$ is always above $g(x)$, on the interval $[a, b]$, then the area of the region between the two curves is found by:

$$\int_a^b (f(x) - g(x)) \, dx$$

In order to determine whether one of the curves is above the other, we can graph them on the calculator. Put the equations $y_1 = 2\ln x$ and $y_2 = \dfrac{x}{2}$ into y = and set the window to $x$min = 2, $x$max = 10, $y$min = 0, and $y$max=5. Then hit GRAPH.

The graph looks like this:

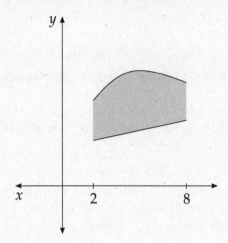

As we can see, the graph of $y = 2\ln x$ is above $y = \dfrac{x}{2}$ on the entire interval, so all we have to do is evaluate the integral $\displaystyle\int_2^8 \left(2\ln x - \dfrac{x}{2}\right)dx =$.

Step 2: We can do the integration one of two ways—on the calculator or analytically.

Calculator: Evaluate $\mathbf{fnint}\left((2\ln x - (x/2)), x, 2, 8\right) = 3.498$

Analytically: $\displaystyle\int_2^8\left(2\ln x - \dfrac{x}{2}\right)dx = 2\int_2^8 \ln x\, dx - \dfrac{1}{2}\int_2^8 x\, dx =$

$= 2(x\ln x - x)\Big|_2^8 - \dfrac{1}{2}\left(\dfrac{x^2}{2}\right)\Big|_2^8 = 18.498 - 15 = 3.498$

By the way, you should have memorized $\displaystyle\int \ln x\, dx = x\ln x - x$, or you can do it as one of the basic Integration By Parts integrals.

This was worth 3 points—1 for setting up the integral, 1 for antidifferentiating correctly, and 1 for evaluating the limits correctly.

(b) Set up, but <u>do not integrate</u>, an integral expression, in terms of a single variable, for the volume of the solid generated when R is revolved about the <u>x-axis</u>.

Step 1: If there are two curves, $f(x)$ and $g(x)$, where $f(x)$ is always above $g(x)$, on the interval $[a, b]$, then the volume of the solid generated when the region is revolved about the x-axis is found by using the method of washers:

$$\pi \int_a^b \left[f(x)\right]^2 - \left[g(x)\right]^2 \, dx$$

Here, we already know that $f(x)$ is above $g(x)$ on the interval, so the integral we need

to evaluate is: $\pi \int_2^8 \left[2\ln x\right]^2 - \left[\dfrac{x}{2}\right]^2 \, dx$

This was worth 2 points—1 for the correct constants and 1 for the correct integral.

(c) Set up, but <u>do not integrate</u>, an integral expression, in terms of a single variable, for the volume of the solid generated when R is revolved about the line $x = -1$.

Step 1: Now we have to revolve the area around a <u>vertical</u> axis. If there are two curves, $f(x)$ and $g(x)$, where $f(x)$ is always above $g(x)$, on the interval $[a, b]$, then the volume of the solid generated when the region is revolved about the y-axis is found by using the method of shells:

$$2\pi \int_a^b x \left[\int(x) - g(x)\right] dx.$$

When we are rotating around a vertical axis, we use the same formula as when we rotate around the y-axis, but we have to account for the shift away from $x = 0$. Here we have a curve that is 1 unit farther away from the line $x = -1$ than it is from the y-axis, so we add 1 to the radius of the shell (For a more detailed explanation of shifting axes, see the unit on Finding the Volume of a Solid of Revolution). This gives us the equation:

$$2\pi \int_2^8 (x+1) \left[2\ln x - \dfrac{x}{2}\right] dx$$

This was worth 3 points—1 for the correct constants, 1 for using the shells method, and 1 for getting the shift correct.

PROBLEM 2. Let f be the function given by $y = f(x) = 2x^4 - 4x^2 + 1$.

(a) Find an equation of the line tangent to the graph at $(-2, 17)$.

In order to find the Equation of a Tangent Line at a particular point we need to take the derivative of the function and plug in the x and y values at that point to give us the slope of the line.

Step 1: The derivative is: $f'(x) = 8x^3 - 8x$. If we plug in $x = -2$, we get:

$f'(-2) = 8(-2)^3 - 8(-2) = -48$. This is the slope m.

Step 2: Now we use the slope-intercept form of the equation of a line, $y - y_1 = m(x - x_1)$, and plug in the appropriate values of x, y, and m.

$y - 17 = -48(x + 2)$. If we simplify this we get $y = -48x - 79$.

This was worth 2 points—1 for finding the slope and 1 for coming up with the correct equation in any form.

(b) Find the x and y-coordinates of the relative maxima and relative minima.

If we want to find the maxima/minima, we need to take the derivative and set it equal to zero. The values that we get are called critical points. We will then test each point to see if it is a maximum or a minimum.

Step 1: We already have the first derivative from part (a), so we can just set it equal to zero: $8x^3 - 8x = 0$.

If we now solve this for x we get:

$$8x(x^2 - 1) = 0 \quad 8x(x+1)(x-1) = 0 \quad x = 0, 1, -1$$

These are our critical points. In order to test if a point is a maximum or a minimum, we usually use the *second derivative test*. We plug each of the critical points into the second derivative. If we get a positive value, the point is a relative minimum. If we get a negative value, the point is a relative maximum. If we get zero, the point is a point of inflection.

Step 2: The second derivative is $f''(x) = 24x^2 - 8$. If we plug in the critical points we get:

$$f''(0) = 24(0)^2 - 8 = -8$$
$$f''(1) = 24(1)^2 - 8 = 16$$
$$f''(-1) = 24(-1)^2 - 8 = 16$$

So $x = 0$ is a relative maximum, and $x = 1, -1$ are relative minima.

Step 3: In order to find the y-coordinates, we plug the x values back into the original equation, and solve.

$$f(0) = 1$$
$$f(1) = -1$$
$$f(-1) = -1$$

and our points are

$(0,1)$ is a relative maximum

$(1,-1)$ is a relative minimum

$(-1,-1)$ is a relative minimum

This was worth 3 points—1 for correctly identifying each critical point.

(c) Find the x-coordinates of the points of inflection.

If we want to find the points of inflection, we set the second derivative equal to zero. The values that we get are the x-coordinates of the points of inflection.

Step 1: We already have the second derivative from part (b), so all we have to do is set it equal to zero and solve for x:

$$24x^2 - 8 = 0 \qquad x^2 = \frac{1}{3} \qquad x = \pm\sqrt{\frac{1}{3}}$$

This was worth 2 points: 1 for correctly identifying each x-coordinate.

(d) Sketch the graph of $f(x)$.

You can graph this function easily using your calculator. Just put the equation into y= and set the window to $x\text{min} = -4$, $x\text{max} = 4$, $y\text{min} = -6$, and $y\text{max} = 6$. Then hit GRAPH.

You should get this:

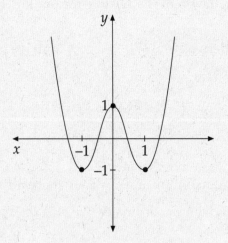

Viewing Window
[–4 x 4] by [–6 x 6]

This was worth 2 points for graphing correctly.

PROBLEM 3: Water is draining at the rate of 48π ft^3/sec from a conical tank whose diameter at its base is 40 feet and whose height is 60 feet.

(a) Find an expression for the volume of water (in ft^3/sec) in the tank in terms of its radius.

The formula for the volume of a cone is: $V = \frac{1}{3}\pi R^2 H$, where R is the radius of the cone, and H is the height. The ratio of the height of a cone to its radius is constant at any point on the edge of the cone, so we also know that $\dfrac{H}{R} = \dfrac{60}{20} = 3$ (Remember that the radius is half the diameter.). If we solve this for H and substitute, we get:

$$H = 3R$$

$$V = \frac{1}{3}\pi R^2 (3R) = \pi R^3$$

This was worth 3 points—1 for the formula of a cone, 1 for the ratio of height to radius, and 1 for the correct final formula.

(b) At what rate (in ft/sec) is the radius of the water in the tank shrinking when the radius is 16 feet?

Step 1: This is a Related Rates question. We now have a formula for the volume of the cone in terms of its radius, so if we differentiate it in terms of t we should be able to solve for the rate of change of the radius $\frac{dR}{dt}$.

We are given that the rate of change of the volume and the radius are, respectively:

$$\frac{dV}{dt} = 48\pi \text{ and } R = 16.$$

Differentiating the formula for the volume, we get: $\frac{dV}{dt} = 3\pi R^2 \frac{dR}{dt}$.

Now we plug in and get: $48\pi = 3\pi 16^2 \frac{dR}{dt}$. Finally, if we solve for $\frac{dR}{dt}$, we get:

$$\frac{dR}{dt} = \frac{1}{16} \text{ ft/sec.}$$

This was worth 2 points — 1 for differentiating correctly and 1 for the correct answer.

(c) How fast (in ft/sec) is the height of the water in the tank dropping at the instant that the radius is 16 feet?

Step 1: This is the same idea as the previous problem, except that we want to solve for $\frac{dH}{dt}$. In order to do this, we need to go back to our ratio of height to radius and solve it for the radius: $\frac{H}{R} = 3$ or $\frac{H}{3} = R$.

Substituting for R in the original equation, we get: $V = \frac{1}{3}\pi \left(\frac{H}{3}\right)^2 H = \frac{\pi H^3}{27}$.

Step 2: Now we need to know what H is when R is 16. Using our ratio:

$$H = 3(16) = 48.$$

Step 3: Now if we differentiate we get: $\frac{dV}{dt} = \frac{\pi H^2}{9} \frac{dH}{dt}$.

Now we plug in and solve:

$$48\pi = \frac{\pi(48)^2}{9}\frac{dH}{dt}$$

$$\frac{dH}{dt} = \frac{3}{16}$$

One should also note that, because $H = 3R$, $\frac{dH}{dt} = 3\frac{dR}{dt}$ in part 2, we merely had to multiply it by 3 to find the answer for part 3.

This was worth 3 points—1 for the new volume formula, 1 for differentiating correctly, and 1 for the correct answer.

PROBLEM 4. Let f be the function given by $f(x) = e^{-4x^2}$.

(a) Find the first four nonzero terms of the power series for $f(x)$ about $x = 0$ and find the general term.

Step 1: If we want to find the power series we need to do the Taylor Series expansion for $f(x)$. The formula for a Taylor Series about the point $x = a$ is:

$$f(a) + f'(a)(x-a) + f''(a)\frac{(x-a)^2}{2!} + f'''(a)\frac{(x-a)^3}{3!} + \ldots + f^{(n)}(a)\frac{(x-a)^n}{n!} + \ldots$$

Here, $a = 0$ and $f(x) = e^{-4x^2}$. You should have memorized the Taylor Series expansion $e^u = 1 + u + \frac{u^2}{2} + \frac{u^3}{3!} + \ldots + \frac{u^n}{n!} + \ldots e^x$, so you don't have to do all of the work here. You can just substitute

$-4x^2$ for u $e^{-4x^2} = 1 + \left(-4x^2\right) + \frac{\left(-4x^2\right)^2}{2} + \frac{\left(-4x^2\right)^3}{3!} = 1 - 4x^2 + 8x^4 - \frac{32x^6}{3}$

Step 2: The general term is: $\dfrac{(-1)^n 2^{2n} x^{2n}}{n!}$

This was worth 3 points—1 for the first and second terms, 1 for the third and fourth terms, and 1 for the general term.

(b) Find the interval of convergence of the power series for $f(x)$ about $x = 0$. Show the analysis that leads to your conclusion.

Step 1: The series for e^{-u^2} converges for $-\infty < u < \infty$. So the series for $f(x) = e^{-4x^2}$ converges if $-\infty < 2x < \infty$. Thus, it converges if $-\infty < x < \infty$.

This was worth 3 points—1 for the convergence interval for e^{-u^2}, 1 for the connection to the series for e^{-4x^2}, and 1 for the correct answer.

(c) Use term-by-term differentiation to show that $f'(x) = -8xe^{-4x^2}$

Step 1: First we differentiate the individual terms of the power series:

$$f(x) = 1 - 4x^2 + 8x^4 - \frac{32x^6}{3} + \dots + \frac{(-1)^n 2^{2n} x^{2n}}{n!}$$

$$f'(x) = 0 - 8x + 32x^3 - 64x^5 + \dots + \frac{(-1)^n 2^{2n}(2n)x^{2n-1}}{n!}$$

$$= -8x + 32x^3 - 64x^5 + \dots + \frac{(-1)^n 2^{2n+1} x^{2n-1}}{(n-1)!}$$

Step 2: Next, we multiply $-8x$ by the individual terms of the power series and compare to the result of step 1.

$$-8xf(x) = 1(-8x) - 4x^2(-8x) + 8x^4(-8x) - \frac{32x^6}{3}(-8x) + \dots + \frac{(-1)^n 2^{2n} x^{2n}}{n!}(-8x)$$

$$= -8x + 32x^3 - 64x^5 + \dots + \frac{(-1)^{n+1} 2^{2n+3} x^{2n+1}}{n!}$$

Step 3: If we substitute $n - 1$ for n, we can make the general terms match exactly.

$$\frac{(-1)^{(n-1)+1} 2^{2(n-1)+3} x^{2(n-1)+1}}{(n-1)!} = \frac{(-1)^n 2^{2n+1} x^{2n-1}}{(n-1)!}$$

This was worth 3 points—1 for differentiating correctly, 1 for the multiplication by $8x$, and 1 for adjusting the general term.

PROBLEM 5. Two particles travel in the xy-plane. For time $t \geq 0$, the position of particle A is given by $x = t + 1$ and $y = (t + 1)^2 - 2t - 2$, and the position of particle B is given by $x = 4t - 2$ and $y = -2t + 2$.

(a) Find the velocity vector for each particle at time $t = 2$.

Step 1: Because velocity is the derivative of position with respect to time, all we have to do is take the derivative of each of the position functions.

Particle A: The position of the x – coordinate is $t + 1$, the velocity is 1.

The position of the y – coordinate is $(t + 1)^2 - 2t - 1$, the velocity is $2(t + 1)(1) - 2 = 2t$

Particle B: The position of the x – coordinate is $4t - 2$, the velocity is 4.

The position of the y – coordinate is $-2t + 2$, the velocity is -2.

Step 2: Now all we have to do is plug in.

At $t = 2$, particle A's velocity vector is $(1, 4)$.

At $t = 2$, particle B's velocity vector is $(4, -2)$.

This was worth 2 points—1 for each correct vector.

(b) Set up an integral expression for the distance traveled by particle A from time $t = 1$ to $t = 3$. Do not evaluate the integral.

Step 1: We are being asked to find the arc length of a curve in parametric form. The

formula is: $\displaystyle\int_a^b \sqrt{\left(\dfrac{dx}{dt}\right)^2 + \left(\dfrac{dy}{dt}\right)^2}\, dt$, where the starting time is a, and the ending time is b.

Plugging into the formula, we get: $\displaystyle\int_1^3 \sqrt{(1)^2 + (2t)^2}\, dt = \int_1^3 \sqrt{1 + 4t^2}\, dt$

This was worth 2 points—1 for realizing that this was an arc length question, 1 for the correct answer.

(c) At what time do the two particles collide? Justify your answer.

Step 1: The particles collide when they have the same x and y-coordinates. If we set the equations for the x-coordinates of the two particles equal we get:

$$t + 1 = 4t - 2 \text{ so } t = 1.$$

At this time, the y-coordinate for particle A is: $(1+1)^2 - 2(1) - 2 = 0$.

The y-coordinate for particle B is: $-2(1) + 2 = 0$. These are the same, so at time $t = 1$, the two particles collide.

This was worth 2 points—1 for setting the coordinates equal to each other, 1 for the correct answer.

(d) In the viewing window provided below, sketch the path of both particles from time $t = 0$ to $t = 4$. Indicate the direction of each particle along its path.

By shifting the calculator into parametric mode, setting the window to match the one below, and inserting the equations for particles A and B into $x_1, y_1, x_2,$ and y_2 respectively, we get:

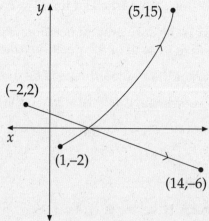

This was worth 2 points—1 for each correct graph.

PROBLEM 6. Let f and g be functions that are differentiable throughout their domains and that have the following properties:

(i) $f(x+y)=f(x)g(y)+g(x)f(y)$

(ii) $\lim\limits_{a\to 0}f(a)=0$

(iii) $\lim\limits_{h\to 0}\dfrac{g(h)-1}{h}=0$

(iv) $f'(0)=1$

(a) Use L'Hopital's rule to show that $\lim\limits_{a\to 0}\dfrac{f(a)}{a}=1$.

Step 1: L'Hopital's rule states that if a function is of the indeterminate form $\dfrac{0}{0}$, then the limit of the derivatives of the numerator is the same as the limit of the original quotient. In other words, if we differentiate the top and bottom of the quotient, we will get the limit. First, we have to show that the quotient gives us an indeterminate form.

$\lim\limits_{a\to 0}\dfrac{f(a)}{a}=\dfrac{0}{0}$ (using property (ii)). Next, we differentiate the top and bottom and take

the limit: $\lim\limits_{a\to 0}\dfrac{f(a)}{a}=\lim\limits_{a\to 0}\dfrac{f'(a)}{1}=f'(0)=1$ (using property (iv)).

This was worth 3 points—1 for stating that the quotient was indeterminate, 1 for using L'Hopital's rule correctly, and 1 for using the properties correctly.

(b) Use the definition of the derivative to show that $f'(x)=g(x)$.

Step 1: The definition of the derivative states: $f'(x)=\lim\limits_{h\to 0}\dfrac{f(x+h)-f(x)}{h}$

If we substitute into the formula, we get:

$f'(x)=\lim\limits_{h\to 0}\dfrac{f(x+h)-f(x)}{h}=\lim\limits_{h\to 0}\dfrac{f(x)g(h)+g(x)f(h)-f(x)}{h}$ (using property (i))

$=\lim\limits_{h\to 0}\dfrac{[g(h)-1]f(x)+g(x)f(h)}{h}=\lim\limits_{h\to 0}\dfrac{[g(h)-1]f(x)}{h}+\dfrac{g(x)f(h)}{h}$

$\lim\limits_{h\to 0}f(x)\dfrac{[g(h)-1]}{h}+g(x)\dfrac{f(h)}{h}=[f(x)](0)+g(x)(1)=g(x)$ (using property (iii))

This was worth 3 points — 1 for setting up the derivative correctly, 1 for evaluating the limits correctly, and 1 for using the properties correctly.

(c) Find $\int \dfrac{g(x)}{f(x)}\, dx$.

Step 1: Using u-substitution, let $u = f(x)$ and $du = f'(x) = g(x)$. Then we have:

$$\int \frac{g(x)}{f(x)}\, dx = \int \frac{du}{u} = \ln|u| + C = \ln|f(x)| + C$$

This was worth 2 points—1 for the u-substitution and 1 for evaluating the integral correctly.

26

APPENDIX

DERIVATIVES AND INTEGRALS THAT YOU SHOULD KNOW

1. $\dfrac{d}{dx}[ku] = k\dfrac{du}{dx}$

2. $\dfrac{d}{dx}[k] = 0$

3. $\dfrac{d}{dx}[uv] = u\dfrac{dv}{dx} + v\dfrac{du}{dx}$

4. $\dfrac{d}{dx}\left[\dfrac{u}{v}\right] = \dfrac{v\dfrac{du}{dx} - u\dfrac{dv}{dx}}{v^2}$

5. $\dfrac{d}{dx}[e^u] = e^u\dfrac{du}{dx}$

6. $\dfrac{d}{dx}[\ln u] = \dfrac{1}{u}\dfrac{du}{dx}$

7. $\dfrac{d}{dx}[\sin u] = \cos u\dfrac{du}{dx}$

8. $\dfrac{d}{dx}[\cos u] = -\sin u\dfrac{du}{dx}$

9. $\dfrac{d}{dx}[\tan u] = \sec^2 u\dfrac{du}{dx}$

10. $\dfrac{d}{dx}[\cot u] = -\csc^2 u\dfrac{du}{dx}$

11. $\dfrac{d}{dx}[\sec u] = \sec u\tan u\dfrac{du}{dx}$

12. $\dfrac{d}{dx}[\csc u] = -\csc u\cot u\dfrac{du}{dx}$

13. $\dfrac{d}{dx}[\sin^{-1} u] = \dfrac{1}{\sqrt{1-u^2}}\dfrac{du}{dx}$

14. $\dfrac{d}{dx}[\tan^{-1} u] = \dfrac{1}{1+u^2}\dfrac{du}{dx}$

15. $\dfrac{d}{dx}[\sec^{-1} u] = \dfrac{1}{|u|\sqrt{u^2-1}}\dfrac{du}{dx}$

---

1. $\displaystyle\int k\,du = ku + C$

2. $\displaystyle\int u^n\,du = \dfrac{u^{n+1}}{n+1} + C; \ n \neq -1$

3. $\displaystyle\int \dfrac{du}{u} = \ln|u| + C$

4. $\displaystyle\int e^u\,du = e^u + C$

5. $\displaystyle\int \sin u\,du = -\cos u + C$

6. $\displaystyle\int \cos u\,du = \sin u + C$

7. $\displaystyle\int \tan u\,du = -\ln|\cos u| + C$

8. $\displaystyle\int \cot u\,du = \ln|\sin u| + C$

9. $\displaystyle\int \sec u\,du = \ln|\sec u + \tan u| + C$

10. $\displaystyle\int \csc u\,du = -\ln|\csc u + \cot u| + C$

11. $\displaystyle\int \sec^2 u\,du = \tan u + C$

12. $\displaystyle\int \csc^2 u\,du = -\cot u + C$

13. $\displaystyle\int \sec u\tan u\,du = \sec u + C$

14. $\displaystyle\int \csc u\cot u\,du = -\csc u + C$

15. $\displaystyle\int \dfrac{du}{\sqrt{a^2-u^2}} = \dfrac{1}{a}\sin^{-1}\dfrac{|u|}{a} + C$

16. $\displaystyle\int \dfrac{du}{a^2+u^2}\,du = \dfrac{1}{a}\tan^{-1}\dfrac{u}{a} + C$

17. $\displaystyle\int \dfrac{du}{u\sqrt{u^2-a^2}} = \dfrac{1}{a}\sec^{-1}\dfrac{u}{a} + C$

PREREQUISITE MATHEMATICS

One of the biggest problems that students have with calculus is that their algebra, geometry, and trigonometry are not solid enough. In calculus, you'll be expected to do a lot of graphing. This requires more than just graphing equations with your calculator. You'll be expected to look at an equation and have a "feel" for what the graph looks like. You'll be expected to factor, combine, simplify, and otherwise rearrange algebraic expressions. You'll be expected to know your formulas for the volume and area of various shapes. You'll be expected to remember trigonometric ratios, their values at special angles, and various identities. You'll be expected to be comfortable with logarithms. And so on. Throughout this book, we spend a lot of time reminding you of these things as they come up, but we thought we should summarize them here at the end.

POWERS

When you multiply exponential expressions with like bases, you add the powers.

$$x^a \cdot x^b = x^{a+b}$$

When you divide exponentiated expressions with like bases, you subtract the powers.

$$\frac{x^a}{x^b} = x^{a-b}$$

When you raise an exponentiated expression to a power, you multiply the powers.

$$\left(x^a\right)^b = x^{ab}$$

When you raise an expression to a fractional power, the denominator of the fraction is the root of the expression, and the numerator is the power.

$$x^{\frac{a}{b}} = \sqrt[b]{x^a}$$

When you raise an expression to the power of zero, you get one.

$$x^0 = 1$$

When you raise an expression to the power of one, you get the expression.

$$x^1 = x$$

When you raise an expression to a negative power, you get the reciprocal of the expression to the absolute value of the power.

$$x^{-a} = \frac{1}{x^a}$$

LOGARITHMS

A logarithm is the power to which you raise a base, in order to get a value. In other words, $\log_b x = a$ means that $b^a = x$. There are several rules of logarithms that you should be familiar with.

When you take the logarithm of the product of two expressions, you add the logarithms.

$$\log(ab) = \log a + \log b$$

When you take the logarithm of the quotient of two expressions, you subtract the logarithms.

$$\log\left(\frac{a}{b}\right) = \log a - \log b$$

When you take the logarithm of an expression to a power, you multiply the logarithm by the power.

$$\log\left(a^b\right) = b \log a$$

The logarithm of 1 is zero.

$$\log 1 = 0$$

The logarithm of its base is 1.

$$\log_b b = 1$$

You cannot take the logarithm of zero or of a negative number.

In calculus, and virtually all mathematics beyond calculus, you will work with natural logarithms. These are logs with base e and are denoted by ln. Thus, you should know the following:

$$\ln 1 = 0$$
$$\ln e = 1$$
$$\ln e^x = x$$
$$e^{\ln x} = x$$

GEOMETRY

The area of a triangle is $\frac{1}{2}(base)(height)$.

The area of a rectangle is $(base)(height)$.

The area of a trapezoid is $\frac{1}{2}(base_1 + base_2)(height)$.

The area of a circle is r^2.

The circumference of a circle is $2 r$.

The Pythagorean Theorem states that the sum of the squares of the legs of a right triangle equals the square of the hypotenuse. This is more commonly stated as $a^2 + b^2 = c^2$ where c equals the length of the hypotenuse.

The volume of a right circular cylinder is $r^2 h$.

The surface area of a right circular cylinder is $2 rh$.

The volume of a right circular cone is $\frac{1}{3}\pi r^2 h$.

The volume of a sphere is $\frac{4}{3}\pi r^3$.

The surface area of a sphere is $4\pi r^2$.

TRIGONOMETRY

Given a right triangle with sides x, y, and r and angle θ below:

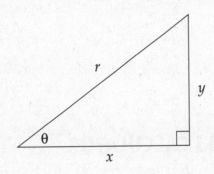

$$\sin\theta = \frac{y}{r} \qquad \csc\theta = \frac{r}{y} \qquad \text{Thus, } \sin\theta = \frac{1}{\csc\theta}$$

$$\cos\theta = \frac{x}{r} \qquad \sec\theta = \frac{r}{x} \qquad \text{Thus, } \cos\theta = \frac{1}{\sec\theta}$$

$$\tan\theta = \frac{y}{x} \qquad \cot\theta = \frac{x}{y} \qquad \text{Thus, } \tan\theta = \frac{1}{\cot\theta}$$

$$\sin 2\theta = 2\sin\theta\cos\theta \qquad \cos 2\theta = 1 - 2\sin^2\theta \qquad \sin^2\theta + \cos^2\theta = 1$$

$$\cos 2\theta = \cos^2\theta - \sin^2\theta \qquad \cos 2\theta = 2\cos^2\theta - 1 \qquad 1 + \tan^2\theta = \sec^2\theta$$

$$\cos^2\theta = \frac{1+\cos 2\theta}{2} \qquad \sin^2\theta = \frac{1-\cos 2\theta}{2} \qquad 1 + \cot^2\theta = \csc^2\theta$$

$$\sin(A + B) = \sin A \cos B + \cos A \sin B$$
$$\sin(A - B) = \sin A \cos B - \cos A \sin B$$
$$\cos(A + B) = \cos A \cos B - \sin A \sin B$$
$$\cos(A - B) = \cos A \cos B + \sin A \sin B$$

You should be able to work in radians and know that $2\pi = 360°$.

You should know the following:

$$\sin 0 = 0 \qquad\qquad \cos 0 = 1 \qquad\qquad \tan 0 = 0$$

$$\sin\frac{\pi}{6} = \frac{1}{2} \qquad \cos\frac{\pi}{6} = \frac{\sqrt{3}}{2} \qquad \tan\frac{\pi}{6} = \frac{1}{\sqrt{3}}$$

$$\sin\frac{\pi}{4} = \frac{1}{\sqrt{2}} \qquad \cos\frac{\pi}{4} = \frac{1}{\sqrt{2}} \qquad \tan\frac{\pi}{4} = 1$$

$$\sin\frac{\pi}{3} = \frac{\sqrt{3}}{2} \qquad \cos\frac{\pi}{3} = \frac{1}{2} \qquad \tan\frac{\pi}{3} = \sqrt{3}$$

$$\sin\frac{\pi}{2} = 1 \qquad\quad \cos\frac{\pi}{2} = 0 \qquad\quad \tan\frac{\pi}{2} = \infty$$

$$\sin \pi = 0 \qquad\qquad \cos \pi = -1 \qquad\quad \tan \pi = 0$$

$$\sin\frac{3\pi}{2} = -1 \qquad \cos\frac{3\pi}{2} = 0 \qquad \tan\frac{3\pi}{2} = \infty$$

$$\sin 2\pi = 0 \qquad\quad \cos 2\pi = 1 \qquad\quad \tan 2\pi = 0$$

USING THE TI-82/83 CALCULATOR

One of the big changes in AP calculus in the last few years has been that ETS now allows students to use graphing calculators on the exam. In fact, there are some questions that can only be solved using a graphing calculator. These are questions that will ask you to find a root, or an extreme point, or the solution to an equation that you can't easily find analytically. Even though you can do most of the exam without a calculator, there are a few calculator functions you should be able to use. We'll teach them to you here.

There are many different calculators on the market (all of them good) which are made by Sharp, Texas Instruments, Casio, and Hewlett Packard, among others. Each has its strengths and weaknesses, but the Texas Instruments series is particularly good for the kinds of calculus problems you're going to run into. The TI-82 is far and away the most popular calculator for calculus students (in fact, you probably use one already) because it has a good size memory, is programmable, and has several handy built-in functions.

Thus, we'll teach you a few facets of the TI-82. If you have a different type of calculator, you should be able to figure out how to convert the process on the TI-82 to the process on your own calculator.

(By the way, if you ever decide to get serious about mathematics, physics, or engineering, you'll need to learn how to use a Hewlett Packard calculator, and you will need to learn how to use either MAPLE or MATHEMATICA on a computer.)

How to Graph

If you want to graph a function on the calculator, push the $Y =$ button on the calculator and type in the equation. For example, suppose you wanted to graph $y = x^3 - 5x^2 + 4x + 2$. Push the button and, on the $Y_1 =$ line, type X^3 – 5X² + 4X + 2. Then push the GRAPH button.

Suppose you wanted to zoom in on a portion of the graph. You have two choices: (1) push the ZOOM button and select a menu choice; or (2) push the WINDOW button and adjust the axes. Experiment with these to see how the graph changes.

TRIG FUNCTIONS

To graph a trig function, make sure that you do two things. First, the calculator must be in radian mode. This is done by pushing the MODE button and selecting "Radian" on the third line. Second, make sure that you use radians for your scale on the coordinate axes. You can do this by pushing the ZOOM button and selecting menu choice 7: ZTRIG.

You can graph a second curve at the same time as the first by pushing the $Y =$ button and, on the $Y_2 =$ line, typing in the second equation. For example, type in $Y_2 = 2x - 1$. See how the calculator draws the graphs one at a time. From these two graphs, you can also find their intersection.

INTERSECTIONS OF CURVES

The calculator has a built-in function that finds the intersection of two curves for you. First, put the two graphs in $Y_1 =$ and $Y_2 =$. Let's use the curves from above:

$$y = x^3 - 5x^2 + 4x + 2 \text{ and } y = 2x - 1.$$

To find the intersection, go to the CALC menu. This is found by pushing the 2ND and TRACE buttons at the same time. Select menu choice 5: INTERSECT. The calculator will then put the cursor (a blinking x) on what it believes to be the first curve. If it's correct, push ENTER. If not, move the cursor to the correct curve, and then push ENTER.

Now the calculator will ask you for the second curve. Follow the same routine as for the first curve, and push ENTER. The calculator will ask you to "guess" the intersection. Move the cursor to the point of intersection that you're trying to find (let's find the middle intersection point) and push ENTER once more. The calculator will tell you the intersection at the bottom of the screen: you should get (1.187, 1.375).

Try this again and find the other two intersections. You should get: (4.388,7.777) and (−0.576,−2.152). Naturally, as with many calculator answers, these answers are not exact, but very close.

EXTREMA

The calculator has built-in functions that will find maxima and minima. Let's use the curve

$$y = x^3 - 5x^2 + 4x + 2$$

and find the local maximum. Go to the CALC menu and select menu choice 4: MAXIMUM. The calculator will ask you for a "lower bound". Move the cursor to the **left** of the point that you believe to be the maximum and push ENTER. Now the calculator will ask you for an "upper bound". Move the cursor to the **right** of the point that you believe to be the maximum and push ENTER again. Now the calculator will ask you for a "guess". Move the cursor to the point that you believe to be the maximum and push ENTER. You should get (0.465,2.879).

Finding the minimum works in much the same way. Go to the CALC menu and select menu choice 3: MINIMUM. Again, the calculator will ask you for a "lower bound". Move the cursor to the **left** of the point that you believe to be the minimum and push ENTER. Now the calculator will ask you for an "upper bound". Move the cursor to the **right** of the point that you believe to be the minimum and push ENTER again. Now the calculator will ask you for a "guess". Move the cursor to the point that you believe to be the minimum and push ENTER. You should get (2.869,−4.065).

ROOTS

Now let's learn to find the roots of an equation. Once again, use the equation

$$y = x^3 - 5x^2 + 4x + 2.$$

To find the roots of this equation, go to the CALC menu and select menu choice 2: ROOT. Then follow the exact procedure we described above involving maxima and minima. You should get (–0.343,0). If you repeat the procedure, you should find that the other root is (3.814,0).

How to Differentiate using the Calculator

You can also find the numerical value of derivatives using the calculator. Suppose you had to find the equation of the tangent line to $y = x^2 + 3x$ at $x = 1.72$. You could find this analytically, but it's easier to make the calculator do the work. Go to the MATH menu and select function 8: NDERIV. Type in the equation, type a comma, and then the value of x at which you want to find the derivative. It should look like this:

$$\text{NDERIV}(X^2 + 3X, X, 1.72).$$

Now push ENTER; you should get 6.44. Now that you have the slope of the tangent line, the rest is easy.

Try another one. Find the derivative of the equation

$$y = \sin^2 x - 2\tan x + 3\ln x \text{ at } x = \frac{\pi}{11}.$$

You should get 8.872 . Wasn't that easy?

How to Integrate using the Calculator

Suppose you wanted to find the area under the curve:

$$y = 2x - x^2 \text{ from } x = \frac{1}{7} \text{ to } x = \frac{9}{7}.$$

You'll need to evaluate the following integral:

$$\int_{\frac{1}{7}}^{\frac{9}{7}} \left(2x - x^2\right) dx.$$

Go to the MATH menu and select function 9: FNINT. Type in the equation, a comma, the lower limit of the integral, another comma, then the upper limit of the integral. It should look like this:

$$\text{FNINT} \left(2X - X^2, X, (1/7), (9/7)\right).$$

Then push ENTER. You should get 0.925.
Here's another example: evaluate the following integral:

$$\int_{\ln 2}^{\ln 3} \sin^3 x \, dx.$$

You should get 0.194.

PROGRAMMING THE TI-82/83

Another useful feature of the calculator is your ability to write programs to simplify some of the numerical work. Not only does ETS not object to your writing programs, they actually publish some suggested ones. Here are two good programs.

The first program performs Newton's Method for you. To create the program first, push the PRGM button. Select menu choice NEW and type in the name NEWTON.

Then push ENTER. Next, type in the following:

$$\text{input } x$$
$$\text{Lbl } 1$$
$$X - Y_1 / NDERIV(Y_1, .001) \rightarrow X$$
$$\text{disp } x$$
$$\text{Pause}$$
$$\text{Goto } 1$$

If you put an equation in the $y =$ under $y_1 = $, the calculator will find roots for you using Newton's Method.

The next program performs Simpson's Rule for you. To create the program, push the PRGM button. Select menu choice NEW and type in the name SIMPSON. Then push ENTER. Next, type in the following:

$$\text{Input } N$$
$$\text{Input } A$$
$$\text{Input } B$$
$$(B - A)/N \rightarrow I$$
$$0 \rightarrow S$$
$$A \rightarrow X$$
$$\text{Lbl } 1$$
$$Y_1 + S \rightarrow S$$
$$X + I \rightarrow X$$
$$4Y_1 + S \rightarrow S$$
$$X + I \rightarrow X$$
$$Y_1 + S \rightarrow S$$
$$\text{If } X < B$$
$$\text{Goto } 1$$
$$S * I / 3 \rightarrow S$$
$$\text{Disp } S$$

Now, put an equation in the Y= under $Y_1 = $ and enter the number of subdivisions you wish to use and the endpoints of the equation.

There are other programs you can learn to write that will also be helpful. We refer you to the calculator's manual for more advice. Your teacher will probably teach you a few more programs as well.

As you can see, the calculator greatly simplifies a lot of the work in calculus. Learn how to use the calculator and you'll have a much easier and more enjoyable time with AP calculus. Good luck!

ABOUT THE AUTHOR

David S. Kahn studied applied mathematics and physics at the University of Wisconsin and has taught courses in calculus, precalculus, algebra, trigonometry, and geometry at the college and high school levels. He has worked as an educational consultant for many years and tutored more students in mathematics than he can count! He has worked for The Princeton Review since 1989, and, in addition to AP calculus, he teaches math and verbal courses for the SAT, SAT II, LSAT, GMAT, and the GRE, trains other teachers, and has now written this book.

ABOUT THIS BOOK

Many of you will be using a college textbook for your AP calculus course. We, too, consulted several college textbooks in the writing of this book. The "default" book was *Calculus and Analytic Geometry*, by Thomas and Finney, 8th <u>ed</u>. In other words, when we weren't sure how to phrase something, or if we wanted to make sure that we were following "standard" calculus, we consulted Thomas. We also used *Calculus with Analytic Geometry*, by Howard Anton, 4th <u>ed</u>., *Differential and Integral Calculus*, by William Granville, earlier editions of Thomas, and a few other classics. Why are we telling you this? Because you may want to consult other books than the one your teacher gave you! All of the above-mentioned books have withstood the test of time, and some of them provide very clear explanations of very difficult subject matter.

NOTES

NOTES

NOTES

NOTES

NOTES

The Princeton Review
Diagnostic Test Form ○ Side 1

1.

YOUR NAME: _____
(Print)   Last   First   M.I.

SIGNATURE: _____ DATE: ___ / ___ / ___

HOME ADDRESS: _____
(Print)   Number and Street

_____
City   State   Zip Code

PHONE NO.: _____
(Print)

IMPORTANT: Please fill in these boxes exactly as shown on the back cover of your test book.

5. YOUR NAME

First 4 letters of last name				FIRST INIT	MID INIT
Ⓐ	Ⓐ	Ⓐ	Ⓐ	Ⓐ	Ⓐ
Ⓑ	Ⓑ	Ⓑ	Ⓑ	Ⓑ	Ⓑ
Ⓒ	Ⓒ	Ⓒ	Ⓒ	Ⓒ	Ⓒ
Ⓓ	Ⓓ	Ⓓ	Ⓓ	Ⓓ	Ⓓ
Ⓔ	Ⓔ	Ⓔ	Ⓔ	Ⓔ	Ⓔ
Ⓕ	Ⓕ	Ⓕ	Ⓕ	Ⓕ	Ⓕ
Ⓖ	Ⓖ	Ⓖ	Ⓖ	Ⓖ	Ⓖ
Ⓗ	Ⓗ	Ⓗ	Ⓗ	Ⓗ	Ⓗ
Ⓘ	Ⓘ	Ⓘ	Ⓘ	Ⓘ	Ⓘ
Ⓙ	Ⓙ	Ⓙ	Ⓙ	Ⓙ	Ⓙ
Ⓚ	Ⓚ	Ⓚ	Ⓚ	Ⓚ	Ⓚ
Ⓛ	Ⓛ	Ⓛ	Ⓛ	Ⓛ	Ⓛ
Ⓜ	Ⓜ	Ⓜ	Ⓜ	Ⓜ	Ⓜ
Ⓝ	Ⓝ	Ⓝ	Ⓝ	Ⓝ	Ⓝ
Ⓞ	Ⓞ	Ⓞ	Ⓞ	Ⓞ	Ⓞ
Ⓟ	Ⓟ	Ⓟ	Ⓟ	Ⓟ	Ⓟ
Ⓠ	Ⓠ	Ⓠ	Ⓠ	Ⓠ	Ⓠ
Ⓡ	Ⓡ	Ⓡ	Ⓡ	Ⓡ	Ⓡ
Ⓢ	Ⓢ	Ⓢ	Ⓢ	Ⓢ	Ⓢ
Ⓣ	Ⓣ	Ⓣ	Ⓣ	Ⓣ	Ⓣ
Ⓤ	Ⓤ	Ⓤ	Ⓤ	Ⓤ	Ⓤ
Ⓥ	Ⓥ	Ⓥ	Ⓥ	Ⓥ	Ⓥ
Ⓦ	Ⓦ	Ⓦ	Ⓦ	Ⓦ	Ⓦ
Ⓧ	Ⓧ	Ⓧ	Ⓧ	Ⓧ	Ⓧ
Ⓨ	Ⓨ	Ⓨ	Ⓨ	Ⓨ	Ⓨ
Ⓩ	Ⓩ	Ⓩ	Ⓩ	Ⓩ	Ⓩ

2. TEST FORM

_____

3. TEST CODE

4. REGISTRATION NUMBER

⓪	Ⓐ Ⓙ	⓪	⓪	⓪	⓪	⓪	⓪	⓪	⓪	⓪
①	Ⓑ Ⓚ	①	①	①	①	①	①	①	①	①
②	Ⓒ Ⓛ	②	②	②	②	②	②	②	②	②
③	Ⓓ Ⓜ	③	③	③	③	③	③	③	③	③
④	Ⓔ Ⓝ	④	④	④	④	④	④	④	④	④
⑤	Ⓕ Ⓞ	⑤	⑤	⑤	⑤	⑤	⑤	⑤	⑤	⑤
⑥	Ⓖ Ⓟ	⑥	⑥	⑥	⑥	⑥	⑥	⑥	⑥	⑥
⑦	Ⓗ Ⓠ	⑦	⑦	⑦	⑦	⑦	⑦	⑦	⑦	⑦
⑧	Ⓘ Ⓡ	⑧	⑧	⑧	⑧	⑧	⑧	⑧	⑧	⑧
⑨		⑨	⑨	⑨	⑨	⑨	⑨	⑨	⑨	⑨

6. DATE OF BIRTH

MONTH	DAY		YEAR	
○ JAN				
○ FEB				
○ MAR	⓪	⓪	⓪	⓪
○ APR	①	①	①	①
○ MAY	②	②	②	②
○ JUN	③	③	③	③
○ JUL		④	④	④
○ AUG		⑤	⑤	⑤
○ SEP		⑥	⑥	⑥
○ OCT		⑦	⑦	⑦
○ NOV		⑧	⑧	⑧
○ DEC		⑨	⑨	⑨

7. SEX
○ MALE
○ FEMALE

SCANTRON® FORM NO. F-592-KIN
© SCANTRON CORPORATION 1989   3289-C553-5
ALL RIGHTS RESERVED.

Begin with number 1 for each new section of the test. Leave blank any extra answer spaces.

SECTION 1

1 Ⓐ Ⓑ Ⓒ Ⓓ Ⓔ	26 Ⓐ Ⓑ Ⓒ Ⓓ Ⓔ	51 Ⓐ Ⓑ Ⓒ Ⓓ Ⓔ	76 Ⓐ Ⓑ Ⓒ Ⓓ Ⓔ
2 Ⓐ Ⓑ Ⓒ Ⓓ Ⓔ	27 Ⓐ Ⓑ Ⓒ Ⓓ Ⓔ	52 Ⓐ Ⓑ Ⓒ Ⓓ Ⓔ	77 Ⓐ Ⓑ Ⓒ Ⓓ Ⓔ
3 Ⓐ Ⓑ Ⓒ Ⓓ Ⓔ	28 Ⓐ Ⓑ Ⓒ Ⓓ Ⓔ	53 Ⓐ Ⓑ Ⓒ Ⓓ Ⓔ	78 Ⓐ Ⓑ Ⓒ Ⓓ Ⓔ
4 Ⓐ Ⓑ Ⓒ Ⓓ Ⓔ	29 Ⓐ Ⓑ Ⓒ Ⓓ Ⓔ	54 Ⓐ Ⓑ Ⓒ Ⓓ Ⓔ	79 Ⓐ Ⓑ Ⓒ Ⓓ Ⓔ
5 Ⓐ Ⓑ Ⓒ Ⓓ Ⓔ	30 Ⓐ Ⓑ Ⓒ Ⓓ Ⓔ	55 Ⓐ Ⓑ Ⓒ Ⓓ Ⓔ	80 Ⓐ Ⓑ Ⓒ Ⓓ Ⓔ
6 Ⓐ Ⓑ Ⓒ Ⓓ Ⓔ	31 Ⓐ Ⓑ Ⓒ Ⓓ Ⓔ	56 Ⓐ Ⓑ Ⓒ Ⓓ Ⓔ	81 Ⓐ Ⓑ Ⓒ Ⓓ Ⓔ
7 Ⓐ Ⓑ Ⓒ Ⓓ Ⓔ	32 Ⓐ Ⓑ Ⓒ Ⓓ Ⓔ	57 Ⓐ Ⓑ Ⓒ Ⓓ Ⓔ	82 Ⓐ Ⓑ Ⓒ Ⓓ Ⓔ
8 Ⓐ Ⓑ Ⓒ Ⓓ Ⓔ	33 Ⓐ Ⓑ Ⓒ Ⓓ Ⓔ	58 Ⓐ Ⓑ Ⓒ Ⓓ Ⓔ	83 Ⓐ Ⓑ Ⓒ Ⓓ Ⓔ
9 Ⓐ Ⓑ Ⓒ Ⓓ Ⓔ	34 Ⓐ Ⓑ Ⓒ Ⓓ Ⓔ	59 Ⓐ Ⓑ Ⓒ Ⓓ Ⓔ	84 Ⓐ Ⓑ Ⓒ Ⓓ Ⓔ
10 Ⓐ Ⓑ Ⓒ Ⓓ Ⓔ	35 Ⓐ Ⓑ Ⓒ Ⓓ Ⓔ	60 Ⓐ Ⓑ Ⓒ Ⓓ Ⓔ	85 Ⓐ Ⓑ Ⓒ Ⓓ Ⓔ
11 Ⓐ Ⓑ Ⓒ Ⓓ Ⓔ	36 Ⓐ Ⓑ Ⓒ Ⓓ Ⓔ	61 Ⓐ Ⓑ Ⓒ Ⓓ Ⓔ	86 Ⓐ Ⓑ Ⓒ Ⓓ Ⓔ
12 Ⓐ Ⓑ Ⓒ Ⓓ Ⓔ	37 Ⓐ Ⓑ Ⓒ Ⓓ Ⓔ	62 Ⓐ Ⓑ Ⓒ Ⓓ Ⓔ	87 Ⓐ Ⓑ Ⓒ Ⓓ Ⓔ
13 Ⓐ Ⓑ Ⓒ Ⓓ Ⓔ	38 Ⓐ Ⓑ Ⓒ Ⓓ Ⓔ	63 Ⓐ Ⓑ Ⓒ Ⓓ Ⓔ	88 Ⓐ Ⓑ Ⓒ Ⓓ Ⓔ
14 Ⓐ Ⓑ Ⓒ Ⓓ Ⓔ	39 Ⓐ Ⓑ Ⓒ Ⓓ Ⓔ	64 Ⓐ Ⓑ Ⓒ Ⓓ Ⓔ	89 Ⓐ Ⓑ Ⓒ Ⓓ Ⓔ
15 Ⓐ Ⓑ Ⓒ Ⓓ Ⓔ	40 Ⓐ Ⓑ Ⓒ Ⓓ Ⓔ	65 Ⓐ Ⓑ Ⓒ Ⓓ Ⓔ	90 Ⓐ Ⓑ Ⓒ Ⓓ Ⓔ
16 Ⓐ Ⓑ Ⓒ Ⓓ Ⓔ	41 Ⓐ Ⓑ Ⓒ Ⓓ Ⓔ	66 Ⓐ Ⓑ Ⓒ Ⓓ Ⓔ	91 Ⓐ Ⓑ Ⓒ Ⓓ Ⓔ
17 Ⓐ Ⓑ Ⓒ Ⓓ Ⓔ	42 Ⓐ Ⓑ Ⓒ Ⓓ Ⓔ	67 Ⓐ Ⓑ Ⓒ Ⓓ Ⓔ	92 Ⓐ Ⓑ Ⓒ Ⓓ Ⓔ
18 Ⓐ Ⓑ Ⓒ Ⓓ Ⓔ	43 Ⓐ Ⓑ Ⓒ Ⓓ Ⓔ	68 Ⓐ Ⓑ Ⓒ Ⓓ Ⓔ	93 Ⓐ Ⓑ Ⓒ Ⓓ Ⓔ
19 Ⓐ Ⓑ Ⓒ Ⓓ Ⓔ	44 Ⓐ Ⓑ Ⓒ Ⓓ Ⓔ	69 Ⓐ Ⓑ Ⓒ Ⓓ Ⓔ	94 Ⓐ Ⓑ Ⓒ Ⓓ Ⓔ
20 Ⓐ Ⓑ Ⓒ Ⓓ Ⓔ	45 Ⓐ Ⓑ Ⓒ Ⓓ Ⓔ	70 Ⓐ Ⓑ Ⓒ Ⓓ Ⓔ	95 Ⓐ Ⓑ Ⓒ Ⓓ Ⓔ
21 Ⓐ Ⓑ Ⓒ Ⓓ Ⓔ	46 Ⓐ Ⓑ Ⓒ Ⓓ Ⓔ	71 Ⓐ Ⓑ Ⓒ Ⓓ Ⓔ	96 Ⓐ Ⓑ Ⓒ Ⓓ Ⓔ
22 Ⓐ Ⓑ Ⓒ Ⓓ Ⓔ	47 Ⓐ Ⓑ Ⓒ Ⓓ Ⓔ	72 Ⓐ Ⓑ Ⓒ Ⓓ Ⓔ	97 Ⓐ Ⓑ Ⓒ Ⓓ Ⓔ
23 Ⓐ Ⓑ Ⓒ Ⓓ Ⓔ	48 Ⓐ Ⓑ Ⓒ Ⓓ Ⓔ	73 Ⓐ Ⓑ Ⓒ Ⓓ Ⓔ	98 Ⓐ Ⓑ Ⓒ Ⓓ Ⓔ
24 Ⓐ Ⓑ Ⓒ Ⓓ Ⓔ	49 Ⓐ Ⓑ Ⓒ Ⓓ Ⓔ	74 Ⓐ Ⓑ Ⓒ Ⓓ Ⓔ	99 Ⓐ Ⓑ Ⓒ Ⓓ Ⓔ
25 Ⓐ Ⓑ Ⓒ Ⓓ Ⓔ	50 Ⓐ Ⓑ Ⓒ Ⓓ Ⓔ	75 Ⓐ Ⓑ Ⓒ Ⓓ Ⓔ	100 Ⓐ Ⓑ Ⓒ Ⓓ Ⓔ

Completely darken bubbles with a No. 2 pencil. If you make a mistake, be sure to erase mark completely. Erase all stray marks.

Begin with number 1 for each new section of the test. Leave blank any extra answer spaces.

SECTION 2

SECTION 3

FOR TPR USE ONLY | V1 | V2 | V3 | V4 | M1 | M2 | M3 | M4 | M5 | M6 | M7 | M8

Free!

Did you know that The Microsoft Network gives you one free month?

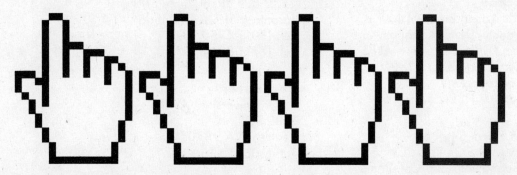

Call us at 1-800-FREE MSN. We'll send you a free CD to get you going.

Then, you can explore the World Wide Web for one month, free. Exchange e-mail with your family and friends. Play games, book airline tickets, handle finances, go car shopping, explore old hobbies and discover new ones. There's one big, useful online world out there. And for one month, it's a free world.

Call **1-800-FREE MSN,** Dept. 3197, for offer details or visit us at **www.msn.com**. Some restrictions apply.

Microsoft Where do you want to go today?®

FIND US...

International

Hong Kong
4/F Sun Hung Kai Centre
30 Harbour Road, Wan Chai,
Hong Kong
Tel: (011)85-2-517-3016

Japan
Fuji Building 40, 15-14
Sakuragaokacho, Shibuya Ku,
Tokyo 150, Japan
Tel: (011)81-3-3463-1343

Korea
Tae Young Bldg, 944-24,
Daechi- Dong, Kangnam-Ku
The Princeton Review- ANC
Seoul, Korea 135-280,
South Korea
Tel: (011)82-2-554-7763

Mexico City
PR Mex S De RL De Cv
Guanajuato 228 Col. Roma
06700 Mexico D.F., Mexico
Tel: 525-564-9468

Montreal
666 Sherbrooke St.
West, Suite 202
Montreal, QC H3A 1E7 Canada
Tel: (514) 499-0870

Pakistan
1 Bawa Park - 90 Upper Mall
Lahore, Pakistan
Tel: (011)92-42-571-2315

Spain
Pza. Castilla, 3 - 5° A, 28046
Madrid, Spain
Tel: (011)341-323-4212

Taiwan
155 Chung Hsiao East Road
Section 4 - 4th Floor,
Taipei R.O.C., Taiwan
Tel: (011)886-2-751-1243

Thailand
Building One, 99 Wireless Road
Bangkok, Thailand 10330
Tel: (662) 256-7080

Toronto
1240 Bay Street, Suite 300
Toronto M5R 2A7 Canada
Tel: (800) 495-7737
Tel: (716) 839-4391

Vancouver
4212 University Way NE,
Suite 204
Seattle, WA 98105
Tel: (206) 548-1100

National (U.S.)

We have over 60 offices around the U.S. and
run courses in over 400 sites. For courses and locations
within the U.S. call 1 (800) 2/Review and you will be
routed to the nearest office.